Lecture Notes in Computer Science   10674

Commenced Publication in 1973
Founding and Former Series Editors:
Gerhard Goos, Juris Hartmanis, and Jan van Leeuwen

More information about this series at http://www.springer.com/series/7410

Helger Lipmaa · Aikaterini Mitrokotsa
Raimundas Matulevičius (Eds.)

Secure IT Systems

22nd Nordic Conference, NordSec 2017
Tartu, Estonia, November 8–10, 2017
Proceedings

 Springer

Editors
Helger Lipmaa
University of Tartu
Tartu
Estonia

Raimundas Matulevičius
University of Tartu
Tartu
Estonia

Aikaterini Mitrokotsa
Chalmers University of Technology
Gothenburg
Sweden

ISSN 0302-9743   ISSN 1611-3349   (electronic)
Lecture Notes in Computer Science
ISBN 978-3-319-70289-6   ISBN 978-3-319-70290-2   (eBook)
https://doi.org/10.1007/978-3-319-70290-2

Library of Congress Control Number: 2017957852

LNCS Sublibrary: SL4 – Security and Cryptology

Printed on acid-free paper

This Springer imprint is published by Springer Nature
The registered company is Springer International Publishing AG
The registered company address is: Gewerbestrasse 11, 6330 Cham, Switzerland

Preface

This volume contains the papers presented at NordSec 2017, the 22nd Nordic Conference on Secure IT Systems. The conference was held during November 8–10, 2017, in Tartu, Estonia.

The NordSec conferences started in 1996 with the aim of bringing together researchers and practitioners in computer security in the Nordic countries, thereby establishing a forum for discussions and cooperation between universities, industry, and computer societies. NordSec addresses a broad range of topics within IT security and privacy and over the years it has developed into an international conference that takes place in the Nordic countries. NordSec is currently a key meeting venue for Nordic university teachers and students with research interests in information security and privacy.

NordSec 2017 received 42 submissions, with all valid submissions receiving three reviews by the Program Committee (PC). After the reviewing phase, 18 papers were accepted for publication and are all included in these proceedings. Furthermore, we had a poster session that encouraged discussions and brainstorming on current topics of information security and privacy.

We were honored to have had three brilliant invited speakers with talks on current topics in information security focusing on machine learning, blockchains, and verifiable computation. More precisely, Dr. Ananth Raghunathan from Google gave a talk on "Security and Privacy Challenges in Machine Learning," Prof. Aggelos Kiayias from the University of Edinburgh gave a talk on "Proof of Stake Blockchain Protocols," and Dr. Dario Fiore from IMDEA Software Institute gave a talk on "Homomorphic Authentication for Computing Securely on Untrusted Machines."

We sincerely thank everyone involved in making this year's instance a success including but not limited to: the authors who submitted their papers, the presenters who contributed to the NordSec program, and the PC members and the additional reviewers for their thorough and very helpful reviews. Last but not least, we sincerely thank the Cybernetica AS company for the support given to the NordSec 2017 conference.

November 2017

Helger Lipmaa
Aikaterini Mitrokotsa
Raimundas Matulevičius

Organization

General Chair

Helger Lipmaa   University of Tartu, Estonia

Program Committee Chairs

Helger Lipmaa   University of Tartu, Estonia
Aikaterini Mitrokotsa   Chalmers University of Technology, Sweden

Program Committee

Tuomas Aura	Aalto University, Finland
Musard Balliu	Chalmers University of Technology, Sweden
Céline Blondeau	Aalto University, Finland
Billy Brumley	Tampere University of Technology, Finland
Sonja Buchegger	KTH Royal Institute of Technology, Sweden
Ahto Buldas	Cybernetica AS, Estonia
Úlfar Erlingsson	Google Brain, Iceland
Simone Fischer-Hübner	Karlstad University, Sweden
Kristian Gjøsteen	Norwegian University of Science and Technology, Norway
Rene Rydhof Hansen	Aalborg University, Denmark
Camilla Hollanti	Aalto University, Finland
Thomas Johansson	Lund University, Sweden
Audun Jøsang	University of Oslo, Norway
Sokratis Katsikas	Norwegian University of Science and Technology, Norway
Martti Lehto	University of Jyväskylä, Finland
Ville Leppänen	University of Turku, Finland
Bei Liang	Chalmers University of Technology, Sweden
Olaf Maennel	Tallinn University of Technology, Estonia
Raimundas Matulevičius	University of Tartu, Estonia
Christian W. Probst	Technical University of Denmark, Denmark
Carla Ràfols	Universitat Pompeu Fabra, Spain
Alejandro Russo	Chalmers University of Technology, Sweden
Berry Schoenmakers	Technical University of Eindhoven, The Netherlands
Carsten Schuermann	IT University of Copenhagen, Denmark
Zheng Yan	Xidian University, China
Bingsheng Zhang	Lancaster University, UK

Additional Reviewers

Daniel Bosk
Alessandro Bruni
Prastudy Fauzi
Wei Feng
Rosario Giustolisi
Oliver Gnilke
Shohreh Hosseinzadeh
Xuyang Jing
Joona Kannisto
Xueqin Liang
Gao Liu

Samuel Marchal
Yoan Miche
Mads C. Olesen
Cesar Pereida Garcia
Sampsa Rauti
Jukka Ruohonen
Razane Tajeddine
Nicola Tuveri
Daren Tuzi
Thomas Zacharias
Daode Zhang

Organization Chair

Raimundas Matulevičius   University of Tartu, Estonia

Steering Committee

Tuomas Aura	Aalto University, Finland (Chair)
Karin Bernsmed	SINTEF ICT, NTNU, Norway
Billy Brumley	Tampere University of Technology, Finland
Bengt Carlsson	Blekinge Institute of Technology, Sweden
Úllfar Erlingsson	Google Inc., Mountain View, USA
Simone Fischer-Huebner	Karlstad University, Sweden
Dieter Gollmann	TUHH Technische Universität Hamburg-Harburg, Germany
Audun Jøsang	University of Oslo, Norway
Stewart Kowalski	Gjøvik University College, Norway
Peeter Laud	Cybernetica AS, Estonia
Helger Lipmaa	University of Tartu, Estonia
Hanne Riis Nielson	Technical University of Denmark, Denmark
Juha Röning	University of Oulu, Finland
Andrei Sabelfeld	Chalmers University of Technology, Sweden
Simin Nadjm-Tehrani	Linköping University, Sweden
Magnus Almgren	Chalmers University of Technology, Sweden
Sonja Buchegger	KTH, Royal Institute of Technology, Sweden

Abstracts of Invited Talks

Homomorphic Authentication for Computing Securely on Untrusted Machines

Dario Fiore

IMDEA Software Institute, Madrid, Spain
dario.fiore@imdea.org

Abstract. Due to phenomena like the ubiquity of the Internet and cloud computing, it is increasingly common to store and process data on third-party machines. In spite of its attractive aspects, this trend raises a number of security concerns, including: how to ensure that the results computed by third parties are correct (integrity) and no unauthorized information is leaked (privacy)? This talk focuses on cryptographic solutions for integrity, and more specifically on the notion of homomorphic authentication. It presents this notion, gives an overview of the state of the art in this area, and covers some of the recent efficient constructions.

Introduction

Due to phenomena like the ubiquity of the Internet and cloud computing, it is increasingly common to store and process data on third-party machines. While this computing trend is undoubtedly successful for its attractive features, it also raises a number of security concerns, such as:

> How to ensure that the results computed by third parties are correct (integrity) and no unauthorized information is leaked (privacy)?

Recent work in cryptography has shown a variety of new cryptographic means for protecting information processed on third-party, untrusted machines. For example, it is widely known that fully homomorphic encryption [6] can solve privacy by allowing one to compute on data that is encrypted. Here, we analyze the problem of guaranteeing the *authenticity of data during computation*, and more specifically we focus on the notion of *homomorphic authentication*.

Homomorphic Authenticators. Akin to standard authentication mechanisms (e.g., digital signatures or message authentication codes), homomorphic authenticators (HAs) allow a user Alice to authenticate a collection of data items $x_1, \ldots, x_n$ using her secret key. The distinguishing feature of HAs is that an untrusted party, without the need of any secret, can use the authenticators on $x_1, \ldots, x_n$ to generate a value $\sigma_{\mathcal{P},y}$ that vouches for the correctness of $y = \mathcal{P}(x_1, \ldots, x_n)$. Finally, a user Bob who is given the tuple $(\mathcal{P}, y, \sigma_{\mathcal{P},y})$ and Alice's verification key can verify the authenticity of y as output of the program $\mathcal{P}$ executed on data authenticated by Alice. In other words, Bob can verify that the server did not tamper with the computation's result and that it used the very same data authenticated by Alice. Alice's verification key can be either secret or public. In the former case, this primitive is known as *homomorphic MACs*, while in the latter case it is known as *homomorphic signatures*.

In terms of security, HAs must be unforgeable. Intuitively, this means that an adversary must not be able to forge a valid authenticator on an incorrect computation's result $y^* \neq \mathcal{P}(x_1, \ldots, x_n)$. In addition to security, HAs are interesting because of two additional properties. The first one is *succinctness*, which says that the authenticators remain short, i.e., much shorter than $\mathcal{P}$'s input size: this means that one can convince Bob about the correctness of a program executed on a huge amount of data by sending him only a very short piece of information. The second interesting property is *composability*, which says that derived authenticators can be used further as inputs to new computations: this means that one can, for example, distribute different subtasks to several untrusted workers, ask each of them to produce a proof of its local task, and use these proofs to create another single proof for the final job (as in the MapReduce approach).

Thanks to these properties, homomorphic authenticators can provide a nice and elegant solution to the problem of ensuring authenticity and integrity of data during computation.

A glance at the state of the art.[1] The notion of homomorphic authentication was first introduced by Desmedt [4] and later reconsidered more formally by Johnson et al. [8]. A more formal definition, as the one depicted above, came only more recently starting with the works of Boneh et al. [1, 2]. Since then, research was mainly devoted towards two fundamental goals: (i) to broaden the class of functionalities that can be computed homomorphically, and (ii) to obtain efficient instantiations. With respect to (i), research has gone far up to the notable result of Gorbunov, Vaikuntanathan and Wichs who showed a scheme that supports boolean circuits of bounded polynomial depth [7]. Yet, the existence of truly fully homomorphic schemes remain an open problem. As far as (ii) is concerned, the problem is less settled as practically efficient instantiations essentially are confined to schemes supporting linear functions. The situation is slightly better in the symmetric-key setting: a fully homomorphic MAC that can deal with all circuits was proposed by Gennaro and Wichs [5] based on FHE, and a simpler, more efficient homomorphic MAC supporting only NC1 circuits has been shown by Catalano and Fiore [3] based on pseudorandom functions.

Talk Overview. This talk begins with an introduction to the notion of homomorphic authentication and an overview of the state of the art. Next, it covers some recent constructions, and finally concludes by discussing some of the main open problems in this research area.

---

[1] This is not meant to be an exhaustive analysis; we only mention a selection of milestones in the area.

Acknowledgements. I would like to thank the program chairs and the entire PC of NordSec 2017 for inviting me to give this talk. I am also grateful to Michael Backes, Manuel Barbosa, Dario Catalano, Rosario Gennaro, Katerina Mitrokotsa, Luca Nizzardo, Elena Pagnin, Valerio Pastro, Raphael Reischuk, Konstantinos Vamvourellis, and Bogdan Warinschi for their fruitful collaboration in this research area.

References

1. Boneh, D., Freeman, D.M.: Homomorphic signatures for polynomial functions. In: Paterson, K.G. (ed.) EUROCRYPT 2011. LNCS, vol. 6632, pp. 149–168. Springer, Berlin (2011)
2. Boneh, D., Freeman, D., Katz, J., Waters, B.: Signing a linear subspace: signature schemes for network coding. In: Jarecki,S., Tsudik, G. (eds.) PKC 2009. LNCS, vol. 5443, pp. 68–87. Springer, Berlin (2009)
3. Catalano, D., Fiore, D.: Practical homomorphic MACs for arithmetic circuits. In: Johansson, T., Nguyen, P.Q. (eds.) EUROCRYPT 2013. LNCS, vol. 7881, pp. 336–352. Springer, Berlin (2013)
4. Desmedt, Y.: Computer security by redefining what a computer is. NSPW (1993)
5. Gennaro, R., Wichs, D.: Fully homomorphic message authenticators. In: Sako, K., Sarkar, P. (eds.) ASIACRYPT 2013, Part II. LNCS, vol. 8270, pp. 301–320. Springer, Berlin (2013)
6. Gentry, C.: Fully homomorphic encryption using ideal lattices. In: 41st ACM STOC, pp. 169–178. ACM Press (2009)
7. Gorbunov, S., Vaikuntanathan, V., Wichs, D.: Leveled fully homomorphic signatures from standard lattices. In: 47th ACM STOC. ACM Press (2015)
8. Johnson, R., Molnar, D., Song, D.X., Wagner, D.: Homomorphic signature schemes. In: Preneel, B. (ed.) CT-RSA 2002. LNCS, vol. 2271, pp. 244–262. Springer, Cham (2002)

Security and Privacy Challenges in Machine Learning

Ananth Raghunathan

Google, Mountain View, CA, USA
pseudorandom@google.com

Abstract. This talk covers many of the security and privacy issues raised by the recent advances in machine learning. In particular, I'll present recent results in protecting the privacy of sensitive training data, and recent attacks enabled by semi-supervised learning and knowledge transfer.

Proof of Stake Blockchain Protocols

Aggelos Kiayias

University of Edinburgh, Edinburgh, UK
akiayias@inf.ed.ac.uk

Abstract. In this talk I will cover recent developments in the design of blockchain protocols focusing on proof of stake based solutions. The talk will overview design challenges and analysis approaches, from both a security and a high performance perspective.

Contents

Applications

Access Control

Emerging Security Areas

Outsourcing Computations

A Server-Assisted Hash-Based Signature Scheme

Ahto Buldas[1], Risto Laanoja[1,2], and Ahto Truu[1,2(✉)]

[1] Tallinn University of Technology, Akadeemia tee 15a, 12618 Tallinn, Estonia
[2] Guardtime AS, A.H. Tammsaare tee 60, 11316 Tallinn, Estonia
ahto.truu@guardtime.com

Abstract. We present a practical digital signature scheme built from a cryptographic hash function and a hash-then-publish digital time-stamping scheme. We also provide a simple proof of existential unforgeability against adaptive chosen-message attack (EUF-ACM) in the random oracle (RO) model.

1 Introduction

All the digital signature schemes in use today (RSA [42], DSA [22], ECDSA [30]) are known to be vulnerable to quantum attacks by Shor's algorithm [46]. While the best current experimental results are still toy-sized [35], it takes a long time for new cryptographic schemes to be accepted and deployed, so it is of considerable interest to look for post-quantum secure alternatives already now. Error-correcting codes, discrete lattices, and multi-variate polynomials have been used as foundations for proposed replacement schemes [4]. However, these are relatively complex structures and new constructions in cryptography, so require significant additional scrutiny before gaining trust.

Hash functions, on the other hand, have been studied for decades and are widely believed to be quite resistant to quantum attacks. The best currently known quantum results against hash functions are using Grover's algorithm [25] to find a pre-image of a given k-bit value in $2^{k/2}$ queries instead of the 2^k queries needed by a classical attacker, and Brassard et al.'s modification [7] to find a collision in $2^{k/3}$ instead of $2^{k/2}$ queries. To counter these attacks, it would be sufficient to deploy hash functions with correspondingly longer outputs when moving from pre-quantum to post-quantum setting.

2 Related Work

The earliest digital signature scheme constructed from hash functions is due to Lamport [19,31]. Merkle [37] introduced two methods for reducing the key sizes, one proposed to him by Winternitz. The Winternitz scheme has subsequently been more thoroughly analyzed and further refined by Even et al. [23],

This research was supported by the European Regional Development Fund through the Estonian smart specialization program NUTIKAS.

H. Lipmaa et al. (Eds.): NordSec 2017, LNCS 10674, pp. 3–17, 2017.
https://doi.org/10.1007/978-3-319-70290-2_1

Dods et al. [21], Buchmann et al. [9], and Hülsing [27]. All of these schemes are *one-time*, and require generation of a new key pair and distribution of a new public key for each message to be signed.

Merkle's arguably most important contribution in [37] was the concept of *hash tree*, which enables a large number of public keys to be represented by a single hash value. With the hash value published, any one of the N public keys can be shown to belong to the tree with a proof consisting of $\log_2 N$ hash values, thus combining N instances of a one-time scheme into an N-*time* scheme. Buldas and Saarepera [16] and Coronado García [17] showed the aggregation to be secure if the hash function used to build the tree is collision resistant. Rohatgi [43] used the XOR-tree construct proposed by Bellare and Rogaway [3] to create a variant of hash tree whose security is based on second pre-image resistance of the hash function instead of collision resistance. Dahmen et al. [18] proposed a similar idea with a more complete security proof.

A drawback of the above hash tree constructs is that the whole tree has to be built at once, which also means all the private keys have to be generated at once. Merkle [38] proposed a certification tree that allows just the root node of the tree to be populated initially and the rest of the tree to be grown gradually as needed. However, to authenticate the lower nodes of the tree, a chain of full-blown one-time signatures (as opposed to a chain of sibling hash values) is needed, unless the protocol is used in an interactive environment where the recipient keeps the public keys already delivered as part of earlier signatures. Malkin et al. [34] and Buchmann et al. [8,11] proposed various multi-level schemes where the keys authenticated by higher-level trees are used to sign roots of lower-level trees to enable the key sets to be expanded incrementally.

Buchmann et al. [10] proposed XMSS, a version of the Merkle signature scheme with improved efficiency compared to previous ones. Hülsing et al. [28] introduced a multi-tree version of it. Hülsing et al. [29] described a modification hardened against so-called multi-target attacks where the adversary will succeed when it can find a pre-image for just one of a large number of target output values of a hash function.

A risk with the N-time schemes is that they are *stateful*: as each of the one-time keys may be used only once, the signer will need to keep track of which keys have already been used. If this state information is lost (for example, when a previous state is restored from a backup), keys may be re-used by accident.

Perrig [39] proposed BiBa which has small signatures and fast verification, but rather large public keys and slow signing. Reyzin and Reyzin [41] proposed the HORS scheme that provides much faster signing than BiBa. These two are not strictly one-time, but so-called *few-time* schemes where a private key can be used to sign several messages, but the security level decreases with each additional use. Bernstein et al. [5] proposed SPHINCS, which combines HORS with XMSS trees to create a *stateless* scheme that uses keys based on a pseudo-random schedule that makes the risk of re-use negligible even without tracking the state.

3   Our Contribution

We propose a signature scheme with a hash function as its sole underlying primitive. At the time of writing, XMSS and SPHINCS are the state of the art in the stateful and stateless hash signature schemes, respectively, so these are what new schemes should be measured against.

XMSS has fast signing and verification, and small signatures, but requires careful management of key state [36]. Our scheme has comparable efficiency, but the private key to be used is determined by signing time, which removes the risk of accidental roll-backs. Also, a single private key can be used to sign multiple messages simultaneously, so no synchronization is required when the scheme is deployed in multi-threaded or multi-processor environments.

SPHINCS has small keys and efficient verification, but quite large signatures and rather expensive signing. Our scheme requires orders of magnitude less computations for signing and produces signatures roughly a tenth the size.

A more general feature is that each signature produced by our scheme is inherently time-stamped. Most other schemes require time-stamping as a separate step after signing to handle key expirations, key revocations, and time-limited signing authority. Due to the time-stamping component, our scheme is necessarily server-assisted. While this may look like a disadvantage, it may in fact be beneficial in enforcing various key usage policies and limiting damage in case of a key leakage. For these reasons, even the technically off-line schemes are usually deployed within on-line frameworks in practice.

4   Preliminaries

Hash Trees. Introduced by Merkle [37], a hash tree is a tree-shaped data structure built using a 2-to-1 hash function $h\colon \{0,1\}^{2k} \to \{0,1\}^k$. The nodes of the tree contain k-bit values. Each node is either a leaf with no children or an internal node with two children. The value x of an internal node is computed as $x \leftarrow h(x_l, x_r)$, where x_l and x_r are the values of the left and right child, respectively. There is one root node that is not a child of any node. We will use $r \leftarrow T^h(x_1, \ldots, x_N)$ to denote a hash tree whose N leaves contain the values $x_1, \ldots, x_N$ and whose root node contains r.

Hash Chains. In order to prove that a value x_i participated in the computation of the root hash r, it is sufficient to present values of all the siblings of the nodes on the unique path from x_i to the root in the tree. For example, to claim that x_3 belongs to the tree shown on the left in Fig. 1, one has to present the values x_4 and $x_{1,2}$ to enable the verifier to compute $x_{3,4} \leftarrow h(x_3, x_4)$, $r \leftarrow h(x_{1,2}, x_{3,4})$, essentially re-building a slice of the tree, as shown on the right in Fig. 1. We will use $x \overset{c}{\leadsto} r$ to denote that the hash chain c links x to r in such a manner.

Intuitively, it seems obvious that if the function h is one-way, the existence of such a chain whose output equals the original r is a strong indication that x was indeed the original input. However, this result was not formally proven until 25 years after the hash tree construct was proposed [16,17].

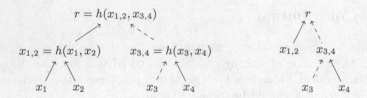

Fig. 1. The hash tree $T^h(x_1, \ldots, x_4)$ and the corresponding hash chain $x_3 \rightsquigarrow r$.

Hash-Then-Publish Time-Stamping. The general idea of time-stamping information by publishing its hash value was used already by Galilei and Hooke in the
XVII century. In more modern cryptographic times, Haber and Stornetta [26]
were the first to propose time-stamping a sequence of records by having each of
them contain the hash of the previous one, in a manner that was later popular-
ized as the *blockchain* structure. Bayer et al. [2] proposed using hash trees to
aggregate the inputs in batches and then linking the roots of the trees instead of
individual records. The most recent results on security bounds of such schemes
are by Buldas et al. [13–15].

5   Description of the Scheme

The principal idea of our signature scheme is to have the signer commit to a
sequence of keys such that each key is assigned a time slot when it can be used
to sign messages and will transition from signing key to verification key once the
time slot has passed.

Signing itself then consists of time-stamping the message-key pair in order to
prove that the signing operation was performed at the correct time. For simplicity
of presentation, we count time in aggregation rounds of the time-stamping service
and use the expression "at time t" to mean "during aggregation round t".

More formally, the classic triple of procedures for key generation, signature
generation, and signature verification [24] is as follows:

Key Generation. To prepare to sign messages at times $1, \ldots, N$, the signer:

1. Generates N signing keys: $(z_1, \ldots, z_N) \leftarrow \mathcal{G}(N, k)$.
   We assume the keys are unpredictable values drawn from $\{0,1\}^k$.
2. Binds each key to its time slot: $x_i \leftarrow h(i, z_i)$ for $i \in \{1, \ldots, N\}$.
3. Computes the public key p by aggregating the key bindings into a hash tree:
   $p \leftarrow T^h(x_1, \ldots, x_N)$.

The resulting data structure is shown in Fig. 2 and its purpose is to be able to
extract hash chains $c_i \leftarrow h(i, z_i) \rightsquigarrow p$ for $i \in \{1, \ldots, N\}$.

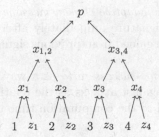

Fig. 2. Computation of public key for $N = 4$.

Signing. To sign message m at time t, the signer:

1. Uses the appropriate key to authenticate the message: $y \leftarrow h(m, z_t)$.
2. Time-stamps the authenticator: $a_t \leftarrow y \rightsquigarrow r_t$.
   Here r_t is the root hash of the aggregation tree built by the time-stamping service for the aggregation round t. We assume the root is committed to in some reliable way, such as broadcasting it to all interested parties, but place no other trust in the service.
3. Outputs the tuple (t, z_t, a_t, c_t), where t is the signing time, z_t is the signing key for time slot t, a_t is the hash chain from the time-stamping service linking the key usage to r_t, and c_t is the hash chain linking the binding of z_t and time slot t to the signer's public key p.

Note that the signature is composed and emitted after the time-stamping step, which makes it safe for the signer to release the key z_t as part of the signature: the aggregation round t has ended and any future uses of the key z_t can no longer be stamped with time t.

Verification. To verify that the message m and the signature $s = (t, z, a, c)$ match the public key p, the verifier:

1. Checks that z was committed as signing key for time t: $h(t, z) \overset{c}{\rightsquigarrow} p$.
2. Checks that m was authenticated with key z at time t: $h(m, z) \overset{a}{\rightsquigarrow} r_t$.

6   Security Proof

Goldwasser et al. [24] proposed a framework for studying security of signature schemes where the attackers have various levels of access to signing oracles and various requirements on what they need to achieve for the attack to be considered successful (and the scheme broken).

As the highest security level, they defined the concept of *existential unforgeability* (EUF) where an attacker should be unable to forge signatures on any messages, even nonsensical ones.

They also defined the *chosen-message attack* where the attacker can submit a number of messages to be signed by the oracle before having to come up with a forged signature on a new message, and in particular, as the one giving the attacker the most power, the *adaptive chosen-message attack* (ACM) where the attacker will receive each signature immediately after submitting the message and can use any information gained from previous signatures to form subsequent messages.

Luby [33] defined the *time-success ratio* as a way to express the resilience of a cryptographic scheme against attacks as the relationship of the probability that the attack will succeed to the computation time the attacker is allowed to spend.

We will now combine these notions to define and prove the security of our signature scheme.

Definition 1. *A signature scheme is S-secure existentially unforgeable against adaptive chosen-message attacks (EUF-ACM), if any T-time adversary, having access to a signer's public key p and to a signing oracle $\mathcal{S}$ to obtain signatures $s_1 \leftarrow \mathcal{S}(m_1), \ldots, s_n \leftarrow \mathcal{S}(m_n)$ on adaptively chosen messages $m_1, \ldots, m_n$, can produce a new message-signature pair (m, s) such that $m \notin \{m_1, \ldots, m_n\}$, but s is a valid signature on m, with probability at most T/S.*

Oracle $\mathcal{S}$ (signing oracle)

Query $\mathsf{Sig}(m, t)$:
    return $h(m, z_t)$
Query $\mathsf{Get}(t)$:
    If $c \geq t$ then:
        return $(z_t, x_t \rightsquigarrow p)$
    else:
        return $\perp$

Oracle $\mathcal{R}$ (repository)

Initialize:
    $c \leftarrow 0$
Query $\mathsf{Put}(r)$:
    $c \leftarrow c + 1$
    $r_c \leftarrow r$
Query $\mathsf{Get}(t)$:
    If $c \geq t$ then:
        return r_t
    else:
        return $\perp$

Fig. 3. The oracles used in the security condition.

To formalize our security assumptions, we introduce three oracles:

We model the publishing of the root hashes of the time-stamping aggregation trees as the oracle $\mathcal{R}$ (Fig. 3, right) that allows each r_t to be published just once.

The signing oracle $\mathcal{S}$ (Fig. 3, left) will compute the message authenticators at any time, but will release only the keys that have already expired for signing (transitioned to verification keys).

We model the hash function h as a random oracle using the *lazy sampling* technique: every time h is queried with a previously unseen input, a new return value is generated by uniform random sampling from $\{0, 1\}^k$; when h is queried with a previously seen input, the same value is returned as last time.

The adversary $\mathcal{A}$ will be interacting with the oracles as shown in Fig. 4 with the goal of producing a forgery.

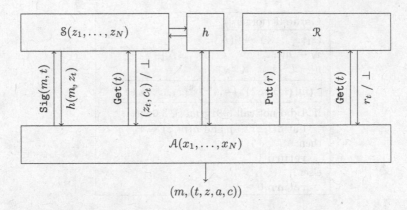

Fig. 4. The adversary's interactions with the oracles.

To model the fact that the signer needs to keep secret only the keys $z_1, \ldots, z_N$, we explicitly initialize the adversary with $x_1, \ldots, x_N$. Note that the verification rule still assumes that the verifier has access only to the signer's public key p, which means the adversary is not limited to presenting hash chains that were actually extracted from $T^h(x_1, \ldots, x_N)$.

Also note that we leave the aggregation process of the time-stamping service fully under the adversary's control; only the repository $\mathcal{R}$ needs to be trusted to operate correctly.

As normally signing message m involves first calling $\mathcal{S}.\mathtt{Sig}(m, t)$, then committing to $\mathcal{R}$ the root of a hash tree that includes the return value, and then calling $\mathcal{S}.\mathtt{Get}(t)$, we formalize the forgery condition by demanding that the adversary can't make the two $\mathcal{S}$ calls in that order:

Definition 2. *The pair (m, s) produced by an adversary is a successful forgery if s is a valid signature on m, but the adversary did not make the calls $\mathcal{S}.\mathtt{Sig}(m, t)$, $\mathcal{S}.\mathtt{Get}(t)$, in that order, for any $t \in \{1, \ldots, N\}$.*

Theorem 1. *Our signature scheme, when instantiated with a hash function $h \colon \{0,1\}^{2k} \to \{0,1\}^k$ indistinguishable from a random oracle, is at least $2^{(k-2)/2}$-secure existentially unforgeable against adaptive chosen-message attacks by any T-time adversary.*

Proof. We will directly show an upper bound on the success probability of the adversary in the forgery game $\mathcal{F}$ (Fig. 5).

Assume that the adversary does not call $\mathcal{S}.\mathtt{Get}(t)$. To win the game $\mathcal{F}$, he must produce t, z, c such that $h(t, z) \overset{c}{\leadsto} p$. For that, the output of the last step of the chain computation must equal the root of the tree $T^h(x_1, \ldots, x_N)$. Let's now consider the inputs to that step. If they equal the corresponding children of the root of the tree, we can repeat the reasoning for the second last step and the corresponding node of the tree, and so on. As we walk a finite chain and

$$\boxed{\begin{array}{l} \textbf{Game } \mathcal{F} \text{ (forgery)} \\ \hline (z_1, \ldots, z_N) \leftarrow \mathcal{G}(N) \\ x_i \leftarrow h(i, z_i) \text{ for } i \in \{1, \ldots, N\} \\ p \leftarrow T^h(x_1, \ldots, x_N) \\ \hline \boxed{(m, (t, z, a, c)) \leftarrow \mathcal{A}^{h, \mathcal{S}, \mathcal{R}}(x_1, \ldots, x_N)} \\ \hline \text{If } \mathcal{A} \text{ did not call } \mathcal{S}.\texttt{Sig}(m, t), \mathcal{S}.\texttt{Get}(t), \\ \quad \text{but } h(t, z) \overset{c}{\leadsto} p \text{ and } h(m, z) \overset{a}{\leadsto} r_t \\ \text{then:} \\ \quad \textbf{return } 1 \\ \text{else:} \\ \quad \textbf{return } 0 \end{array}}$$

Fig. 5. The forgery game.

simultaneously traverse a finite tree from the root towards leaves, one of the following events must eventually happen:

1. We run out of the chain at the same time we run out of the tree. This means the adversary has found t and z such that $x_i = h(t, z)$ for some $i \in \{1, \ldots, N\}$. If $i \neq t$, then the adversary has found a second pre-image for the x_i originally computed as $h(i, z_i)$. With h being a random oracle, the probability of a T-time adversary achieving that for any given i is $\leq T/2^k$. If $i = t$, then the adversary may have found a second pre-image for x_t, with probability $\leq T/2^k$, or may have guessed z_t, also with probability $\leq T/2^k$. Thus the total probability of $h(t, z)$ matching a leaf of the tree is $\pi_{A,1} \leq (N+1)T/2^k$.
2. We run out of the chain before we run out of the tree. This means $h(t, z)$ matches one of the internal nodes of the tree, say x. This can be the case in two ways:
   (a) The left child of x contains t and the adversary uses the right child of x as z. The probability of any given node having the given value t is $1/2^k$. As there are $N - 1$ candidate nodes and N possible values of t, the total probability is $\leq (N - 1)N/2^k$.
   (b) The adversary has found a second pre-image for x. The probability of a T-time adversary achieving that for any given node is $\leq T/2^k$. As the adversary has $N - 1$ nodes as potential targets for such a hit, the total probability is $\leq (N - 1)T/2^k$.
   Thus the total probability of $h(t, z)$ matching an internal node of the tree is $\pi_{A,2} \leq (N - 1)(N + T)/2^k$.
3. We run out of the tree before we run out of the chain. This means that the adversary has found a pre-image for one of the $2N$ values $\{1, z_1, \ldots, N, z_N\}$. The probability of that is $\pi_{A,3} \leq 2NT/2^k$.
4. We encounter a hash step where the output of the step equals an internal node in the tree, say x, but the inputs of the step do not match the children of x. This means the adversary has found a second pre-image for x. The probability of that is $\pi_{A,4} \leq (N - 1)T/2^k$.

So, the total success probability of a T-time adversary who does not call $\mathcal{S}.\mathsf{Get}(t)$ is $\pi_A \leq \pi_{A,1} + \pi_{A,2} + \pi_{A,3} + \pi_{A,4} \leq (N+1)T/2^k + (N-1)(N+T)/2^k + 2NT/2^k + (N-1)T/2^k < (N^2 + 5NT)/2^k$.

Assume now that the adversary does call $\mathcal{S}.\mathsf{Get}(t)$. Then we can, without loss of generality, also assume that

- he calls $\mathcal{S}.\mathsf{Get}(t)$ only after committing r_t, as before that $\mathcal{S}.\mathsf{Get}(t)$ would always return $\perp$, which would provide no useful information;
- he calls $\mathcal{S}.\mathsf{Get}(t)$ only once, as all additional calls to $\mathcal{S}.\mathsf{Get}(t)$ would return the same result, which would provide no new information;
- he never calls $\mathcal{S}.\mathsf{Sig}(m,t)$, as he is not allowed to call $\mathcal{S}.\mathsf{Sig}(m,t)$ before calling $\mathcal{S}.\mathsf{Get}(t)$ according to the security condition, but after calling $\mathcal{S}.\mathsf{Get}(t)$ he already has z_t and can compute $h(m, z_t)$ directly with no need to call the signing oracle any more.

Finally, we can also assume that in order to win the game $\mathcal{F}$, the adversary must produce m and a such that $h(m, z_t) \overset{a}{\rightsquigarrow} r_t$. Indeed, if the adversary wins the game with $h(m, z) \overset{a}{\rightsquigarrow} r_t$ where $z \neq z_t$, then he has not used the information gained from the $\mathcal{S}.\mathsf{Get}(t)$ call and thus could not have done any better than without the call, a case we have already analyzed.

Let H be the set of h-calls $y \leftarrow h(x_1, x_2)$ the adversary made before committing r_t. As the adversary is T-time, we have $|H| \leq T$. Consider now the h-calls to be made during the computation of $h(m, z_t) \overset{a}{\rightsquigarrow} r_t$:

1. If all the calls are in H, then the adversary must have called $h(m, z_t)$ before committing r_t and thus also before learning z_t from the call to $\mathcal{S}.\mathsf{Get}(t)$. This means that the adversary guessed z_t. The probability of a T-time adversary achieving that is $\pi_{B,1} \leq T/2^k$.
2. If none of the calls are in H, then there are two possibilities:
   (a) The value r_t was not returned from any of the calls in H. This means the adversary was able to find a pre-image of r_t after committing it, the probability of which is $\leq T/2^k$.
   (b) The value r_t was returned by some call in H. Since the chain a is computed entirely using calls not in H, the inputs of the final step of the computation represent a second pre-image of r_t. The probability of a T-time adversary achieving that is also $\leq T/2^k$.
   Thus the total probability of the adversary finding a chain entirely outside of H is $\pi_{B,2} \leq 2T/2^k$.
3. Some, but not all of the calls are in H. Let's examine, among the calls that are not in H, the one made last during the computation of the chain. Let it be $y \leftarrow h(x_1, x_2)$. Again, there are two possibilities:
   (a) The value y was not returned from any of the calls in H. However, the next step in a is already a call in H. This means that y is among the inputs of calls in H and the adversary was able to find a pre-image of it. The probability of the adversary achieving that for any given y is $\leq T/2^k$. As there are $2|H|$ possible values of y, the total probability is $\leq 2|H|T/2^k$.
   (b) The value y was returned by some call in H. Since the call $y \leftarrow h(x_1, x_2)$ is not in H, the adversary must have found a second pre-image of y. The total probability of that over all available values of y is $\leq |H|T/2^k$.

Thus the probability of the adversary finding a chain entering into H is $\pi_{B,3} \leq 3|H|T/2^k \leq 3T^2/2^k$.

Hence the total success probability of a T-time adversary who calls $\mathcal{S}.\mathtt{Get}(t)$ is $\pi_B \leq \pi_{B,1} + \pi_{B,2} + \pi_{B,3} \leq T/2^k + 2T/2^k + 3T^2/2^k = (3T + 3T^2)/2^k$.

Summary. If the adversary does not call $\mathcal{S}.\mathtt{Get}(t)$, he can win the forgery game $\mathcal{F}$ with probability $\pi_A < (N^2 + 5NT)/2^k$. If he does call $\mathcal{S}.\mathtt{Get}(t)$, he can win with probability $\pi_B \leq (3T + 3T^2)/2^k$. Overall, he can win with probability $\pi = \max(\pi_A, \pi_B)$.

Since generating the N keys $z_1, \ldots, z_N$ and making the $2N - 1$ calls to h to compute $x_1, \ldots, x_N$ and $T^h(x_1, \ldots, x_N)$ is something the signers are expected to do routinely, we can assume that $N \ll T$. Already with $N < T/10$, we have $\pi_A < (N^2 + 5NT)/2^k < (T^2/100 + T^2/2)/2^k < T^2/2^k$. With $T > 10N \geq 10$, we have $T^2 > 3T$ and thus $\pi_B \leq (3T + 3T^2)/2^k < 4T^2/2^k$.

Therefore, $\pi = \max(\pi_A, \pi_B) < 4T^2/2^k$, or $T^2/\pi > 2^{k-2}$. As $\pi \leq 1$, we also have $(T/\pi)^2 \geq T^2/\pi$, which yields the claim $T/\pi > 2^{(k-2)/2}$.

7   Practical Considerations

Key Generation. In the description of the scheme we assumed that the signing keys $z_1, \ldots, z_N$ are unpredictable values drawn from $\{0,1\}^k$, but left unspecified how they might be generated in practice. Obviously they could be generated as independent truly random values, but this would be rather expensive and also would necessitate keeping a large number of secret values over a long time. It would be more practical to generate them pseudo-randomly from a single random seed s. There are several known ways of doing that:

– Iterated hashing: $z_N \leftarrow s$, $z_{i-1} \leftarrow h(z_i)$ for $i \in \{2, \ldots, N\}$.
  This idea of generating a sequence of one-time keys from a single seed is due to Lamport [32] and has also been used in the TESLA protocol by Perrin et al. [40]. Implemented this way, our scheme would also bear some resemblance to the Guy Fakes protocol by Anderson et al. [1]. Note that the keys have to be generated in reverse order, otherwise the earlier keys released as signature components could be used to derive the later ones that are still valid for signing. To be able to use the keys in the direct order, the signer would have to either remember them all, re-compute half of the sequence on average, or implement a traversal algorithm such as the one proposed by Schoenmakers [45].
– Counter hashing: $z_i \leftarrow h(s, i)$.
  With a hash function behaving as a random oracle, this scheme would generate keys indistinguishable from truly random values, but there does not appear to be much research on the security of practical hash functions when used in this mode.
– Counter encryption: $z_i \leftarrow E_s(i)$.
  The signing keys are generated by encrypting their indices with a symmetric block cipher using the seed as the encryption key. This is equivalent to

using the block cipher in the counter mode as first proposed by Diffie and Hellman [20]. The security of this mode is extensively studied and well understood for all common block ciphers. Another benefit of this approach is that it can be implemented using standard hardware security modules where the seed is kept in a protected storage and the encryption operations are performed in a security-hardened environment.

Time-Stamping. As already mentioned, we side-step the key state management problems [36] common for most N-time signing schemes by making the signing keys not one-time, but *time-bound* instead. This in turn raises the issue of clock synchronization.

We first note that even when the signer's local clock is running fast, premature key release is easy to prevent by having the signer verify the time-stamp on $h(m, z_t)$ before releasing z_t. This is how the condition $c \geq t$ of the signing oracle $\mathcal{S}$ in Fig. 3 should be implemented in practice.

The next issue is that the signer needs to select the key z_t before computing $h(m, z_t)$ and submitting it to time-stamping. If, due to clock drift or network latency, the time in the time-stamp received does not match t, the signature can't be composed. To counter clock drift and stable latency, the signer can first time-stamp a dummy value and use the result to compare its local clock to that of the time-stamping service.

To counter network jitter, the signer can compute the message authenticators $h(m, z_{t'})$ for several consecutive values of t', submit all of them in parallel, and compose the signature using the components whose t' matches the time t in the time-stamps received. Buldas et al. [12] have shown that with careful scheduling the latency can be made stable enough for this strategy even in an aggregation network with world-wide scale.

Finally, we note that time-stamping services operating in discrete aggregation rounds are particularly well suited for use in our scheme, as they only return time-stamps once the round is closed, thus eliminating the risk that a fast adversary could still manage to acquire a suitable time-stamp after the signer has released a key.

Efficiency. In the following estimates, we assume the use of SHA-256, a common 256-bit hash function. On small inputs, a moderate laptop can perform about a million SHA-256 evaluations per second. We also assume a signing key sequence containing one key per second for a year, or a total of a bit less than 32 million, or roughly 2^{25} keys.

Using the techniques described above, generation of N signing keys takes N applications of either a hash function or a symmetric block cipher. Binding them into a public key takes $2N - 1$ hashing operations. Thus, the key generation in our example takes about 100 seconds.

The resulting public key consists of just one hash value. In the private key, only the seed s has to be kept secret. The signing keys $z_1, \ldots, z_N$ can be erased once the public key has been computed, an then re-generated as needed for signing. The hash tree $T^h(x_1, \ldots, x_N)$ presents a space-time trade-off. It may be kept (in regular unprotected storage, as it contains no sensitive information), taking up $2N - 1$ nodes, or about 1 GB, and then the key authentication hash chains can be just read from the tree with no additional computations needed. Alternatively, one can use a hash tree traversal algorithm, such as the one proposed by Szydlo [47], to keep only $3 \log_2 N$ nodes of the tree and spend $2 \log_2 N$ hash function evaluations per chain extraction, assuming all chains are extracted consecutively.

The size of the signature (t, z_t, a_t, c_t) is dominated by the two hash chains. The key authentication chain consists of $\log_2 N$ hash values, for a total of about 800 B for our 1-year key sequence. The time-stamping chain consists of $\log_2 M$ hash values, where M is the number of requests received by the time-stamping service in the round t. Assuming the use of the KSI service described in [12] under its theoretical maximum load of 2^{50} requests, this adds about 1 600 B. Thus we can expect signatures of less than 3 kB.

As the verification means re-computing the hash chains, it amounts to less than a hundred hash function evaluations.

8 Conclusions and Outlook

We have presented a simple and efficient digital signature scheme built from a hash function and a hash-then-publish time-stamping scheme. Considering that the existence of hash functions is a necessary pre-condition for the existence of digital signatures [44], one could argue our scheme is based on minimal assumptions. However, there is still much room for improvement in both theoretical and practical aspects.

Current security proofs are given in the random oracle model and in the classical setting. It would be desirable to prove the security also in the standard model and in the quantum setting, in particular taking into account the effects of quantum-oracle access to the hash function [6] and possible quantum interactions between the aggregation and the hash chain extraction phases of time-stamping, as these are all under the adversary's control.

It would also be good to reduce, or at least defer, the key generation costs, perhaps by adopting some of the incremental tree generation approaches, and to develop a version of the scheme suitable for personal signing devices like smart cards and USB dongles. These devices, in addition to having significantly less memory and computational power, also lack several functional qualities of the full-sized computers: they are powered on only intermittently, and do not have on-board real-time clocks or independent network communication capabilities.

References

1. Anderson, R.J., Bergadano, F., Crispo, B., Lee, J.-H., Manifavas, C., Needham, R.M.: A new family of authentication protocols. Oper. Syst. Rev. **32**(4), 9–20 (1998)

2. Bayer, D., Haber, S., Stornetta, W.S.: Improving the efficiency and reliability of digital time-stamping. In: Capocelli, R., De Santis, A., Vaccaro, U. (eds.) Sequences II, Proceedings. LNCS, vol. 9056, pp. 329–334. Springer, Heidelberg (1992). doi:10.1007/978-1-4613-9323-8_24

3. Bellare, M., Rogaway, P.: Random oracles are practical: a paradigm for designing efficient protocols. In: ACM CCS 1993, Proceedings, pp. 62–73. ACM (1993)

4. Bernstein, D.J., Buchmann, J.A., Dahmen, E. (eds.): Post-Quantum Cryptography. Springer, Heidelberg (2009). doi:10.1007/978-3-540-88702-7

5. Bernstein, D.J., et al.: SPHINCS: practical stateless hash-based signatures. In: Oswald, E., Fischlin, M. (eds.) EUROCRYPT 2015. LNCS, vol. 9056, pp. 368–397. Springer, Heidelberg (2015). doi:10.1007/978-3-662-46800-5_15

6. Boneh, D., Dagdelen, Ö., Fischlin, M., Lehmann, A., Schaffner, C., Zhandry, M.: Random oracles in a quantum world. In: Lee, D.H., Wang, X. (eds.) ASIACRYPT 2011. LNCS, vol. 7073, pp. 41–69. Springer, Heidelberg (2011). doi:10.1007/978-3-642-25385-0_3

7. Brassard, G., Høyer, P., Tapp, A.: Quantum cryptanalysis of hash and claw-free functions. In: Lucchesi, C.L., Moura, A.V. (eds.) LATIN 1998. LNCS, vol. 1380, pp. 163–169. Springer, Heidelberg (1998). doi:10.1007/BFb0054319

8. Buchmann, J.A., Coronado García, L.C., Dahmen, E., Döring, M., Klintsevich, E.: CMSS – an improved Merkle signature scheme. In: Barua, R., Lange, T. (eds.) INDOCRYPT 2006. LNCS, vol. 4329, pp. 349–363. Springer, Heidelberg (2006). doi:10.1007/11941378_25

9. Buchmann, J.A., Dahmen, E., Ereth, S., Hülsing, A., Rückert, M.: On the security of the Winternitz one-time signature scheme. IJACT **3**(1), 84–96 (2013)

10. Buchmann, J.A., Dahmen, E., Hülsing, A.: XMSS - a practical forward secure signature scheme based on minimal security assumptions. In: Yang, B.-Y. (ed.) PQCrypto 2011. LNCS, vol. 7071, pp. 117–129. Springer, Heidelberg (2011). doi:10.1007/978-3-642-25405-5_8

11. Buchmann, J.A., Dahmen, E., Klintsevich, E., Okeya, K., Vuillaume, C.: Merkle signatures with virtually unlimited signature capacity. In: Katz, J., Yung, M. (eds.) ACNS 2007. LNCS, vol. 4521, pp. 31–45. Springer, Heidelberg (2007). doi:10.1007/978-3-540-72738-5_3

12. Buldas, A., Kroonmaa, A., Laanoja, R.: Keyless signatures' infrastructure: how to build global distributed hash-trees. In: Nielson, H.R., Gollmann, D. (eds.) NordSec 2013. LNCS, vol. 8208, pp. 313–320. Springer, Heidelberg (2013). doi:10.1007/978-3-642-41488-6_21

13. Buldas, A., Laanoja, R.: Security proofs for hash tree time-stamping using hash functions with small output size. In: Boyd, C., Simpson, L. (eds.) ACISP 2013. LNCS, vol. 7959, pp. 235–250. Springer, Heidelberg (2013). doi:10.1007/978-3-642-39059-3_16

14. Buldas, A., Laanoja, R., Laud, P., Truu, A.: Bounded pre-image awareness and the security of hash-tree keyless signatures. In: Chow, S.S.M., Liu, J.K., Hui, L.C.K., Yiu, S.M. (eds.) ProvSec 2014. LNCS, vol. 8782, pp. 130–145. Springer, Cham (2014). doi:10.1007/978-3-319-12475-9_10

15. Buldas, A., Niitsoo, M.: Optimally tight security proofs for hash-then-publish time-stamping. In: Steinfeld, R., Hawkes, P. (eds.) ACISP 2010. LNCS, vol. 6168, pp. 318–335. Springer, Heidelberg (2010). doi:10.1007/978-3-642-14081-5_20

16. Buldas, A., Saarepera, M.: On provably secure time-stamping schemes. In: Lee, P.J. (ed.) ASIACRYPT 2004. LNCS, vol. 3329, pp. 500–514. Springer, Heidelberg (2004). doi:10.1007/978-3-540-30539-2_35

17. Coronado García, L.C.: Provably secure and practical signature schemes. Ph.D. thesis, Darmstadt University of Technology, Germany (2005)
18. Dahmen, E., Okeya, K., Takagi, T., Vuillaume, C.: Digital signatures out of second-preimage resistant hash functions. In: Buchmann, J.A., Ding, J. (eds.) PQCrypto 2008. LNCS, vol. 5299, pp. 109–123. Springer, Heidelberg (2008). doi:10.1007/978-3-540-88403-3_8
19. Diffie, W., Hellman, M.E.: New directions in cryptography. IEEE Trans. Inf. Theor. 22(6), 644–654 (1976)
20. Diffie, W., Hellman, M.E.: Privacy and authentication: an introduction to cryptography. Proc. IEEE 67(3), 397–427 (1979)
21. Dods, C., Smart, N.P., Stam, M.: Hash based digital signature schemes. In: Smart, N.P. (ed.) Cryptography and Coding 2005. LNCS, vol. 3796, pp. 96–115. Springer, Heidelberg (2005). doi:10.1007/11586821_8
22. ElGamal, T.: A public key cryptosystem and a signature scheme based on discrete logarithms. IEEE Trans. Inf. Theor. 31(4), 469–472 (1985)
23. Even, S., Goldreich, O., Micali, S.: On-line/Off-line digital signatures. J. Cryptol. 9(1), 35–67 (1996)
24. Goldwasser, S., Micali, S., Rivest, R.L.: A digital signature scheme secure against adaptive chosen-message attacks. SIAM J. Comput. 17(2), 281–308 (1988)
25. Grover, L.K.: A fast quantum mechanical algorithm for database search. In: 28th ACM STOC, Proceedings, pp. 212–219. ACM (1996)
26. Haber, S., Stornetta, W.S.: How to time-stamp a digital document. J. Cryptol. 3(2), 99–111 (1991)
27. Hülsing, A.: W-OTS+ – shorter signatures for hash-based signature schemes. In: Youssef, A., Nitaj, A., Hassanien, A.E. (eds.) AFRICACRYPT 2013. LNCS, vol. 7918, pp. 173–188. Springer, Heidelberg (2013). doi:10.1007/978-3-642-38553-7_10
28. Hülsing, A., Rausch, L., Buchmann, J.A.: Optimal parameters for $XMSS^{MT}$. In: Cuzzocrea, A., Kittl, C., Simos, D.E., Weippl, E., Xu, L. (eds.) CD-ARES 2013. LNCS, vol. 8128, pp. 194–208. Springer, Heidelberg (2013). doi:10.1007/978-3-642-40588-4_14
29. Hülsing, A., Rijneveld, J., Song, F.: Mitigating multi-target attacks in hash-based signatures. In: Cheng, C.-M., Chung, K.-M., Persiano, G., Yang, B.-Y. (eds.) PKC 2016. LNCS, vol. 9614, pp. 387–416. Springer, Heidelberg (2016). doi:10.1007/978-3-662-49384-7_15
30. Johnson, D., Menezes, A., Vanstone, S.A.: The elliptic curve digital signature algorithm (ECDSA). Int. J. Inf. Secur. 1(1), 36–63 (2001)
31. Lamport, L.: Constructing digital signatures from a one way function. Technical report, SRI International, Computer Science Laboratory (1979)
32. Lamport, L.: Password authentification with insecure communication. Commun. ACM 24(11), 770–772 (1981)
33. Luby, M.: Pseudorandomness and Cryptographic Applications. Princeton University Press, Princeton (1996)
34. Malkin, T., Micciancio, D., Miner, S.: Efficient generic forward-secure signatures with an unbounded number of time periods. In: Knudsen, L.R. (ed.) EUROCRYPT 2002. LNCS, vol. 2332, pp. 400–417. Springer, Heidelberg (2002). doi:10.1007/3-540-46035-7_27
35. Martín-López, E., Laing, A., Lawson, T., Alvarez, R., Zhou, X.-Q., O'Brien, J.L.: Experimental realization of Shor's quantum factoring algorithm using qubit recycling. Nat. Photonics 6(11), 773–776 (2012)

36. McGrew, D., Kampanakis, P., Fluhrer, S., Gazdag, S.-L., Butin, D., Buchmann, J.A.: State management for hash-based signatures. In: Chen, L., McGrew, D., Mitchell, C. (eds.) SSR 2016. LNCS, vol. 10074, pp. 244–260. Springer, Cham (2016). doi:10.1007/978-3-319-49100-4_11

37. Merkle, R.C.: Secrecy, authentication and public key systems. Ph.D. thesis, Stanford University (1979)

38. Merkle, R.C.: A digital signature based on a conventional encryption function. In: Pomerance, C. (ed.) CRYPTO 1987. LNCS, vol. 293, pp. 369–378. Springer, Heidelberg (1988). doi:10.1007/3-540-48184-2_32

39. Perrig, A.: The BiBa one-time signature and broadcast authentication protocol. In: ACM CCS 2001, Proceedings, pp. 28–37. ACM (2001)

40. Perrig, A., Canetti, R., Tygar, J.D., Song, D.: The TESLA broadcast authentication protocol. CryptoBytes 5(2), 2–13 (2002)

41. Reyzin, L., Reyzin, N.: Better than BiBa: short one-time signatures with fast signing and verifying. In: Batten, L., Seberry, J. (eds.) ACISP 2002. LNCS, vol. 2384, pp. 144–153. Springer, Heidelberg (2002). doi:10.1007/3-540-45450-0_11

42. Rivest, R.L., Shamir, A., Adleman, L.M.: A method for obtaining digital signatures and public-key cryptosystems. Commun. ACM 21(2), 120–126 (1978)

43. Rohatgi, P.: A compact and fast hybrid signature scheme for multicast packet authentication. In: ACM CCS 1999, Proceedings, pp. 93–100. ACM (1999)

44. Rompel, J.: One-way functions are necessary and sufficient for secure signatures. In: 22nd ACM STOC, Proceedings, pp. 387–394. ACM (1990)

45. Schoenmakers, B.: Explicit optimal binary pebbling for one-way hash chain reversal. In: Grossklags, J., Preneel, B. (eds.) FC 2016. LNCS, vol. 9603, pp. 299–320. Springer, Heidelberg (2017). doi:10.1007/978-3-662-54970-4_18

46. Shor, P.W.: Polynomial-time algorithms for prime factorization and discrete logarithms on a quantum computer. SIAM Rev. 41(2), 303–332 (1999)

47. Szydlo, M.: Merkle tree traversal in log space and time. In: Cachin, C., Camenisch, J.L. (eds.) EUROCRYPT 2004. LNCS, vol. 3027, pp. 541–554. Springer, Heidelberg (2004). doi:10.1007/978-3-540-24676-3_32

Outsourcing of Verifiable Attribute-Based Keyword Search

Go Ohtake[1]([✉]), Reihaneh Safavi-Naini[2], and Liang Feng Zhang[3]

[1] Japan Broadcasting Corporation, Tokyo, Japan
ohtake.g-fw@nhk.or.jp
[2] University of Calgary, Calgary, Canada
[3] ShanghaiTech University, Shanghai, China

Abstract. In integrated broadcast-broadband services, viewers receive content via the airwaves as well as additional content via the Internet. The additional content can be personalized by using the viewing histories of each viewer. Viewing histories however contain private data that must be handled with care. A verifiable attribute-based keyword search (VABKS) scheme allows data users (service providers), whose attributes satisfy a policy that is specified by the data owner (viewer), to securely search and access stored data in a malicious cloud server, and verify the correctness of the operations by the cloud server. VABKS, however, requires data owners who have computationally weak terminals, such as television sets, to perform heavy computations due to the attribute-based encryption process. In this paper, we propose a new VABKS scheme where such heavy computations are outsourced to a cloud server and hence the data owner is kept as light as possible. Our scheme is provably secure against two malicious cloud servers in the random oracle model: one performing the attribute-based encryption process, and the other performing the keyword search process on the encrypted data. We implement our scheme and the previous VABKS scheme and show that our scheme significantly reduces the computation cost of the data owner.

1 Introduction

The user's history of interaction with a service provider provides a valuable source of information with which service providers can construct a user profile and offer personalized services that match the profile. This data however is private and users who are *data owners* must be able to control access to it. On the other hand, data owners likely do not have sufficient computational resources to store and control access to data, so these operations have to be delegated to a cloud server that in general cannot be trusted. In this paper, we focus on a particular type of service history and propose an architecture that allows the user's history data to be stored in the cloud server and makes it available to the service providers that the user approves, while keeping the user computation at an acceptable level by outsourcing part of it to the cloud server. This architecture and approach can be extended to other broadcast services such as online games and entertainment services.

© Springer International Publishing AG 2017
H. Lipmaa et al. (Eds.): NordSec 2017, LNCS 10674, pp. 18–35, 2017.
https://doi.org/10.1007/978-3-319-70290-2_2

Integrated broadcast-broadband services: Integrated broadcast-broadband services [5,7,8,16] allow viewers to view content via the airwaves and, simultaneously, additional content via the Internet. The additional content can be used to personalize broadcasts and provides opportunities for electronic commerce. To make the personalization of the broadcast and services effective, viewers must share their viewing preferences with the service provider. Viewing histories are a rich source of data for service providers to learn about the viewers' interests. Their data however could reveal sensitive personal information about a viewer and so must be handled with care. Ideally, viewers want to share their viewing histories with service providers that pass certain criteria, including being trustworthy or having a high rating based on customer reviews. *Attribute-based encryption (ABE)* [2,14] enables a viewer to specify a policy when encrypting their viewing history at the user terminal and store them in a cloud server, and only service providers whose attributes satisfy the policy can decrypt it.

Attribute-based keyword search: Attribute-based keyword search (ABKS) [11,12, 15] provides attribute-based access control for data and the ability to search for keywords in the encrypted domain. This allows service providers to acquire the desired viewing histories from the cloud server while viewer privacy is protected. Most of the existing ABKS schemes assume that the cloud server that performs the keyword search is *honest-but-curious*. Zheng, Xu, and Ateniese [18] proposed *verifiable attribute-based keyword search (VABKS)* that allows the cloud server to be *malicious* and gave a generic construction of VABKS that is based on an ABE scheme, an ABKS scheme, a digital signature scheme, and a Bloom filter and that enables one to verify the correctness of the keyword search result. However, the computational cost of this construction, in particular, the index generation algorithm, is rather high for weak user terminals such as television sets. This is because the index generation algorithm uses an ABE scheme and an ABKS scheme many times, which imposes heavier loads than those of conventional public key encryption schemes such as RSA and ElGamal encryption.

1.1 Our Contribution

In this paper, we propose a new VABKS scheme where the heavy computations for a data owner (viewer) are outsourced to a cloud server, while being able to verify the correctness of the computation. Our scheme is provably secure against two malicious cloud servers in the random oracle model: one performing the attribute-based encryption process, and the other performing the keyword search process on the encrypted data. We upgrade the model of VABKS in [18] with more security requirements to accommodate the additional outsourcing of the attribute-based encryption process. This outsourcing process makes our scheme more efficient than the VABKS scheme in [18].

Concretely, we follow the generic construction of [18] and use an ABE scheme and an ABKS scheme whose computations can be verifiably outsourced. This allows us to construct a VABKS scheme with a verifiably outsourceable computation (See Fig. 1). For an ABE scheme, we will use the construction of Ohtake, Safavi-Naini, and Zhang [10] that requires the data owner to only perform

ElGamal encryption and outsources the heavy computations of the ABE scheme to the cloud server. However, to the best of our knowledge, no outsourcing scheme has been proposed for ABKS. Hence, to make an ABKS scheme whose computations can be outsourced and be compatible with the ABE construction, we start by defining the model of secure outsourcing of ABKS to a malicious cloud server and then give a construction based on the ABKS scheme in [15], which is provably secure in the random oracle model. After that, we define the model of VABKS outsourcing scheme for the generic construction of VABKS [18] and show our construction of VABKS that uses the outsourceable ABE scheme [10] and our outsourceable ABKS scheme.

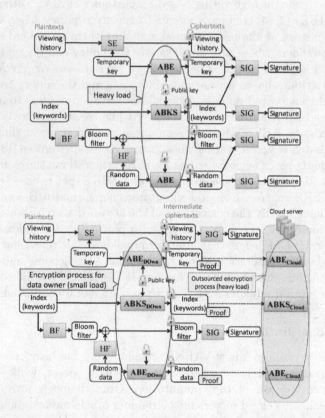

Fig. 1. Overview of index generation algorithms of VABKS scheme [18] (the upper half of this figure) and our VABKS outsourcing scheme (the lower half of this figure). SE denotes a symmetric key encryption scheme, BF denotes a Bloom filter generation algorithm, HF denotes a hash function, SIG denotes a digital signature scheme, ABE_{DOwn} and $ABKS_{DOwn}$ are part of the encryption processes of ABE and ABKS performed by the data owner, and ABE_{Cloud} and $ABKS_{Cloud}$ are the encryption processes of ABE and ABKS that are outsourced to a cloud server.

As shown in Fig. 1, the generic construction of VABKS [18] has four kinds of signature for verifying the correctness of search results from a cloud server, whereas our VABKS outsourcing scheme has three kinds of *proofs*, generated simply by using hash function, and two kinds of signature. In the VABKS scheme [18], a data user can verify that the cloud server faithfully performed the keyword search operation by using the signatures that validate the integrity of the ciphertexts of the data, temporary keys for encrypting data, keywords, Bloom filter, and random data for masking the Bloom filter. In contrast, in our VABKS outsourcing scheme, a data user can verify the correctness of the outsourced ABE/ABKS encryption process by using the proofs that validate the integrity of the ciphertexts of the temporary keys for encrypting the data, keywords, and random data for masking the Bloom filter. Namely, both signatures and proofs are used only for validating the integrity of the ciphertexts. Hence, the proofs for verifying the outsourced encryption data can also be used for verifying the search results and the number of signatures can be reduced by replacing some of them with proofs, which means that our scheme is more efficient in terms of the signing costs of the data owner.

Finally, we compare our VABKS outsourcing scheme with the previous VABKS schemes (which do not outsource the encryption processes to a cloud server) in terms of the computation cost for the index generation algorithm and show that our scheme is the most efficient VABKS scheme for the data owner. Outsourcing computations to the cloud server increases the total cost of the data owner and the cloud server. However, the cloud server usually has higher performance CPUs compared with those of the user terminal, so the total processing time for the index generation algorithm in our scheme will be smaller than that of the previous VABKS scheme. We also implement the index generation algorithms of our scheme and one of the previous VABKS schemes on a PC and show that our scheme requires only half the processing time of the previous one because it outsources part of the encryption process of ABE and ABKS to the cloud server.

1.2   Related Work

Attribute-based encryption and its outsourcing: Sahai and Waters [13] proposed the first ABE scheme as an extension of identity-based encryption (IBE). ABE schemes can be classified into (i) key-policy ABE (KP-ABE) [4] and (ii) ciphertext-policy ABE (CP-ABE) [2,14]. In KP-ABE, a ciphertext is associated with a set of attributes, and a private key is associated with a policy. In CP-ABE, a private key is associated with a set of attributes, and a ciphertext is associated with a policy. A ciphertext can be decrypted by a user whose attributes satisfy the policy that is attached to the ciphertext. In this paper, we consider CP-ABE.

Green, Hohenberger, and Waters [3] first considered outsourcing of ABE. They proposed an outsourcing scheme for ABE decryption with the goal of minimizing the users' decryption cost. After that, several outsourcing schemes for ABE encryption were proposed [6,9,17], but they all assumed *honest* or *honest-but-curious* cloud servers. In contrast, Ohtake, Safavi-Naini, and Zhang [10] proposed an outsourcing scheme of ABE encryption for when the cloud server is *malicious*. In this paper, we use the same approach as taken in [10] and extend it to include a search functionality.

Attribute-based keyword search: ABKS combines ABE with a keyword search functionality. Several ABKS schemes have been proposed [11,12,15], but they all assume the cloud server that performs keyword search to be *honest-but-curious*. In contrast, Zheng, Xu, and Ateniese [18] proposed the first VABKS scheme that assumes that the cloud server performing the keyword search is *malicious* and proposed a generic construction of VABKS that enables anyone to verify the result of the keyword search. However, the generic construction imposes high computation costs on the data owner for the index generation algorithm. In this paper, we modify the generic construction by using the idea of ABE outsourcing in [10] and construct a more efficient VABKS scheme than that in [18].

2   Preliminaries

Definition 1 *(Access structure* [1]). *Let $\mathcal{P} = \{P_1, P_2, ..., P_n\}$ be a set of parties. $\mathbb{A} \subseteq 2^{\mathcal{P}} \backslash \{\emptyset\}$, a collection of non-empty subsets of $\mathcal{P}$, is a monotone access structure if $B \in \mathbb{A}$ and $B \subseteq C$, then $C \in \mathbb{A}$ ($\forall B, C$). The sets in $\mathbb{A}$ are called authorized sets, and the sets not in $\mathbb{A}$ are called unauthorized sets.*

In our context, attributes play the role of parties in the secret sharing scheme. Thus, the access structure $\mathbb{A}$ contains the authorized sets of attributes.

Definition 2 *(Linear secret sharing schemes (LSSS)* [14]). *A secret-sharing scheme Π over a set of ℓ parties $\mathcal{P}$ is called linear (over $\mathbb{Z}_p$) if,*

1. *The shares of the parties form a vector of length ℓ over $\mathbb{Z}_p$.*
2. *There exists a matrix M with ℓ rows and n columns, called the share-generating matrix for Π, and a function ρ which maps each row of the matrix to an associated party. That is, for $i = 1, ..., \ell$, the value $\rho(i)$ is the party associated with row i. If we consider a column vector $v = (s, r_2, ..., r_n)$, where $s \in \mathbb{Z}_p$ is the secret to be shared, and $r_2, ..., r_n \in \mathbb{Z}_p$ are randomly chosen, then Mv is the vector of ℓ shares of the secret s according to Π. The share $(Mv)_i$ belongs to party $\rho(i)$.*

It is shown in [1] that every linear secret sharing scheme having the above definition also enjoys the *linear reconstruction* property, defined as follows: Suppose that Π is a LSSS for the access structure $\mathbb{A}$. Let $S \in \mathbb{A}$ be any authorized set, and let $I \subset \{1, 2, ..., \ell\}$ be defined as $I = \{i : \rho(i) \in S\}$. Then, there exist constants $\{w_i \in \mathbb{Z}_p\}_{i \in I}$ such that, if $\{\lambda_i\}$ are valid shares of any secret s according to Π, then $\sum_{i \in I} w_i \lambda_i = s$. Furthermore, it is shown in [1] that these constants $\{w_i\}$ can be found in polynomial time to the size of the share-generating matrix M.

Definition 3 *(Bilinear maps)*. *Let $\mathbb{G}$ and $\mathbb{G}_T$ be two multiplicative cyclic groups of prime order p. Let g be a generator of $\mathbb{G}$ and $e : \mathbb{G} \times \mathbb{G} \to \mathbb{G}_T$ be a bilinear map that has the following properties: (Bilinearity) $e(u^a, v^b) = e(u, v)^{ab}$ for all $u, v \in \mathbb{G}$ and $a, b \in \mathbb{Z}_p$, (Non-degeneracy) $e(g, g) \neq 1$.*

We say that $\mathbb{G}$ is a bilinear group if the group operation in $\mathbb{G}$ and the bilinear map $e : \mathbb{G} \times \mathbb{G} \rightarrow \mathbb{G}_T$ are both efficiently computable. Notice that the map e is symmetric, since $e(g^a, g^b) = e(g, g)^{ab} = e(g^b, g^a)$.

3   System Model

The system model of our VABKS outsourcing scheme is shown in Fig. 2. There are five entities: the trusted authority (TA), a data owner (DOwn), two untrusted cloud servers (Cloud 1 and Cloud 2) with a public storage that can be written to by only Cloud 1 and is accessible to public (including Cloud 2), and a data user (DUsr). TA issues secret keys to DUsrs according to their attributes. DOwn performs a basic encryption of their data files as well as the related keywords (index), and uses Cloud 1 to perform the remaining computation of generating attribute-based ciphertexts for the files and the keywords. Cloud 1 stores the result in a public storage. When DUsr wants to files that are attached to a particular keyword, they create a token for the keyword search and send it to Cloud 2, who will perform the search over the encrypted keywords and gives the corresponding encrypted files to DUsr, iff the set of attributes of DUsr satisfies the policy attached to the encrypted keywords. DUsr can verify the correctness of the search result and can decrypt the data with their secret key. We assume that Cloud 1 and Cloud 2 are *malicious* and may not follow the protocol.

In the case of integrated broadcast-broadband services, DOwn is a viewer that holds their viewing history and DUsr is a service provider that uses it for

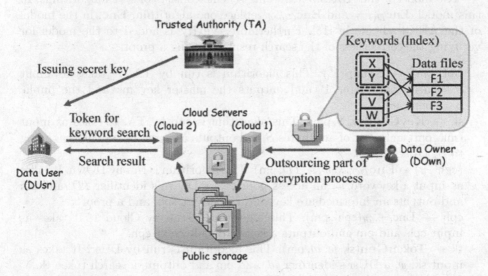

Fig. 2. System model of our VABKS outsourcing scheme. The data file F1 is related to the keywords X and V, the data file F2 is related to the keywords Y , and the data file F3 is related to the keywords X and W. The keywords X and Y are encrypted with access policy 1, and the keywords V and W are encrypted with access policy 2.

their service. The index generation for the viewing history imposes a heavy load on DOwn, so part of the above process is outsourced to Cloud 1.

4   ABKS Outsourcing Scheme

Our ABKS outsourcing scheme is based on the idea of ABE outsourcing in [10]. The generic construction of the VABKS scheme in [18] uses an ABKS scheme as a building block. In particular, the index generation algorithm of the VABKS scheme uses the keyword encryption algorithm of ABKS many times, which imposes a heavy load on DOwn. Therefore, we outsource part of the encryption process to a cloud server. First, we define the model and security of the ABKS outsourcing scheme by modifying those of the ABKS scheme in [18]. After that, we describe our construction. To achieve expressive policy settings by using a LSSS and provable security in the random oracle model, our scheme is based on the ABKS scheme in [15].

4.1   Model of ABKS Outsourcing Scheme

The model of the ABKS outsourcing scheme is based on the model of the ABKS scheme in [18], where a cloud server can see if an encrypted keyword, called a "keyword ciphertext", corresponds to the search token from DUsr. Multiple keyword ciphertexts are attached to an encrypted data file, but the data file is outside the scope of this model.

The model of the ABKS outsourcing scheme consists of seven algorithms. In this model, $\mathbf{Enc}_{DOwn}$ and $\mathbf{Enc}_{Cloud}$ replace one algorithm, $\mathbf{Enc}$, in the model of the ABKS scheme in [18]. Furthermore, $\mathbf{Verify}$ is added to the model for verifying the correctness of the search result by using a proof.

- $(\mathsf{mk}, \mathsf{pm}) \leftarrow \mathbf{Setup}(1^\ell)$: This algorithm is run by TA. It takes as input a security parameter 1^ℓ and outputs the master key mk and the public parameter pm.
- $\mathsf{sk} \leftarrow \mathbf{KeyGen}(\mathsf{mk}, S, \mathsf{pm})$: This algorithm is run by TA. It takes as input $(\mathsf{mk}, \mathsf{pm})$ and a set of attributes S and outputs a secret key sk corresponding to S. DUsr gets sk.
- $(\mathsf{cph}', \pi) \leftarrow \mathbf{Enc}_{DOwn}(w, \mathbb{A}, ID, \mathsf{pm})$: This algorithm is run by DOwn. It takes as input a keyword w, an access policy $\mathbb{A}$, a DOwn's identifier ID, and pm and outputs an intermediate keyword ciphertext cph' and a proof π.
- $\mathsf{cph} \leftarrow \mathbf{Enc}_{Cloud}(\mathsf{cph}', \mathsf{pm})$: This algorithm is run by Cloud 1. It takes as input cph' and pm and outputs a keyword ciphertext cph.
- $\mathsf{tk} \leftarrow \mathbf{TokenGen}(\mathsf{sk}, w, id, \mathsf{pm})$: This algorithm is run by DUsr. It takes as input sk, w, a DUsr's identifier id, and pm and outputs a search token tk.
- $(\{0, 1\}, \mathsf{aux}) \leftarrow \mathbf{Search}(\mathsf{cph}, \mathsf{tk}, ID, \mathsf{pm})$: This algorithm is run by Cloud 2. It takes as input cph, tk, ID, and pm and outputs $(1, \mathsf{aux})$ if (i) S satisfies $\mathbb{A}$ and (ii) cph and tk correspond to the same keyword, where aux is auxiliary data for verifying the search result; otherwise, it outputs $(0, \mathsf{aux})$.

– $\{0, 1\} \leftarrow$ **Verify**(cph, tk, aux, π, pm): This algorithm is run by DUsr. It takes as input cph, tk, aux, π, and pm and outputs 1 if π is a valid proof of cph; otherwise, it outputs 0.

4.2 Security Definition of ABKS Outsourcing Scheme

We define the security of the ABKS outsourcing scheme by modifying the security of the ABKS scheme in [18]. The model of the ABKS outsourcing scheme in Sect. 4.1 is created by the combination of those of the ABKS scheme in [18] and the ABE outsourcing scheme in [10], so we consider the security of the ABKS outsourcing scheme based on those in [10,18]. We assume that DOwn is *honest*, DUsr is *honest-but-curious*, Cloud 1 is *malicious*, and Cloud 2 is *honest-but-curious*, as in the security models of the ABE outsourcing scheme in [10] and ABKS scheme in [18]. We define three kinds of security as follows (we omit the formal security definitions because of limited space and will show them in the full version of this paper):

(1) *Selective security against chosen-keyword attack:* Without being given any matching search token, an adversary (Cloud 1) cannot infer any information about the plaintext keyword of an intermediate keyword ciphertext in the selective security model. This security requirement is a modification of the one in [18]: in the challenge phase of our scheme, an adversary gets a tuple of an intermediate keyword ciphertext and a proof. This modification is due to outsourcing the encryption process to Cloud 1.

(2) *Keyword secrecy:* The probability that an adversary (Cloud 2) learns the keyword from the intermediate keyword ciphertext and search tokens is negligibly more than the probability of a correct random keyword guess. This security requirement is a modification of the one in [18]: in the challenge phase of our scheme, an adversary gets a tuple of an intermediate keyword ciphertext and a proof as well as a related token. This modification is due to outsourcing the encryption process to Cloud 1 (Note that Cloud 1 and Cloud 2 might collude with each other).

(3) *Unforgeability:* Given an intermediate keyword ciphertext, an adversary (Cloud 1) cannot create a keyword ciphertext that corresponds to a different keyword from the original one. This is a security requirement based on [10], not in [18].

The above security definitions assume that the adversaries that break the ABKS outsourcing scheme are Cloud 1 and Cloud 2, and they collude with each other. Intuitively, (1) and (3) are the securities related to the encryption process outsourced to Cloud 1, and (2) is the security related to the search process run by Cloud 2. Namely, (1) ensures that Cloud 1 cannot get any information on the keyword from the intermediate keyword ciphertexts that DOwn sent to it, (2) ensures that Cloud 2 cannot get any information on the keyword from the search tokens that DUsr sent to it, and (3) ensures that Cloud 1 cannot modify a keyword ciphertext by using the intermediate ciphertexts that DOwn sent to it.

4.3   Our ABKS Outsourcing Scheme

To reduce the computation cost for DOwn, we use the idea of [10] to outsource part of the keyword encryption process to Cloud 1. Our scheme is based on the ABKS scheme in [15], which achieves expressive policy settings by using a LSSS and is likely provably secure in the random oracle model (although there is no security proof provided in [15]).

Setup(1^ℓ): TA selects a bilinear group $\mathbb{G}$ of prime order p, a bilinear map $e : \mathbb{G} \times \mathbb{G} \to \mathbb{G}_T$, a generator g of $\mathbb{G}$, hash functions $H : \{0,1\}^* \to \mathbb{G}$, $H' : \{0,1\}^* \times \mathbb{G}_T \times \{0,1\}^* \to \{0,1\}^*$, $H'' : \{0,1\}^* \to \{0,1\}^*$, and a message authentication code function $F : \mathbb{G}_T \times \{0,1\}^* \to \{0,1\}^m$, where m is the length of the message authentication code. It also chooses random numbers $\alpha, a \in \mathbb{Z}_p$. TA sets the public parameter and the master key as $\mathsf{pm} = \langle g, g^a, e(g,g)^\alpha, F(\cdot,\cdot), H(\cdot), H'(\cdot,\cdot,\cdot), H''(\cdot) \rangle$ and $\mathsf{mk} = \alpha$.

KeyGen(mk, S, pm): DUsr sends its identifier id and a set of its attributes S to TA. TA chooses a random value $t \in \mathbb{Z}_p$, creates $K = g^\alpha g^{at}$, $L = g^t$, and $\{K_x = H(x)^t\}_{x \in S}$, and adds an entry (id, g^{at}) to the user list. TA sets a secret key as $\mathsf{sk} = \langle K, L, \{K_x\}_{x \in S} \rangle$ and sends it to DUsr.

Enc$_{DOwn}$(w, (M, ρ), ID, pm): This is an algorithm for encrypting a keyword w with a policy (M, ρ). Here, M is an access matrix and ρ is a function that associates the rows of M with attributes. DOwn randomly chooses $s, y_2, ..., y_n$, $\beta_1, \beta_2, ..., \beta_n \in \mathbb{Z}_p$ and sets a column vector $v = (s + \beta_1, y_2 + \beta_2, ..., y_n + \beta_n) \in \mathbb{Z}_p^n$. It then calculates $C' = g^s$ and $\hat{k} = e(g,g)^{\alpha s} e(g, H(w))^s$. After that, it chooses a random bit string $\hat{t}$ and sets the index of the keyword w: $idx(w) = (\hat{t}, F(\hat{k}, \hat{t}))$. After that, it calculates a proof $\pi = H'(ID, \hat{k}, H''(\mathsf{pos}_M))$. Here, ID is an identifier of DOwn, and pos_M is a string including all of the positions of 1 in the access matrix M. For example, if the matrix M is

$$M = \begin{pmatrix} 0 & 1 \\ 1 & 1 \end{pmatrix},$$

then $\mathsf{pos}_M = \{(1,2), (2,1), (2,2)\}$. For $1 \le i \le \ell$, let J_i be a set $J_i = \{j : M_{ij} = 1(1 \le j \le n)\}$. DOwn calculates $E_i = g^{a \sum_{j \in J_i} \beta_j}$ and sets an intermediate ciphertext as $\mathsf{cph'} = \langle C', (E_i)_{1 \le i \le \ell}, v, (M, \rho), idx(w) \rangle$. It sends $(ID, \mathsf{cph'}, \pi)$ to Cloud 1.

Enc$_{Cloud}$($\mathsf{cph'}$, pm): For $1 \le i \le \ell$, Cloud 1 calculates $\lambda_i = M_i v$, where M_i is the row vector corresponding to the ith row of M. In addition, it chooses random numbers $r_1, ..., r_\ell \in \mathbb{Z}_p$. It then calculates

$$C_i = \frac{g^{a\lambda_i} H(\rho(i))^{-r_i}}{E_i}, \quad D_i = g^{r_i} \quad (1 \le i \le \ell)$$

and sets a ciphertext as $\mathsf{cph} = \langle C', (C_i, D_i)_{1 \le i \le \ell}, (M, \rho), idx(w) \rangle$. Cloud 1 stores (ID, cph, π) in a public database.

TokenGen(sk, w, id, pm): DUsr chooses a random value $u \in \mathbb{Z}_p$ and computes $q_u = g^u$. It sends id and q_u to TA. Then, TA retrieves g^{at} according to id, generates $q_{id} = g^{at} q_u^\alpha$, and sends it to DUsr. DUsr calculates

$T_q(w) = H(w)q_{id}^{1/u}$, $L' = L^{1/u}$, and $K'_x = K_x^{1/u}$ ($\forall x \in S$), and sets a search token as $\mathsf{tk} = \langle T_q(w), L', \{K'_x\}_{x \in S} \rangle$. It sends the token to Cloud 2.

Search(cph, tk, ID, pm): Cloud 2 outputs $(0, \bot)$ if S does not satisfy (M, ρ). Otherwise, let $I \subset \{1, 2, ..., \ell\}$ be defined as $I = \{i : \rho(i) \in S\}$ and $\{\mu_i \in \mathbb{Z}_p\}_{i \in I}$ be a set of constants such that if $\{\lambda_i\}$ are valid shares of any secret s according to M, then $\sum_{i \in I} \lambda_i \mu_i = s$. Then, Cloud 2 calculates

$$aux_1 = \prod_{i \in I} \left(e(C_i, L')e(D_i, K'_{\rho(i)}) \right)^{\mu_i}, \quad k = \frac{e(C', T_q(w))}{aux_1}.$$

Cloud 2 sets $\mathsf{aux} = (aux_1, ID)$. If $F(k, \hat{t}) = F(\hat{k}, \hat{t})$, Cloud 2 outputs $(1, \mathsf{aux})$. Otherwise, it outputs $(0, \mathsf{aux})$.

Verify(cph, tk, aux, π pm): DUsr outputs 0 if $\mathsf{aux} = \bot$. Otherwise, it outputs 1 if

$$\pi = H'\left(ID, \frac{e(C', T_q(w))}{aux_1}, H''(\mathsf{pos}_M) \right).$$

Otherwise, it outputs 0.

Remark 1. DOwn can verify the correctness of a tuple of (ID, cph, π) stored in a public database if DOwn keeps the random number s that is generated in the Enc_{DOwn} algorithm. Namely, even if Cloud 1 creates another tuple of $(ID, \mathsf{cph}^*, \pi^*)$ from scratch and stores it in the public database, DOwn can detect the attack (although this requires DOwn to check the status of the database periodically).

Theorem 1. *The security of the above ABKS outsourcing scheme is as follows:*

- *It is selectively secure against chosen-keyword attack in the random oracle model if the decisional q-parallel DBHE assumption holds.*
- *It has keyword secrecy in the random oracle model if H is a one-way hash function.*
- *It is unforgeable in the random oracle model if H is collision-resistant.*

We omit the proof of Theorem 1 because of limited space (We will show it in the full version of this paper).

5 VABKS Outsourcing Scheme

The generic construction of VABKS in [18] is composed of ABE, ABKS, digital signatures, and a Bloom filter. However, ABE and ABKS imposes heavier loads than those of conventional public key encryption schemes such as RSA and ElGamal encryption. In particular, the keyword encryption process using ABE and ABKS is performed by DOwn, whose device (e.g. television set) might have a low-performance CPU. To reduce the burden on such devices, we make it so that part of the encryption process of ABE and ABKS by DOwn is outsourced to Cloud 1.

5.1   Model of VABKS Outsourcing Scheme

Let $\mathsf{FS} = \{\mathsf{F}_1, ..., \mathsf{F}_n\}$ be a set of data files. Let KG_j $(1 \leq j \leq l)$ be a set of keywords (called keyword group) that will be encrypted with an access policy $\mathbb{A}_j$. That is, $\mathbb{A}_1$, ..., $\mathbb{A}_l$ are assigned to KG_1, ..., KG_l, respectively. Let $\mathsf{KG} = \{\mathsf{KG}_1, ..., \mathsf{KG}_l\}$. For each keyword w, let $\mathsf{MP}(w)$ be the set of identifiers of data files that contain keywords w. Let $\mathsf{MP} = \{\mathsf{MP}(w)|w \in \bigcup_{i=1}^{l} \mathsf{KG}_i\}$. Let $\mathsf{D} = (\mathsf{KG}, \mathsf{MP}, \mathsf{FS})$ denote data files with keywords and identifiers.

The VABKS outsourcing scheme consists of the following seven algorithms. In this model, **BuildIndex**$_{DOwn}$ and **BuildIndex**$_{Cloud}$ replace one algorithm, **BuildIndex**, in the model of the VABKS scheme in [18].

- $(\mathsf{mk}, \mathsf{pm}) \leftarrow \mathbf{Init}(1^\ell)$: This algorithm is run by TA. It takes as input a security parameter 1^ℓ and outputs the master key mk and the public parameter pm.
- $\mathsf{sk} \leftarrow \mathbf{KeyGen}(\mathsf{mk}, S, \mathsf{pm})$: This algorithm is run by TA. It takes as input $(\mathsf{mk}, \mathsf{pm})$ and a set of attributes S for DUsr and outputs a secret key sk corresponding to S.
- $(\mathsf{Au}', \mathsf{Index}', \mathsf{D}'_{\mathsf{cph}}) \leftarrow \mathbf{BuildIndex}_{DOwn}(\{\mathbb{A}\}_l, \{\mathbb{A}'\}_n, \mathsf{D}, \mathsf{pm})$: This algorithm is run by DOwn. It takes as input a set of access policies $\{\mathbb{A}\}_l = \{\mathbb{A}_1, ..., \mathbb{A}_l\}$ for encrypting the l keyword groups $\mathsf{KG}_1; ..., \mathsf{KG}_l$ respectively, a set of access policies $\{\mathbb{A}'\}_n = \{\mathbb{A}'_1, ..., \mathbb{A}'_n\}$ for encrypting the n data files $\mathsf{FS}_1, ..., \mathsf{FS}_n$ respectively, data D, and pm. It outputs intermediate auxiliary information Au', an intermediate index ciphertext Index' that includes encrypted keywords related to data files, and an intermediate data ciphertext $\mathsf{D}'_{\mathsf{cph}}$ that includes encrypted data files.
- $(\mathsf{Au}, \mathsf{Index}, \mathsf{D}_{\mathsf{cph}}) \leftarrow \mathbf{BuildIndex}_{Cloud}(\mathsf{Au}', \mathsf{Index}', \mathsf{D}'_{\mathsf{cph}}, \mathsf{pm})$: This algorithm is run by Cloud 1. It takes as input Au', Index', $\mathsf{D}'_{\mathsf{cph}}$, and pm and outputs auxiliary information Au, an index ciphertext Index, and data ciphertext $\mathsf{D}_{\mathsf{cph}}$.
- $\mathsf{tk} \leftarrow \mathbf{TokenGen}(\mathsf{sk}, w, \mathsf{pm})$: This algorithm is run by DUsr. It takes as input sk, a keyword w, and pm and outputs a search token tk.
- $(\mathsf{rslt}, \mathsf{proof}) \leftarrow \mathbf{SearchIndex}(\mathsf{Au}, \mathsf{Index}, \mathsf{D}_{\mathsf{cph}}, \mathsf{tk}, \mathsf{pm})$: This algorithm is run by Cloud 2. It takes as input Au, Index, $\mathsf{D}_{\mathsf{cph}}$, tk, and pm and outputs a search result rslt and a proof proof.
- $\{0,1\} \leftarrow \mathbf{Verify}(\mathsf{sk}, w, \mathsf{tk}, \mathsf{rslt}, \mathsf{proof}, \mathsf{pm})$: This algorithm is run by DUsr. It takes as input sk, w, tk, rslt, proof, and pm and outputs 1 if $(\mathsf{rslt}, \mathsf{proof})$ is valid and 0 otherwise.

5.2   Security Definition of VABKS Outsourcing Scheme

The security of the VABKS outsourcing scheme is a modification of the security of the VABKS scheme in [18]. We assume that DOwn is *honest*, DUsr is *honest-but-curious*, and Cloud 1 and Cloud 2 are *malicious*. Note that Cloud 2 is assumed to be honest-but-curious in the security definition of ABKS outsourcing scheme in Sect. 4.2. This is because an ABKS scheme, where a cloud server who performs keyword search is assumed to be honest-but-curious, can be used as one of the building blocks to construct a VABKS scheme (See [18]). We define four kinds of security as follows (we omit the formal security definitions because of limited space and will show them in the full version of this paper):

(1) *Data secrecy:* Given encrypted keywords and search tokens, an adversary (Cloud 2) cannot learn any information about the data files. This definition can be formalized as a chosen-plaintext security game, where two challenges $D_0 = (\mathsf{KG}, \mathsf{MP}, \mathsf{FS}_0)$ and $D_1 = (\mathsf{KG}, \mathsf{MP}, \mathsf{FS}_1)$ correspond to the same KG and MP, and $|\mathsf{FS}_0| = |\mathsf{FS}_1|$. This security requirement is the same as that in [18].

(2) *Selective security against chosen-keyword attack:* Same as the *selective security against chosen-keyword attack* of ABKS (See Sect. 4.2).

(3) *Keyword secrecy:* Same as the *keyword secrecy* of ABKS (See Sect. 4.2).

(4) *Verifiability:* If an adversary (Cloud 1) illegally modified a keyword ciphertext or an adversary (Cloud 2) returns an incorrect search result, it can be detected by Dusr with an overwhelming probability. This security definition is a modification of verifiability in [18]: in the setup phase of our scheme, an adversary gets a tuple of an intermediate index ciphertext and an intermediate data ciphertext as well as intermediate auxiliary information. This modification is due to outsourcing the encryption process to Cloud 1.

In the above security definitions, we assume that the adversaries that attempt to break the VABKS outsourcing scheme are Cloud 1 and Cloud 2, and they collude with each other. Intuitively, (2) and (4) are related to the encryption process outsourced to Cloud 1, and (1), (3), and (4) are related to the search process run by Cloud 2. Namely, (1) ensures that Cloud 2 can get no information of the data file from the keyword ciphertexts in the public storage and the search tokens that DUsr sent to it, (2) ensures that Cloud 1 can get no information of the keyword from the intermediate keyword ciphertexts that DOwn sent to it, (3) ensures that Cloud 2 can get no information of the keyword from the search tokens that DUsr sent to it, and (4) ensures that Cloud 1 cannot modify the keyword ciphertext and Cloud 2 cannot modify the search result, by using the intermediate ciphertexts that DOwn sent to it.

5.3 Our VABKS Outsourcing Scheme

We construct our VABKS outsourcing scheme by combining the generic construction of VABKS in [18] with the idea of outsourcing attribute-based encryption processes to Cloud 1. Concretely, we replace ABE.Enc in the **BuildIndex** algorithm in [18] by $\mathsf{Encrypt}_u$ of the ABE outsourcing scheme in [10] and ABKS.Enc by Enc_{DOwn} of the ABKS outsourcing scheme in Sect. 4.3. These outsourcing processes reduce the computation cost for DOwn. In addition, four kinds of signature are generated in the **BuildIndex** algorithm in [18], which imposes a heavy load on DOwn. Therefore, we try to reduce these signatures as much as possible by using a proof in the ABE outsourcing scheme and the ABKS outsourcing scheme.

Init(1^ℓ): Given a security parameter ℓ, TA chooses k universal hash functions H'_1, ..., H'_k, which are used to construct an m-bit Bloom filter.

Let $H : \{0,1\}^\ell \to \{0,1\}^m$ be a secure pseudorandom generator, SE be a secure symmetric encryption scheme, Sig be a secure signature scheme, ABEO be the ABE outsourcing scheme in [10], and ABKSO be the ABKS outsourcing scheme in Sect. 4.3. TA executes (ABEO.pm, ABEO.mk) $\leftarrow$ ABEO.Setup(1^ℓ) and (ABKSO.pm, ABKSO.mk) $\leftarrow$ ABKSO.Setup(1^ℓ). It sets the public parameter and the master key as pm = (ABEO.pm, ABKSO.pm, $H_1', ..., H_k'$) and mk = (ABEO.mk, ABKSO.mk).

KeyGen(mk, S, pm): TA runs ABEO.sk $\leftarrow$ ABEO.KeyGen(ABEO.pm, ABEO.mk, S) and ABKSO.sk $\leftarrow$ ABKSO.KeyGen (ABKSO.pm, ABKSO.mk, S), sets sk = (ABEO.sk, ABKSO.sk), and sends sk to DUsr over an authenticated private channel.

BuildIndex$_{DOwn}$($\{\mathbb{A}\}_l, \{\mathbb{A}'\}_n$, D, pm): DOwn runs (Sig.sk, Sig.pk) $\leftarrow$ Sig.KeyGen(1^ℓ), keeps Sig.sk private, and makes Sig.pk public. Given D = (KG = $\{KG_1, ..., KG_l\}$, MP = $\{MP(w) \mid w \in \bigcup_{i=1}^{l} KG_i\}$, FS = $\{F_1, ..., F_n\}$), DOwn performs the following actions:

1. Encrypt each data file with hybrid encryption: $\forall F_j \in$ FS, generate an intermediate ciphertext cph$'_{F_j}$ = (cph$'_{sk_j}$, cph$_{SE_j}$) by running SE.sk$_j \leftarrow$ SE.KeyGen(1^ℓ), cph$_{SE_j}$ $\leftarrow$ SE.Enc(SE.sk$_j$, F_j), and (cph$'_{sk_j}$, π_{sk_j}) $\leftarrow$ ABEO.Encrypt$_u$(pm, ID$_{DOwn}$, $\mathbb{A}'_j$, SE.sk$_j$), where ID$_{DOwn}$ is an identifier of DOwn. In addition, generate $\sigma_{SE_j} \leftarrow$ Sig.Sign (Sig.sk, cph$_{SE_j}$)
2. Encrypt each keyword: Given KG_i, $1 \leq i \leq l$, for each $w \in KG_i$, run (cph$'_w$, π_w) $\leftarrow$ ABKSO.Enc$_{DOwn}$(w, $\mathbb{A}_i$, ID$_{DOwn}$, pm), and set MP(cph$'_w$) = $\{$ID$_{cph'_{F_j}}$|ID$_{F_j} \in$ MP(w)$\}$, where ID$_{F_j}$ and ID$_{cph'_{F_j}}$ are identifiers for identifying data file F_j and intermediate data ciphertext cph$'_{F_j}$, respectively.
3. Generate a Bloom filter for each group KG_i: Let BF$_i \leftarrow$ BFGen($\{H_1', ..., H_k'\}$, KG_i), (cph$'_{BF_i}$, π_{BF_i}) $\leftarrow$ ABEO.Encrypt$_u$(pm, ID$_{DOwn}$, $\mathbb{A}_i$, M) for some randomly chosen M from the message space of ABEO, compute BF$'_i = H$(M) $\oplus$ BF$_i$, and generate $\sigma_{BF_i} \leftarrow$ Sig.Sign(Sig.sk, BF$'_i$).
4. Let Au$'$ = (ID$_{DOwn}$, cph$'_{BF_1}$, ..., cph$'_{BF_l}$, σ_{BF_1}, ..., σ_{BF_l}, σ_{SE_1}, ..., σ_{SE_n}, π_{sk_1}, ..., π_{sk_n}, $\{\pi_w | w \in \bigcup_{i=1}^{l} KG_i\}$, π_{BF_1}, ..., π_{BF_l}, BF$'_1$, ..., BF$'_l$), Index$'$ = ($\{$cph$'_w | w \in \bigcup_{i=1}^{l} KG_i\}$, $\{$MP(cph$'_w$)$| w \in \bigcup_{i=1}^{l} KG_i\}$), and D$'_{cph}$ = ($\{$cph$'_{F_j}|F_j \in$ FS$\}$).

BuildIndex$_{Cloud}$(Au$'$, Index$'$, D$'_{cph}$, pm): Cloud 1 runs cph$_{sk_j}$ $\leftarrow$ ABEO.Encrypt$_c$(pm, cph$'_{sk_j}$), cph$_w \leftarrow$ ABKSO.Enc$_{Cloud}$(cph$'_w$, pm), and cph$_{BF_i}$ $\leftarrow$ ABEO.Encrypt$_c$(pm, cph$'_{BF_i}$). Set MP(cph$_w$) = $\{$ID$_{cph_{F_j}}$ $\mid$ ID$_{cph'_{F_j}} \in$ MP(cph$'_w$)$\}$. Let Au = (ID$_{DOwn}$, cph$_{BF_1}$, ..., cph$_{BF_l}$, σ_{BF_1}, ..., σ_{BF_l}, σ_{SE_1}, ..., σ_{SE_n}, π_{sk_1}, ..., π_{sk_n}, $\{\pi_w | w \in \bigcup_{i=1}^{l} KG_i\}$, π_{BF_1}, ..., π_{BF_l}, BF$'_1$, ..., BF$'_l$), Index = ($\{$cph$_w | w \in \bigcup_{i=1}^{l} KG_i\}$, $\{$MP(cph$_w$)$| w \in \bigcup_{i=1}^{l} KG_i\}$), and D$_{cph}$ = ($\{$cph$_{F_j}|F_j \in$ FS$\}$).

TokenGen(sk, w, pm): DUsr generates a search token tk $\leftarrow$ ABKSO.TokenGen(ABKSO.sk, w, ID$_{DUsr}$, pm), where ID$_{DUsr}$ is an identifier of DUsr.

SearchIndex(Au, Index, D$_{cph}$, tk, pm): Let rslt and proof initially be empty sets. Cloud 2 enumerates $\prod_i = \{$cph$_w | w \in KG_i\}, 1 \leq i \leq l$, which are the keyword ciphertexts with the same access control policy.

- For each $\mathsf{cph}_w \in \prod_i$, it runs $(\gamma, \mathsf{aux}) \leftarrow \mathsf{ABKSO.Search}(\mathsf{cph}_w, \mathsf{tk}, \mathsf{ID}_{DOwn}, \mathsf{pm})$. If $\gamma = 0$, it continues to process the next keyword ciphertext in $\prod_i$. If $\gamma = 1$, it adds the tuple $(\mathsf{ID}_{DOwn}, \mathsf{cph}_w, \{\mathsf{cph}_{\mathsf{F}_j} | \mathsf{ID}_{\mathsf{cph}_{\mathsf{F}_j}} \in \mathsf{MP}(\mathsf{cph}_w)\})$ to rslt and $(\pi_w, \mathsf{cph}_{\mathsf{BF}_i})$ to proof.
- If there is no $\gamma = 1$ after processing all cph_w in $\prod_i$, it adds $(\mathsf{BF}'_i, \mathsf{cph}_{\mathsf{BF}_i}, \sigma_{\mathsf{BF}_i}, \pi_{\mathsf{BF}_i})$ to proof.

Verify$(\mathsf{sk}, w, \mathsf{tk}, \mathsf{rslt}, \mathsf{proof}, \mathsf{pm})$: DUsr verifies the search result from Cloud 2 as follows:

1. For $i = 1, ..., l$, it verifies that the cloud indeed returned the correct result for each keyword group i as follows:

Case 1: If $(\mathsf{ID}_{DOwn}, \mathsf{cph}_w, \{\mathsf{cph}_{\mathsf{F}_j} | \mathsf{ID}_{\mathsf{cph}_{\mathsf{F}_j}} \in \mathsf{MP}(\mathsf{cph}_w)\}) \in \mathsf{rslt}$, meaning that a keyword ciphertext cph_w exists which corresponds to the same access control policy as specified by $\mathsf{cph}_{\mathsf{BF}_i}$ and having the same keyword specified by tk, it runs $(\gamma, \mathsf{aux}) \leftarrow \mathsf{ABKSO.Search}(\mathsf{cph}_w, \mathsf{tk}, \mathsf{ID}_{DOwn}, \mathsf{pm})$ and $\gamma' \leftarrow \mathsf{ABKSO.Verify}(\mathsf{cph}_w, \mathsf{tk}, \mathsf{aux}, \pi_w, \mathsf{pm})$ to verify whether cph_w matches tk and is not modified. If either $\gamma = 0$ or $\gamma' = 0$, it returns 0. Otherwise, it runs $\{\mathsf{SE.sk}_j / \perp\} \leftarrow \mathsf{ABEO.Dec}(\mathsf{cph}_{\mathsf{sk}_j}, \pi_{\mathsf{sk}_j}, \mathsf{ABEO.sk}, \mathsf{pm}, \mathsf{ID}_{DOwn})$. If the output is $\perp$, it returns 0. Otherwise, it runs $\gamma'' \leftarrow \mathsf{Sig.Verify}(\mathsf{Sig.pk}, \sigma_{\mathsf{SE}_j}, \mathsf{cph}_{\mathsf{SE}_j})$. If $\gamma'' = 0$, it returns 0. Otherwise, it continues to $i = i + 1$.

Case 2: If $(\mathsf{BF}'_i, \mathsf{cph}_{\mathsf{BF}_i}, \sigma_{\mathsf{BF}_i}, \pi_{\mathsf{BF}_i}) \in \mathsf{proof}$ meaning that there is no matching keyword ciphertext, it continues to verify the integrity of the masked Bloom filter by running $\gamma' \leftarrow \mathsf{Sig.Verify}(\mathsf{Sig.pk}, \sigma_{\mathsf{BF}_i}, \mathsf{BF}'_i)$. If $\gamma' = 0$, it returns 0; otherwise, it executes the following:

- If DUsr is authorized, compute $\{\mathsf{M} / \perp\} \leftarrow \mathsf{ABEO.Dec}(\mathsf{cph}_{\mathsf{BF}_i}, \pi_{\mathsf{BF}_i}, \mathsf{ABEO.sk}, \mathsf{pm}, \mathsf{ID}_{DOwn})$. If the output is $\perp$, return 0; otherwise, $\mathsf{BF}_i = H(\mathsf{M}) \oplus \mathsf{BF}'_i$. Execute $\delta \leftarrow \mathsf{BFVerify}(\{H'_1, ..., H'_k\}, \mathsf{BF}_i, w)$ to check whether w is present in the keyword group as represented by BF_i.
  - If $\delta = 0$, meaning that w is not present in the keyword group as represented by BF_i, continue to $i = i + 1$.
  - If $\delta = 1$, download $\prod_i = \{\mathsf{ID}_{DOwn}, (\mathsf{cph}_w, \pi_w) | w \in \mathsf{KG}_i\}$ from Cloud 2. Then, run $(\tau, \mathsf{aux}) \leftarrow \mathsf{ABKSO.Search}(\mathsf{cph}_w, \mathsf{tk}, \mathsf{ID}_{DOwn}, \mathsf{pm})$ and $\tau' \leftarrow \mathsf{ABKSO.Verify}(\mathsf{cph}_w, \mathsf{tk}, \mathsf{aux}, \pi_w, \mathsf{pm})$ by enumerating cph_w in $\{\mathsf{cph}_w | w \in \mathsf{KG}_i\}$. If there exists some $\tau' = 0$ after processing all cph_w (meaning that the ciphertext is modified), return 0; otherwise, if there exists some $\tau = 1$ after processing all cph_w (meaning that there exists cph_w that matches tk), return 0; otherwise, continue to $i = i + 1$.
- If DUsr is unauthorized, continue to $i = i + 1$ because $\mathsf{cph}_{\mathsf{BF}_i}$ cannot be decrypted.

Case 3: If neither case happens, return 0.

2. Return 1 if all tuples in the search result have been successfully verified, and 0 otherwise.

Theorem 2. *The security of our VABKS outsourcing scheme is as follows:*

- *It achieves data secrecy if* ABEO *and* SE *are secure against chosen-plaintext attack.*
- *It is selectively secure against chosen-keyword attack if* ABEO *is secure against chosen-plaintext attack,* H *is a secure pseudorandom generator, and* ABKSO *is selectively secure against chosen-keyword attack.*
- *It achieves keyword secrecy if* ABEO *is secure against chosen-plaintext attack,* H *is a secure pseudorandom generator, and* ABKSO *achieves keyword secrecy.*
- *It achieves verifiability if* Sig, ABEO, *and* ABKSO *are unforgeable.*

We omit the proof of Theorem 2 for lack of space (We will show it in the full version of this paper).

6 Comparison

Table 1 compares our VABKS outsourcing scheme with the previous VABKS schemes in terms of the computation cost for DOwn. In this table, $M_{\mathbb{G}}$ and $M_{\mathbb{G}_T}$ denote the computation cost of one modular exponentiation in $\mathbb{G}$ and $\mathbb{G}_T$, respectively, P denotes the computation cost of one pairing over an elliptic curve, ℓ denotes the number of attributes in a policy (also the number of rows of the access matrix), l denotes the number of keyword groups, n_w denotes the number of keywords in all of the keyword groups, and n denotes the number of data files. ABE.Enc and ABKS.Enc respectively denote the computation costs of one ABE encryption and one ABKS encryption. ZXA-1 denotes the VABKS scheme that uses the CP-ABE scheme in [2] and the CP-ABKS scheme in [18] in the generic construction in [18]. ZXA-2 denotes the VABKS scheme that uses the CP-ABE scheme in [14] and the CP-ABKS scheme in [15] in the generic construction in [18]. Ours denotes the VABKS outsourcing scheme in Sect. 5.3. DOwn runs ABE $n + l$ times and ABKS n_w times, so the total encryption costs of ZXA-1, ZXA-2, and Ours are $((2\ell + 1)n + (2\ell + 1)l + (2\ell + 4)n_w)M_{\mathbb{G}} + (n + l)M_{\mathbb{G}_T}$, $((3\ell+1)n+(3\ell+1)l+(3\ell+1)n_w)M_{\mathbb{G}}+(n+l+2n_w)M_{\mathbb{G}_T}+n_wP$, and $((\ell+1)n+(\ell+1)l+(\ell+1)n_w)M_{\mathbb{G}}+(n+l+2n_w)M_{\mathbb{G}_T}+n_wP$, respectively. For example, in the case of $\ell = 5$, $n = 10$, $l = 1$, and $n_w = 30$, the total costs of ZXA-1, ZXA-2, and Ours are $541M_{\mathbb{G}} +11M_{\mathbb{G}_T}$, $656M_{\mathbb{G}} + 71M_{\mathbb{G}_T} + 30P$, and $246M_{\mathbb{G}} + 71M_{\mathbb{G}_T} + 30P$, respectively. ZXA-2 and Ours require n_w pairing computations, so their costs are higher than that of ZXA-1. However, while ZXA-1 is provably secure in the generic group model, ZXA-2 and Ours are provably secure in the random oracle model. Therefore, under this more realistic assumption, Ours is the most efficient VABKS scheme in terms of the encryption cost for DOwn.

ZXA-1, ZXA-2, and Ours use $2l + n_w + 1$, $2l + n_w + 1$, and $l + n$ signatures, respectively. For example, in the case of $n = 10$, $l = 1$, and $n_w = 30$, ZXA-1, ZXA-2, and Ours use 33, 33, and 11 signatures, respectively. Hence, Ours uses fewer signatures. On the other hand, in Ours, some of the signatures are replaced by proofs and the number of proofs is $n + l + n_w$. In the case of $n = 10$, $l = 1$, and $n_w = 30$, the number of proofs is 41, which is larger than the number

Table 1. Comparison of our VABKS outsourcing scheme and previous VABKS schemes.

	ZXA-1	ZXA-2	Ours
ABE.Enc (for DOwn)	$(2\ell+1)M_{\mathbb{G}}+M_{\mathbb{G}_T}$	$(3\ell+1)M_{\mathbb{G}}+M_{\mathbb{G}_T}$	$(\ell+1)M_{\mathbb{G}}+M_{\mathbb{G}_T}$
ABE.Enc (for Cloud 1)	-	-	$(3\ell)M_{\mathbb{G}}$
ABKS.Enc (for DOwn)	$(2\ell+4)M_{\mathbb{G}}$	$(3\ell+1)M_{\mathbb{G}}+2M_{\mathbb{G}_T}+P$	$(\ell+1)M_{\mathbb{G}}+2M_{\mathbb{G}_T}+P$
ABKS.Enc (for Cloud 1)	-	-	$(3\ell)M_{\mathbb{G}}$
Num. of signatures	$2l+n_w+1$	$2l+n_w+1$	$l+n$
Num. of proofs	0	0	$n+l+n_w$
Policy structure	access tree	LSSS	LSSS
Security model	generic group model	random oracle model	random oracle model

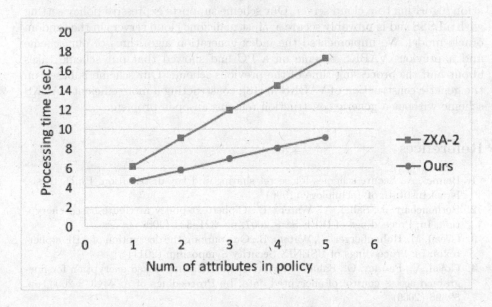

Fig. 3. Experimental results. (CPU: Intel Core i7-4790 (3.60GHz))

of signatures in ZXA-1 and ZXA-2. However, a proof can be generated simply by using a hash function at a smaller computation cost than that of a digital signature scheme. This means that Ours is the most efficient VABKS scheme in terms of the signing cost for DOwn.

We implemented the index generation algorithms of Ours and ZXA-2 on a PC having the following specifications. CPU: Intel Core i7-4790 (3.60 GHz), Memory: 8 GB, OS: CentOS 7.2, and Browser: Firefox 38.3.0. The algorithms were mainly written in JavaScript; the pairing computations were written in C/C++. We considered a viewing history that consisted of 13 records including 54 related keywords (index) and five possible attributes for DUsr. We measured the processing time of the index generation algorithms as the average of 100 trials. In Fig. 3, the horizontal axis denotes the number of attributes in the policy, and the vertical axis denotes the processing time (seconds). For simplicity, we used a ciphertext policy consisting of only AND-gates. It can be seen that Ours costs much less than ZXA-2 and the processing time depends on the number of attributes in the policy. In particular, for five attributes, Ours takes about 9 s, while ZXA-2 takes about 18 s. Therefore, our scheme significantly reduces the computation cost for DOwn.

7   Conclusion

We proposed a VABKS outsourcing scheme where the computation cost for a data owner can be significantly reduced by outsourcing part of the index generation algorithm to a cloud server. Our scheme supports expressive policy setting with a LSSS and is provably secure against malicious cloud servers in the random oracle model. We implemented the index generation algorithms of our scheme and a previous VABKS scheme on a PC and showed that our scheme takes about half the processing time of the previous scheme. Our scheme is based on the generic construction of VABKS in [18]; constructing a more efficient VABKS scheme without a generic construction remains an open problem.

References

1. Beimel, A.: Secure schemes for secret sharing and key distribution. Ph.D. thesis, Israel Institute of Technology (1996)
2. Bethencourt, J., Sahai, A., Waters, B.: Ciphertext-policy attribute-based encryption. In: Proceedings of IEEE S&P 2007, pp. 321–334 (2007)
3. Green, M., Hohenberger, S., Waters, B.: Outsourcing the decryption of ABE ciphertexts. In: Proceedings of USENIX Security Symposium (2011)
4. Goyal, V., Pandey, O., Sahai, A., Waters, B.: Attribute-based encryption for fine-grained access control of encrypted data. In: Proceedings of ACMCCS 2006, pp. 89–98 (2006)
5. ETSI: TS 102 796: hybrid broadcast broadband TV, V1.3.1
6. Hohenberger, S., Waters, B.: Online/Offline attribute-based encryption. In: Proceedings of PKC 2014, pp. 293–310 (2014)
7. NHK, Hybridcast (in Japanese). http://www.nhk.or.jp/hybridcast/online/
8. KBS: Icon (in Korean). http://icon.kbs.co.kr/site/main/main.php
9. Li, J., Jia, C., Li, J., Chen, X.: Outsourcing encryption of attribute-based encryption with MapReduce. In: Proceedings of ICICS 2012, pp. 191–201 (2012)

10. Ohtake, G., Safavi-Naini, R., Zhang, L.: Outsourcing scheme of ABE encryption secure against malicious adversary. In: Proceedings of ICISSP 2017, pp. 71–82 (2017)
11. Shi, J., Lai, J., Li, Y., Deng, R., Weng, J.: Authorized keyword search on encrypted data. In: Proceedings of ESORICS 2014, pp. 419–435 (2014)
12. Sun, W., Yu, S., Lou, W., Hou, Y., Li, H.: Protecting your right: attribute-based keyword search with fine-grained owner-enforced search authorization in the cloud. In: Proceedings of IEEE Infocom 2014, pp. 226–234 (2014)
13. Sahai, A., Waters, B.: Fuzzy identity-based encryption. In: Proceedings of Eurocrypt 2005, pp. 457–473 (2005)
14. Waters, B.: Ciphertext-policy attribute-based encryption: an expressive, efficient, and provably secure realization. eprint 2008/290
15. Wang, C., Li, W., Li, Y., Xu, X.: A ciphertext-policy attribute-based encryption scheme supporting keyword search function. In: Proceedings of CSS 2013, pp. 377–386 (2013)
16. BBC: YouView: extraordinary TV for everyone. http://www.youview.com/
17. Zhou, Z., Huang, D.: Efficient and secure data storage operations for mobile cloud computing. eprint 2011/185
18. Zheng, Q., Xu, S., Ateniese, G.: VABKS: verifiable attribute-based keyword search over outsourced encrypted data. In: Proceedings of IEEE Infocom 2014, pp. 522–530 (2014)

Privacy Preservation

Is RCB a Leakage Resilient Authenticated Encryption Scheme?

Farzaneh Abed[1], Francesco Berti[2(✉)], and Stefan Lucks[1]

[1] Bauhaus-Universität Weimar, Weimar, Germany
{farzaneh.abed,stefan.lucks}@uni-weimar.de
[2] ICTEAM/ELEN/Crypto Group, Université catholique de Louvain,
Louvain-la-neuve, Belgium
francesco.berti@uclouvain.be

Abstract. Leakage resilient cryptography wants to provide security against side channel attacks. In this paper, we present several issues of the RCB block cipher mode, proposed by Agrawal et al. in [2]. RCB is the first Leakage Resilient Authenticated Encryption (AE) scheme ever presented. In particular, we present a forgery attack that breaks the INT-CTXT security which is a fundamental requirement in the design of AE schemes.

Keywords: Authenticated encryption · Leakage resilience · Block cipher · Attack

1 Introduction

One of the main issues of modern cryptography is the vulnerability of cryptosystem implementations against side channel attacks. To thwart this kind of attack, countermeasures such as masking [19], shuffling [25] and noise addition [12] have been proposed. For constrained devices, which are likely exposed to side-channel attacks, those countermeasures are quite expensive.

Leakage Resilient Cryptography. A complementary approach initiated with high hopes [9,14] is to design "leakage resilient" schemes. The goal is to maintain a certain level of security even when the implementation leaks to the adversary some information about the internal values computed by the algorithm. During the last decade, many methods have been proposed, yet few did focus on symmetric cryptography, or block cipher based schemes.

There have been a handful of proposals for leakage resilient encryption schemes (such as [16]), message authentication codes (MACs) (such as [16,22]) leakage resilient pseudorandom generators (PRGs) or stream ciphers (such as [9, 24,26,27]) and pseudorandom functions (PRFs) (such as [10,18]). Although in some cases the underlying primitives have looser requirements (e.g. "weak prf"), all those proposals can be naturally instantiated using a block cipher. They often use a rekeying scheme, introduced by Borst [7].

© Springer International Publishing AG 2017
H. Lipmaa et al. (Eds.): NordSec 2017, LNCS 10674, pp. 39–52, 2017.
https://doi.org/10.1007/978-3-319-70290-2_3

To the best of our knowledge, the problem of leakage resilient authenticated encryption remained untouched until Agrawal et al. [2] proposed RCB. Later, other schemes appeared: Berti et al. [6] proposed DTE and DCE, Dobraunig et al. [8] proposed ISAP, and Barwell et al. [4] proposed SIVAT in three independent works.

The RCB *Mode.* Agrawal et.al [2] proposed the RCB mode, which is based on the OCB mode [20], a well-known Authenticated Encryption scheme. The security of the OCB mode has been proved in the black-box model (i.e., without leakage) [11,20]. Agrawal et al. [2] enhanced OCB using a rekeying schemes [13]. With other minor modifications, explained in Sect. 3, this generates RCB. Using rekeying makes sure that the block cipher is never called twice under the same key. This is the core of the claims that RCB is leakage resilient.

Our Contribution and Results. In this paper we analyze security of RCB. As it will turn out, RCB suffers from some issues. In fact:

1. Neither RCB nor OCB are nonce-misuse resistant. Moreover they cannot face the release of unverified plaintexts (INT-RUP), see [1,3].[1]
2. RCB requires sender and receiver to synchronize counters and to keep them continually synchronized, otherwise an interactive resynchronization is required (cf. [2, Figure 2]).
3. RCB is vulnerable to Denial-of-Service (DoS) attacks,
4. RCB is not secure to be used in full-duplex communication.
5. The privacy provided by RCB fails in many practical side channel settings.
6. RCB fails to provide secure authenticated encryption since it cannot provide INT-CTXT security.

We stress that none of the issues 2–4, 6 applies to OCB, when used with nonces. Issues 1–4 might be acceptable trade-offs for leakage resilient schemes. Issue 2 is an intentional design decision made by RCB's creators [2]. Issue 5 is a general problem for block cipher based leakage resilient cryptography (we will discuss this in Sect. 5). Issue 6, the lack of secure authentication is damning for any Authenticated Encryption scheme. We presents explicit attacks to prove these issues. Apart from issue 5, all attacks do not use Side-Channel leakages, so they are in the black-box model.

Outlook. We start fixing fixing the notions and the general definitions in Sect. 2. Section 3 provides an overview over the RCB scheme, and Sect. 4 describes our attacks with regard to Issue 1–4 and 6. Section 5 discusses the privacy of RCB when the adversary has also access to leakage (issue 5), and at the end we conclude providing an idea to face issue 6.

2   Preliminaries and Notions

Notations. We denote the set of all n-bits strings with $\{0,1\}^n$. If x is a string, we denote with $|x|$, its bit length. Given two strings x, y, we denote with $x \oplus y$

---

[1] The authors of OCB did never claim nonce misuse resistance, but [2] made such claims for RCB.

their bitwise exclusive-OR (XOR) and with $x\|y$ the padding of the two strings x, y. Given a string $m \in \mathcal{M}$, we denote with $m = (m_1, ..., m_l)$ the parsing of the string m in l strings with $|m_1| = ... = |m_{l-1}| = b$ (the size b will be clear by the context) and $|m_l| \leq b$ ($m = m_1\|m_2\|...\|m_l$). Given a string x with $|x| \leq b$, we denote with $x0^b$ the string $x\|0^{b-|x|}$ (with $0^{b-|x|}$ we denote the string of length $b - |x|$ bit, whose bits are all 0). Given a string x with $|x| \geq b$, we denote with $\lfloor x \rfloor_\tau$ the string obtained from x keeping only the first τ bits of x and dropping the others. With $x \leftarrow \mathcal{X}$, we denote that an element x is taken uniformly random from the set $\mathcal{X}$.

If an algorithm Alg has many arguments, we may write part of them as subscripts or superscripts, that is $\text{Alg}(x_1, x_2, x_3, x_4) = \text{Alg}_{x_1}^{x_2, x_3}(x_4)$.

We denote adversary with $\mathcal{A}$. An adversary is an efficient Turing machine interacting with some oracles, and we denote with $\mathcal{A}^{\mathcal{O}_1, ..., \mathcal{O}_n} \rightarrow 1$ the event that the adversary $\mathcal{A}$ interacts with the n oracles $\mathcal{O}_1, ..., \mathcal{O}_n$ and outputs 1. Adversaries have a bounded running time and are allowed to ask a certain number of queries to the oracles they are granted access to.

The algorithm RCB uses a block cipher E, which is a pseudorandom function $\{0, 1\}^b \longmapsto \{0, 1\}^b$. The *block length* of the block cipher E is b bits. A *pseudorandom function* is a family of functions (indexed by a key $k \in \mathcal{K}$) which has the property that the outputs of a random instance of the family should be "computationally indistinguishable" from those of a random function, even if the inputs, for these outputs, are chosen by the adversary. This means that every adversary behaves in the same way if it has access to an oracle implemented with a random instance of the family or with a random function.

For authenticated encryption we follow Namprempre et al. [15].

Definition 1 (Authenticated Encryption (AE)). *A nonce-based authenticated encryption (AE) scheme is a tuple $\Pi = (\mathcal{K}, \text{Enc}, \text{Dec})$ of a deterministic encryption algorithm $\text{Enc} : \mathcal{K} \times \mathcal{N} \times \mathcal{H} \times \mathcal{M} \rightarrow \mathcal{C} \times \mathcal{T}$, and a deterministic decryption algorithm $\text{Dec} : \mathcal{K} \times \mathcal{N} \times \mathcal{H} \times \mathcal{C} \times \mathcal{T} \rightarrow \mathcal{M} \cup \{\bot\}$. It uses an associated non-empty key space $\mathcal{K}$, a non-empty nonce space $\mathcal{N}$. We denote with $\mathcal{H}$, $\mathcal{M}$, $\mathcal{C} \subseteq \{0, 1\}^*$, the header space, message space, and ciphertext space, respectively. The tag space $\mathcal{T} = \{0, 1\}^\tau$ for a fixed $\tau \geq 0$. If given a tuple (n, h, c, τ), $\text{Dec}_k(n, h, c, \tau)$ returns a plaintext $m \neq \bot$, we say that the tuple (n, h, c, τ) is valid, otherwise it is invalid.*

We require Enc and Dec to be the inverse of each other. This is formalized by the following two properties:

- *Correctness*: if $\text{Enc}_k^{n,h}(m) = (c, \tau)$, then $\text{Dec}_k^{n,h}(c, \tau) = m$.
- *Tidiness*: if $\text{Dec}_k^{n,h}(c, \tau) = m \neq \bot$, then $\text{Enc}_k^{n,h}(m) = (c, \tau)$.

3 General Overview of RCB

Agrawal et al. [2] proposed a new symmetric leakage resilient authenticated encryption scheme called RCB. It uses a block cipher E with b bits block length to encrypt and authenticate the message, but in every call the block cipher E is used with a different key provided by the re-keying scheme G. The scheme uses a counter ctr and a master key k^*, so that every round key $k_i := G(k^*, \text{ctr}_i)$ is chosen randomly and it is independent from the others keys. The rekeying scheme is supposed not to leak any information about the master key k^*, while the block cipher E may leak some information about its key k_i. Both sender and receiver need to synchronize the counter. To do this, at the beginning, both sender and the receiver initialize the values of their counters ctr to 1, and, if the synchronization is lost, they need to perform a resynchronization process.

As shown in Fig. 1, RCB first parses a message m into l blocks, with $|m_1| = ... = |m_{l-1}| = b$ and $|m_l| \leq b$. Then it encrypts these message blocks to $c = (c_1, ..., c_l)$ (with $|c_i| = |m_i| \; \forall i = 1, ..., l$) using E. The output is the pair $(\text{ctr}, c, \tau) \in \{0,1\}^{\tau_s}$ for some tag size $\tau_s \leq b$ (see Algorithm 1).

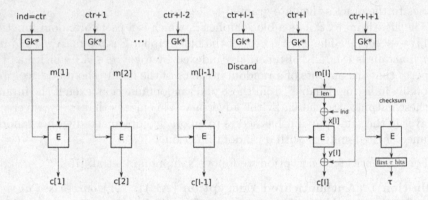

Fig. 1. A schematic view of the RCB structure.

As shown in Algorithm 1, the first $l - 1$ blocks, which are always full b-bit blocks, are encrypted in lines 4–7. These lines are skipped if the message is only a single block one. The last message block m_l, which can be smaller than b bits, is encrypted in lines 8–13, and the authentication tag is computed in lines 14–17.

To decrypt, given a pair (Ind, c, τ), RCB first computes m, then its own authentication tag τ', and returns the original message m if $\tau' = \tau$, else it returns $\bot$.

The authors of RCB claimed the following securities for their construction:

– Privacy under chosen plaintext attacks (IND-CPA), even with leakage, and
– authenticity against chosen-message existential forgery attacks (INT-CTXT) even with leakage.

They also claimed that RCB is robust, meaning it can provide security in both

Algorithm 1. RCB encryption.

1: **state** long-term key k^*, counter ctr   ($* k^*$ is constant, ctr always increases $*$)
2: **input** message $m = (m_1, \ldots, m_l)$
3: Ind $\leftarrow$ ctr
4: **for** $i \in \{1, \ldots, l-1\}$ **do**
5:      $k_i \leftarrow G_{k^*}(\text{ctr})$
6:      ctr $\leftarrow$ ctr $+ 1$
7:      $c_i \leftarrow \mathsf{E}_{k_i}(m_i)$
8: ctr $\leftarrow$ ctr $+ 1$ ($*$ skip one value $*$)
9: $k_l \leftarrow G_{k^*}(\text{ctr})$
10: ctr $\leftarrow$ ctr $+ 1$
11: $x \leftarrow |m_l| \oplus (\text{ctr} + l + 1)$
12: $y \leftarrow \mathsf{E}_{k_l}(x)$
13: $c_l \leftarrow y \oplus m_l$
14: $s \leftarrow m_1 \oplus \cdots \oplus m_{l-1} \oplus (c_l 0^b) \oplus y$ ($*$ checksum $*$)
15: ctr $\leftarrow$ ctr $+ 2$
16: $k_{l+1} \leftarrow G_{k^*}(\text{ctr})$
17: $\tau \leftarrow \lfloor \mathsf{E}_{k_{l+1}}(s) \rfloor_{\tau_s}$ [first τ_s bits]
18: **return** $(\mathsf{Ind}, \overbrace{(c_1, \cdots, c_l)}^{c}, \tau)$

- nonce-misuse setting, and
- decryption misuse or release of unverified plaintexts (RUP) setting.

We show, presenting our attacks, that none of the above claims is correct for the RCB construction. Thus, we prove that the RCB construction is totally insecure and that it is broken even in the black-box model, therefore it should not be considered as a secure AE scheme.

4   Attacks on RCB

In this section, we show some attacks which violate the securities claimed in [2]. We start each subsection presenting the property of the attack that breaks, and then we explain the attack itself. We first present the most important attack, the forgery attack which breaks the ciphertext integrity (INT-CTXT), then the other attacks.

Below, we assume Alice uses RCB to send authenticated and encrypted messages to the receiver Bob. We assume that Alice and Bob have shared a secret key k^* and synchronized the counter ctr. The adversary $\mathcal{A}$ is trying to attack Alice and Bob.

4.1   Forgery Attack

Definition 2 (INTCTXT Security [5]). *Let $\Pi = (\mathcal{K}, \mathsf{Enc}, \mathsf{Dec})$ be a nonce-based AE scheme, $k \leftarrow \mathcal{K}$, and let $\mathcal{A}$ be a computationally bounded adversary with access to an encryption and a decryption oracle. We suppose that $\mathcal{A}$ never* *queries the decryption oracle with outputs of the encryption one. Then, the INT-CTXT advantage of $\mathcal{A}$ with respect to Π is defined as*

$$\mathbf{Adv}_{\Pi}^{INT\text{-}CTXT}(\mathcal{A}) := \left| \Pr\left[\mathcal{A}^{\mathsf{Enc}_k(\cdot,\cdot,\cdot),\mathsf{Dec}_k(\cdot,\cdot,\cdot,\cdot)} \Rightarrow 1 \right] - \Pr\left[\mathcal{A}^{\mathsf{Enc}_k(\cdot,\cdot,\cdot),\perp(\cdot,\cdot,\cdot,\cdot)} \Rightarrow 1 \right] \right|,$$

where $\perp (\cdot,\cdot,\cdot,\cdot)$ is an oracle that always outputs $\perp$ for every input. Furthermore, we define $\mathbf{Adv}_{\Pi}^{INT\text{-}CTXT}(q,\ell,t)$ as the maximum advantage over all INT-CTXT adversaries $\mathcal{A}$ against Π that run in time at most t, and make at most q queries of total length ℓ to the available oracles.

The idea of the attack is to use one valid ciphertext to produce another valid ciphertext. To do this, we prevent Bob from receiving the valid ciphertext with the aim that his counter will not change. We prove that this message will contain all the information that an adversary $\mathcal{A}$ needs to forge a 2 blocks message. We first start presenting a forgery attack using only an encryption of a 5 blocks message, we then generalize the attack.

Attack on 5 Blocks Message:

1. Alice and Bob share the same counter $\mathsf{ctr} = \mathsf{Ind}$ at the beginning.
2. An adversary $\mathcal{A}$ chooses arbitrary message blocks m_1, m_2 in $\{0,1\}^b$.
3. He then chooses another arbitrary message block $m_2' \in \{0,1\}^{b'}$ with $b' \leq b$.
4. As a third message block, he chooses $m_3 = |m_2'| \oplus \mathsf{Ind}$.
5. Now, he computes $m_4 = m_1 \oplus (m_2'0^b)$.
6. As a last message block, an adversary $\mathcal{A}$ chooses arbitrary message m_5 in $\{0,1\}^{b''}$ with $b'' \leq b$.
7. Now, an adversary $\mathcal{A}$ asks for the encryption of $m = (m_1, ..., m_5)$.
8. Alice encrypts the message m and obtains corresponding ciphertexts and the tag $(\mathsf{Ind}, (c_1, ..., c_5), \tau)$.
9. In this phase, an adversary $\mathcal{A}$ sets $c_2' := c_3 \oplus m_2'$.
10. He then sends $(\mathsf{Ind}, (c_1, c_2'), \lfloor c_4 \rfloor_{\tau_s})$ to Bob.
11. Bob accepts this message as a valid one.

In our attack, we assume that the counter ctr is a known value. This hypothesis is sound, since to get the actual value of ctr, it is enough to see the last encryptions made by Alice $(\mathsf{Ind}, c_1, ..., c_{l'}, \tau)$. Now Alice's counter is $\mathsf{Ind} + l' + 2$.

We argue that $(\mathsf{Ind}, (c_1, c_2'), \lfloor c_4 \rfloor_{\tau_s})$ is the legitimate encryption of the message (m_1, m_2') with initial counter $\mathsf{ctr} = \mathsf{Ind}$, therefore Bob accepts it as a valid one.

Firstly, it is easy to see that c_1 is the encryption of m_1.

Secondly, the ciphertext block c_3 is the encryption of $|m_2'| \oplus \mathsf{Ind}$ under the key $G_{k^*}(\mathsf{Ind} + 2)$, and the ciphertext block c_2' has been chosen as $c_3 \oplus m_2'$, as required by lines 8–13 of Algorithm 1.

Thirdly, as can be seen, m_4 has been chosen as $m_1 \oplus m_2'0^b$, and the tag is the encryption of $m_1 \oplus m_2'0^b$ under the key $G_{k^*}(\mathsf{Ind} + 3)$.

Finally, note that, if the last message block m_2' of the forged message is a b-bit block, then the ciphertext is valid, because the tag is correct. Otherwise, it is valid if the last $b - b'$ bits of y are 0, where $y := \mathsf{E}_{k_4}(|m_2'| \oplus (\mathsf{Ind} + 2 + 1))$. In this case, the success probability of the attack is about $2^{-b+b'}$.

Generalized Attack on Message ≥ 4:

1. Initially, Alice and Bob share the same counter $\text{ctr} = \text{Ind}$.
2. An adversary $\mathcal{A}$ chooses $a \geq 4$.
3. He chooses arbitrary message blocks $m_1, \ldots, m_{a-4}, m_{a-3}$ in $\{0,1\}^b$.
4. He chooses another arbitrary message block $m'_{a-3} \in \{0,1\}^{b'}$ with $b' \leq b$.
5. As a next, he chooses $m_{a-2} = |m'_{a-3}| \oplus (\text{Ind} + a - 2 + 1)$.
6. Now, the adversary $\mathcal{A}$ computes $m_{a-1} = \left(\bigoplus_{i=1}^{a-4} m_i \right) \oplus m'_{a-3}$.
7. As a last message, the adversary $\mathcal{A}$ chooses arbitrary message m_a in $\{0,1\}^{b''}$ with $b'' \leq b$.
8. In this step, the adversary $\mathcal{A}$ asks for the encryption of the message $m = (m_1, \ldots, m_a)$.
9. Alice encrypts the message $m = (m_1, \ldots, m_a)$ and gets the corresponding ciphertext and the tag $(\text{Ind}, (c_1, \ldots, c_a), \tau)$.
10. The adversary $\mathcal{A}$ sets $c'_{a-3} := c_{a-3} \oplus m'_{a-3}$.
11. He now sends $(\text{Ind}, (c_1, \ldots, c_{a-4}, c'_{a-3}), \lfloor c_{a-1} \rfloor_{\tau s})$ to Bob.
12. Bob accepts this message as a valid one.

The validity of the attack derives from the previous observations.

Decryption-misuse: Since RCB can not provide INT-CTXT security in the black-box model, it can neither provide this security in the RUP setting (introduced by Andreeva et al. [3]). As a result, RCB in not robust in terms of nonce-misuse and decryption-misuse settings.

4.2   Attacks on Misuse Resistance

Misuse-resistance Agrawal et al. [2] claimed that RCB is nonce-misuse resistance because it does not have the nonce requirement. From our point of view, the counter ctr behaves as a nonce, since it is used in a way that prevents its reuse. Moreover, if the counter ctr is reused then the nonce misuse security (Definition 3) will fall.

Definition 3 (MRAE (Misuse resistance Security. [21]). *Let $\Pi = (\mathcal{K}, \mathsf{Enc}, \mathsf{Dec})$ be a nonce-based AE scheme, as defined in Definition 1. Then, the MRAE advantage of $\mathcal{A}$ with respect to Π is defined as*

$$\mathbf{Adv}_\Pi^{\mathrm{MRAE}}(\mathcal{A}) := \left| \Pr\left[\mathcal{A}^{\mathsf{Enc}_k(\cdot,\cdot,\cdot,\cdot), \mathsf{Dec}_k(\cdot,\cdot,\cdot,\cdot,\cdot)} \Rightarrow 1 \right] - \Pr\left[\mathcal{A}^{\$(\cdot,\cdot,\cdot,\cdot), \perp(\cdot,\cdot,\cdot,\cdot,\cdot)} \Rightarrow 1 \right] \right|,$$

where $\$(\cdot,\cdot,\cdot,\cdot)$ is an oracle which outputs random values at every query of length $|\mathsf{Enc}(\cdot,\cdot,\cdot,\cdot)|$, and $\perp (\cdot,\cdot,\cdot,\cdot,\cdot)$ is an oracle that outputs always $\perp$ for every input. Furthermore, we define $\mathbf{Adv}_\Pi^{\mathrm{MRAE}}(q, \ell, t)$ as the maximum advantage over all MRAE adversaries $\mathcal{A}$ against Π that run in time at most t, and make at most q queries of total length ℓ to the available oracles. If $\mathbf{Adv}_\Pi^{\mathrm{MRAE}}(q, \ell, t) \leq \epsilon$ we say that Π is (q, t, l, ϵ)-nonce misuse resistant.

An example of nonce-misuse resistant AE scheme is SCT presented by Peyrin and Seurin [17].

As we show below, when the counters repeat (i.e., if nonces are misused), then there is a simple attack on RCB as follow:

1. An adversary $\mathcal{A}$ knows the current value of Alice's counter, ctr = Ind.
2. He chooses three different message blocks $m_1^1, m_1^2, m_2^2 \in \{0,1\}^b$.
3. Alice encrypts message $m^1 = (m_1^1, m_2^1)$ obtaining $(\mathsf{Ind}, (c_1^1, c_2^1), \tau^1)$.
4. Now, an adversary $\mathcal{A}$ resets the Alice's counter to ctr = Ind (nonce-reuse!)
5. Then Alice encrypts $m^2 = (m_1^2[= m_1^1 \oplus e], m_2^2 = m_2^1)$ (with $e \in \{0,1\}^b, e \neq 0...0$) obtaining $(\mathsf{Ind}, (c_1^2, c_2^2)\tau^2)$ (Due to the nonce reuse, the value ctr is the same as in step 3.)
6. An adversary $\mathcal{A}$ can easily distinguish the real encryption RCB from the random oracle $\$(\cdot,\cdot,\cdot)$, In fact if $c_1^2 = c_1^1 \oplus e$, Alice is using real encryption with overwhelming probability, otherwise she is using the random oracle.

The success probability of an adversary $\mathcal{A}$ to correctly distinguish the two oracles is $1 - 2^{-b}$ because it may happen that $\$(m_1^2) = c_1^2$.

4.3   A Denial-of-Service (DoS) Attack

In general, an AE scheme is *correct*, if decrypting a ciphertext obtained by a genuine encryption oracle, it recovers the original message (Definition 1). We explain two attacks on the correctness of the RCB algorithm. For this purpose, we show that an adversary $\mathcal{A}$ can tamper the counter in order to deny the service and to make Bob reject a valid ciphertext. Our first DoS attack works as follow:

1. Alice's counter current value is ctr = Ind.
2. Alice chooses a message m to encrypt and obtains (Ind, c, τ). Her new counter value is now ctr = Ind + a for some $a > 0$.
3. Alice chooses another message m' to encrypt and obtains $(\mathsf{Ind} + a + 2, c', \tau')$. Her counter value is incremented to ctr = Ind + $a + a' + 4$ for some $a' > 0$.
4. Now, an adversary $\mathcal{A}$ forwards $(\mathsf{Ind} + a + 2, c', \tau')$ to Bob.
5. If $a + 2$ does not exceed a pre-defined threshold,[2] then Bob decrypts $(\mathsf{Ind} + a + 2, c', \tau')$ obtaining m'. Bob's new counter is now ctr = Ind + $a + a' + 4$.
6. As the final step, an adversary $\mathcal{A}$ forwards (Ind, c, τ) to Bob. Because Ind < Ind + $a + a' + 4$, Bob aborts to decrypt (Ind, c, τ). Therefore, he cannot recover the original message m, and he only performs a resynchronization.

Our second DoS attack does not even require Alice to encrypt two messages:

1. Alice's counter current value is ctr = Ind.
2. Alice chooses message m to encrypt and obtains (Ind, c, τ). Her new counter value is now ctr = Ind + $a + 2$ for a certain $a > 0$.
3. An adversary $\mathcal{A}$ chooses c' and τ' on his own desire and sends $(\mathsf{Ind}, c', \tau')$ to Bob to get the message.
4. Bob decrypts $(\mathsf{Ind}, c', \tau')$ to $\perp$. Bob's new counter value is ctr = Ind + $a' + 2$ for a certain $a' > 0$.[3]

---

[2] Else, Alice and Bob would perform interactive resynchronization [2, Fig. 2].
[3] Bob must increase the counter, even if the message turns out to be invalid. Otherwise, Bob would use the same internal key more than once, thus destroying the main purpose of using RCB, namely its claimed leakage-resilience.

5. As a final step, an adversary $\mathcal{A}$ forwards (Ind, c, τ) to Bob. Because $\mathsf{Ind} < \mathsf{Ind} + a' + 2$, Bob aborts instead of decrypting (Ind, c, τ). Therefore, he is not able to recover the original message m, and he only performs a resynchronization.

4.4 Attack on Full-Duplex Communication

Contrary to many symmetric schemes, it is not possible for both Alice and Bob to use the same key to communicate with each other. In fact, if they use the same master key k^*, then the adversary $\mathcal{A}$ can destroy the privacy of the communication in this way:

1. Alice and Bob share the same initial counter $\mathsf{ctr} = \mathsf{Ind}$.
2. An adversary $\mathcal{A}$ chooses $m_1, m_2 \in \{0,1\}^b$ with $m_1 \neq m_2$ and asks Alice to encrypt it.
3. Alice encrypts (m_1, m_2) and gets $(\mathsf{Ind}, (c_1, c_2), \tau)$. Bob should not receive this message.
4. An adversary $\mathcal{A}$ chooses another message $m' = (m_1, m_1)$ and asks Bob to encrypt it.
5. Bob encrypts the message m' and gets $(\mathsf{Ind}, (c'_1, c'_2), \tau')$.
6. Since $c_1 = c'_1$ in the first case, an adversary $\mathcal{A}$ is able to distinguish between the real encryption RCB and a random function.

It is easy to evade this attack using two independent keys, one for messages from Alice to Bob, and the other one for messages from Bob to Alice. One can also use two different counters ctr_{AB} for communication from Alice to Bob and ctr_{BA} for communications from Bob to Alice, but the set of values taken by ctr_{AB} and that of those taken by ctr_{BA} must have no intersection.

5 Privacy by RCB

In many practical leakage settings, algorithm RCB fails to provide privacy [15]. Actually, as we mentioned before, this does not contradict the security claims made by Agrawal et al. [2], but it is related to a more general problem for block cipher based (authenticated) encryption.

One of the many implications of the semantic security (Defintion 4) is the following: given two messages m_0, m_1 with $|m_0| = |m_1|$, if Alice encrypts either of two messages, any adversary $\mathcal{A}$ should not be able to decide which message has been encrypted, otherwise, the privacy is gone.

Definition 4 (IND-CPA Security). *Let $\Pi = (\mathcal{K}, \mathsf{Enc}, \mathsf{Dec})$ be an AE scheme, as defined in Definition 1. Let $\mathcal{A}$ be a computationally bounded adversary. Then, the IND-CPA (*Indistinguishibility under a Chosen Plaintext Attack*) advantage of $\mathcal{A}$ is defined as*

$$\mathbf{Adv}_{\Pi}^{IND\text{-}CPA}(\mathcal{A}) = \left| \Pr\left[\mathcal{A}^{\mathsf{Enc}_k(\cdot,\cdot,\cdot)} \Rightarrow 1 \right] - \Pr\left[\mathcal{A}^{\$(\cdot,\cdot,\cdot)} \Rightarrow 1 \right] \right|,$$

where the probabilities are taken over a key $k \leftarrow \mathcal{K}$ and the random coins of $\mathcal{A}$. Here $\$(\cdot, \cdot, \cdot)$ *is an oracle that for every input (n, h, m) outputs a random string whose length* $|\$(n, h, m)| = |\mathsf{Enc}_k(n, h, m)|$. *Moreover the oracle* $\$(\cdot, \cdot, \cdot)$ *keeps track of its queries in order to answer in the same way if the same query is asked again. Furthermore, we define* $\mathbf{Adv}_\Pi^{IND\text{-}CPA}(q, \ell, t)$ *as the maximum advantage over all IND-CPA adversaries $\mathcal{A}$ against Π that run in time at most t, and make at most q queries of total length ℓ to the available oracles.*

In leakage resilient schemes, we consider this security notion in the presence of the leakage function L.

Leakage. Let $\mathcal{L}$ be the leaking space. A *leakage function* is a function $L : \mathcal{K} \times \mathcal{N} \times \mathcal{H} \times \mathcal{M} \longmapsto \mathcal{L}$. The leakage function can be any function.

The *Leakage privacy* or *Indistinguishibility under a Chosen Plaintext Attack with Leakage* (IND-CPA-L) is defined as the following in [2]:

Definition 5 (IND-CPA-L Security). *Let $\Pi = (\mathcal{K}, \mathsf{Enc}, \mathsf{Dec})$ be an AE scheme, as defined in Definition 1, and L be the leakage function of Enc. Let $\mathcal{A}$ be a computationally bounded adversary. Then, the IND-CPA-L advantage of $\mathcal{A}$ is defined as*

$$\mathbf{Adv}_\Pi^{IND\text{-}CPA\text{-}L}(\mathcal{A}) = \left| \Pr\left[\mathcal{A}^{\mathsf{Enc}_k(\cdot,\cdot,\cdot), \mathsf{Enc}_k^L(\cdot,\cdot,\cdot)} \Rightarrow 1 \right] - \Pr\left[\mathcal{A}^{\$(\cdot,\cdot,\cdot), \mathsf{Enc}_k^L(\cdot,\cdot,\cdot)} \Rightarrow 1 \right] \right|,$$

where the probabilities are taken over a key $k \leftarrow \mathcal{K}$ and the random coins of $\mathcal{A}$. Here $\$(\cdot, \cdot, \cdot)$ *is an oracle that for every input (n, h, m) outputs a random string whose length* $|\$(n, h, m)| = |\mathsf{Enc}_k(n, h, m)|$. *Moreover, the oracle* $\$(\cdot, \cdot, \cdot)$ *keeps track of its queries in order to answer in the same way if the same query is asked again. The oracle* $\mathsf{Enc}_k^L(\cdot, \cdot, \cdot)$ *is an oracle that on input (n, h, m) outputs the leakage of the algorithm* Enc_k. *Furthermore, we define* $\mathbf{Adv}_\Pi^{IND\text{-}CPA\text{-}L}(q, \ell, t)$ *as the maximum advantage over all IND-CPA-L adversaries $\mathcal{A}$ against Π that run in time at most t, and make at most q queries of total length ℓ to the available oracles.*

We now employ an attack on RCB showing how it is possible to violate the privacy of RCB in terms of distinguishing the encryption of two messages m_0 and m_1, with $|m_0| = |m_1|$.

Recall line 7 in Algorithm 1:

$$c_i \leftarrow \mathsf{E}_{k_i}(m_i).$$

If any information about m_i leaks, then the privacy of RCB will completely be gone.

As it turns out, typical side channels allow an adversary to gather information about the messages m_i, the outputs c_i and the inner computation. Therefore, it is very difficult to obtain security of privacy against side channel adversaries. See Fig. 2 for a power analysis trail. The "peaks" may, e.g., reveal information about the Hamming weight of the data currently processed, including the Hamming weight of m_i at the beginning and the Hamming weight of c_i at the end.

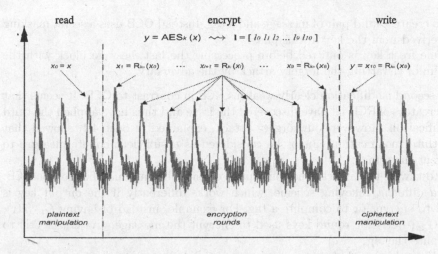

Fig. 2. A leakage trail for a block cipher encryption, computing $c_i = E_{k_i}(m_i)$ [23].

This means that the adversary will be able to distinguish what message m_i is being encrypted.

Implicitly, RCB assumes that information about the round key k_i may leak, but also that information about m_i must not leak. In particular Agrawal et al. [2] relates the privacy of RCB to the privacy of the block cipher implementation. Thus, if a side channel trace, such as the one in Fig. 2 leaks information about m_i, then the privacy of the block cipher implementation (one-time leakage resiliency) is low, therefore the privacy claimed for RCB becomes invalid.

We need to stress that maintaining a good privacy security for block cipher based leakage resilient cryptosystems seems to be an unsolved issue because it seem impossible to have that every block cipher evaluation does not leak any information about the message. For example, for CPA-security Pereira et al. [16], give a security notion where the security of a multiple block cipher calls is related to the security of a single block cipher call (the quantification of the security of a single block cipher call remains an open problem).

6 Conclusion

What went wrong? And can one repair RCB? RCB has been derived from the well-established OCB mode. OCB is neither leakage resilient nor misuse resistant, but it provides secure authenticated encryption in the black-box model (without leakage). On the other hand, RCB is not even secure in the black-box model.

In [2] Agrawal et al. list the following modifications to turn OCB into RCB:

1. Instead of masking the input and output of the block cipher, they change the key of the block cipher.
2. The starting counter, which is not secret, is XORed to the input of the block cipher E during processing the last block of the message to prevent adversary

to create a valid pair of message and a tag. Instead OCB uses a secret masking derived from the key in this phase.

3. One fresh key is omitted before processing the last message block with the aim of thwarting the forgery attack by the adversary.

The second modification clearly weakens RCB in contrast to OCB. It seems that the creators of RCB [2] have discovered the issue and thus have applied the third modification. However, our forgery attack, explained in Sect. 4.1, showed that the third modification cannot be considered as a sufficient countermeasure to prevent the attack.

To thwart the attack, one could propose the following modification on RCB: use a different rekeying scheme, which works differently if the output key is used to encrypt or to compute a tag. For example, instead of having $G_{k^*}(\text{ctr})$, use $G_{k^*}(0, \text{ctr})$ for round keys used to encrypt the message, and $G_{k^*}(1, \text{ctr})$ to compute the tag.

We conjecture that this would prevent black-box forgery attacks such as ours, but it would not solve any of the other issues.

Summary. This work described several attacks on RCB leakage resilient Authenticated Encryption scheme. RCB is not robust, neither against nonce misuse, nor release of unverified plaintexts. The most important issue of RCB is the forgery attack in the black-box model, which is a fundamental requirement for any AE schemes. Moreover, its requirement to maintain synchronized counters between sender and receiver opens the door to Denial-of-Service attacks. In addition, RCB cannot be securely used for full-duplex communication. Finally, its privacy succumb to many practical side channel attacks.

In spite of all these negative results, we give credit to Agrawal et al. [2] to be the first, to the best of our knowledge, to present a leakage-resilient authenticated encryption scheme, and we hope that our solution can help them to improve their construction.

Acknowledgments. Farzaneh Abed was supported by the Simple Scry project with Cisco.

References

1. Abed, F., Fluhrer, S.R., Forler, C., List, E., Lucks, S., McGrew, D.A., Wenzel, J.: Pipelineable on-line encryption. In: Cid, C., Rechberger, C. (eds.) FSE 2014. LNCS, vol. 8540, pp. 205–223. Springer, Heidelberg (2015). doi:10.1007/978-3-662-46706-0_11
2. Agrawal, M., Bansal, T.K., Chang, D., Chauhan, A.K., Hong, S., Kang, J., Sanadhya, S.K.: RCB: leakage-resilient authenticated encryption via re-keying. J. Supercomput. 1–26. Springer, Heidelberg (2016). doi:10.1007/s11227-016-1824-6
3. Andreeva, E., Bogdanov, A., Luykx, A., Mennink, B., Mouha, N., Yasuda, K.: How to securely release unverified plaintext in authenticated encryption. In: Advances in Cryptology -ASIACRYPT 2014–20th International Conference on the Theory and Application of Cryptology and Information Security, Kaoshiung, Taiwan, R.O.C., 7–11 December 2014, Proceedings, Part I, pp. 105–125 (2014)

4. Barwell, G., Martin, D.P., Oswald, E., Stam, M.: Authenticated encryption in the face of protocol and side channel leakage. IACR Cryptology ePrint Archive 2017, 68 (2017)
5. Bellare, M., Namprempre, C.: Authenticated encryption: relations among notions and analysis of the generic composition paradigm. In: Okamoto, T. (ed.) ASI-ACRYPT 2000. LNCS, vol. 1976, pp. 531–545. Springer, Heidelberg (2000). doi:10.1007/3-540-44448-3_41
6. Berti, F., Koeune, F., Pereira, O., Peters, T., Standaert, F.-X.: Leakage-resilient and misuse-resistant authenticated encryption. IACR Cryptol. ePrint Arch. **2016**, 996 (2016)
7. Borst, J.: Block ciphers: design, analysis and side-channel analysis. Ph.D. thesis, KULeuven, Belgium (2001)
8. Dobraunig, C., Eichlseder, M., Mangard, S., Mendel, F., Unterluggauer, T.: ISAP - towards side-channel secure authenticated encryption. IACR Trans. Symmetric Cryptol. **2017**(1), 80–105 (2017)
9. Dziembowski, S., Pietrzak, K.: Leakage-resilient cryptography. In: 49th Annual IEEE Symposium on Foundations of Computer Science, FOCS 2008, 25–28 October 2008, Philadelphia, PA, USA, pp. 293–302 (2008)
10. Faust, S., Pietrzak, K., Schipper, J.: Practical leakage-resilient symmetric cryptography. In: Prouff, E., Schaumont, P. (eds.) CHES 2012. LNCS, vol. 7428, pp. 213–232. Springer, Heidelberg (2012). doi:10.1007/978-3-642-33027-8_13
11. Krovetz, T., Rogaway, P.: The software performance of authenticated-encryption modes. In: Joux, A. (ed.) FSE 2011. LNCS, vol. 6733, pp. 306–327. Springer, Heidelberg (2011). doi:10.1007/978-3-642-21702-9_18
12. Mangard, S.: Hardware countermeasures against DPA - a statistical analysis of their effectiveness. In: Okamoto, T. (ed.) CT-RSA 2004. LNCS, vol. 2964, pp. 222–235. Springer, Heidelberg (2004). doi:10.1007/978-3-540-24660-2_18
13. Medwed, M., Petit, C., Regazzoni, F., Renauld, M., Standaert, F.-X.: Fresh re-keying II: securing multiple parties against side-channel and fault attacks. In: Prouff, E. (ed.) CARDIS 2011. LNCS, vol. 7079, pp. 115–132. Springer, Heidelberg (2011). doi:10.1007/978-3-642-27257-8_8
14. Micali, S., Reyzin, L.: Physically observable cryptography. In: Naor, M. (ed.) TCC 2004. LNCS, vol. 2951, pp. 278–296. Springer, Heidelberg (2004). doi:10.1007/978-3-540-24638-1_16
15. Namprempre, C., Rogaway, P., Shrimpton, T.: Reconsidering generic composition. In: Nguyen, P.Q., Oswald, E. (eds.) EUROCRYPT 2014. LNCS, vol. 8441, pp. 257–274. Springer, Heidelberg (2014). doi:10.1007/978-3-642-55220-5_15
16. Pereira, O., Standaert, F.-X., Vivek, S.: Leakage-resilient authentication and encryption from symmetric cryptographic primitives. In: Proceedings of the 22nd ACM SIGSAC Conference on Computer and Communications Security, Denver, CO, USA, 12–16 October 2015, pp. 96–108 (2015)
17. Peyrin, T., Seurin, Y.: Counter-in-tweak: authenticated encryption modes for tweakable block ciphers. In: Robshaw, M., Katz, J. (eds.) CRYPTO 2016. LNCS, vol. 9814, pp. 33–63. Springer, Heidelberg (2016). doi:10.1007/978-3-662-53018-4_2
18. Pietrzak, K.: A leakage-resilient mode of operation. In: Joux, A. (ed.) EURO-CRYPT 2009. LNCS, vol. 5479, pp. 462–482. Springer, Heidelberg (2009). doi:10.1007/978-3-642-01001-9_27
19. Rivain, M., Emmanuel, P.: Provably secure higher-order masking of AES. IACR Cryptol. ePrint Arch. **2010**, 441 (2010)

20. Rogaway, P., Bellare, M., Black, J.: OCB: a block-cipher mode of operation for efficient authenticated encryption. ACM Trans. Inf. Syst. Secur. **6**(3), 365–403 (2003)
21. Rogaway, P., Thomas, S.: Deterministic authenticated-encryption: a provable-security treatment of the key-wrap problem. IACR Cryptol. ePrint Arch. **2006**, 221 (2006)
22. Schipper, J.H.: Leakage resilient authentication, master thesis, Utrecht university, The Netherlands (2010)
23. Standaert, F.-X.: Directory authorities specifications from the tor project. http://perso.uclouvain.be/fstandae/PUBLIS/96_slides.pdf. Invited talk at SKEW 2011
24. Standaert, F.-X., Pereira, O., Yu, Y., Quisquater, J.-J., Yung, M., Oswald, E.: Leakage resilient cryptography in practice. In: Towards Hardware-Intrinsic Security - Foundations and Practice, pp. 99–134 (2010)
25. Veyrat-Charvillon, N., Medwed, M., Kerckhof, S., Standaert, F.-X.: Shuffling against side-channel attacks: a comprehensive study with cautionary note. In: Wang, X., Sako, K. (eds.) ASIACRYPT 2012. LNCS, vol. 7658, pp. 740–757. Springer, Heidelberg (2012). doi:10.1007/978-3-642-34961-4_44
26. Yu, Y., Standaert, F.-X.: practical leakage-resilient pseudorandom objects with minimum public randomness. In: Dawson, Ed (ed.) CT-RSA 2013. LNCS, vol. 7779, pp. 223–238. Springer, Heidelberg (2013)
27. Yu, Y., Standaert, F.-X., Pereira, O., Yung, M.: Practical leakage-resilient pseudorandom generators. In: Proceedings of the 17th ACM Conference on Computer and Communications Security, CCS 2010, Chicago, Illinois, USA, 4–8 October 2010, pp. 141–151 (2010)

Practical and Secure Searchable Symmetric Encryption with a Small Index

Ryuji Miyoshi[1], Hiroaki Yamamoto[1(✉)], Hiroshi Fujiwara[1], and Takashi Miyazaki[2]

[1] Faculty of Engineering, Shinshu University, 4-17-1 Wakasato,
Nagano 380-8553, Japan
{yamamoto,fujiwara}@cs.shinshu-u.ac.jp
[2] National Institute of Technology, Nagano College, 716 Tokuma,
Nagano 381-8550, Japan
miya@nagano-nct.ac.jp

Abstract. From a view point of information security, researches on an encrypted search system have been done intensively. Such search systems are called searchable symmetric encryption (SSE). The main part of SSE is an encrypted index which affects security and efficiency. Until now many SSE schemes have been proposed, but most of them uses a random oracle to achieve both adaptive security and an optimal index size. The index size of adaptively secure SSE schemes without a random oracle can be much larger. In this paper, we propose a new SSE scheme which satisfies adaptive security in the standard model and has an optimal index size. Furthermore the index of our scheme consists of Bloom filters and simple arrays, that is, arrays of integers. Since Bloom filters are also implemented by an array of integers, the structure of the index is simple. Thus, unlike other SSE schemes with an optimal index size, the size does not depend on a security parameter.

Keywords: Searchable symmetric encryption · Keyword search · Bloom filter

1 Introduction

1.1 Backgrounds

In recent years, many services have come to be performed through a network. In such an information society, protection of personal data and confidential data has been a very important subject. Also in information retrieval, the importance of the security is increasing. For example, in a remote storage service and an email service, there is a case where a server administrator differs from the owner of data. In this case, the server administrator is assumed to be trusted. However, putting this assumption on the administrator who is the 3rd person is not enough from a viewpoint of security. In order to use a server more safely, a user stores data in an encrypted form. By such an encryption, the server administrator

© Springer International Publishing AG 2017
H. Lipmaa et al. (Eds.): NordSec 2017, LNCS 10674, pp. 53–69, 2017.
https://doi.org/10.1007/978-3-319-70290-2_4

can access only encrypted data, but he cannot know the contents. A problem is that a search becomes difficult by encryption. A trivial method to resolve a search problem is that a user downloads all encrypted data and searches data by decrypting it. However it is clear that this method is impractical. Therefore a technique for searching efficiently the encrypted data on the server is desired. Up to now, researches on efficient and secure search techniques for encrypted data, which is called *searchable symmetric encryption* (SSE), have been actively done under such a background [1, 4–8, 11–14, 16, 18, 19, 22–32].

1.2   Our Contributions

The main part of SSE schemes is an encrypted index and a number of SSE schemes use an inverted index which consists of keyword-document pairs in order to realize an encrypted index. Since a keyword is usually encrypted by a pseudo-random function, the size of the index depends on a security parameter. Therefore the size becomes larger in proportion to the security parameter. In this paper, we propose a new SSE scheme which is constructed using Bloom filters and integer arrays. Our index does not depend on a security parameter. Therefore we can realize a smaller index. To do this, we design a secure mapping from an encrypted keyword to a keyword ID using a Bloom filter. Namely, the Bloom filter does not leak any information about an encrypted keyword. Using this mapping, we can realize an inverted index of keyword-document pairs with a simple array of integers. Since Bloom filters are also implemented by an array of integers, the whole index can be implemented by only integer arrays. Our scheme uses two rounds of communication at search phase. By this, however, we can design an adaptively secure scheme in the standard model and achieve an almost optimal index size which does not depend on a security parameter. We summarize the features of our scheme. Here let D be the set of documents and $\mathcal{K}(D)$ be the set of distinct keywords.

- For any keyword w, the search time is $O(\log m + n_w)$ and the communication complexity is $O(n_w)$. Here, $m = |\mathcal{K}(D)|$, that is, the number of distinct keywords, n_w is the number of documents containing w. In addition, the number of rounds of communication is 2.
- The encrypted index can be constructed using arrays of integers and the size is $O(m \log m + N)$, where $N = \sum_{w \in \mathcal{K}(D)} n_w$, that is, the number of keyword-document pairs. In fact, if we use a 4-byte integer, then the size is at most $m \log m + 12m + 8N$ bytes. This is bounded by $9N$ from above if m is small enough compared to N. Thus the size is almost optimal and does not depend on a security parameter. The size of adaptive SSE schemes proposed previously depends on a security parameter, which is typically 128 or more bits, because of a pseudo-random function. Therefore previous SSE schemes need *at least* $16N$ bytes.
- The proposed scheme is secure against adaptive attacks in the standard model. Most of the existing schemes which achieves an optimal size is proved to be adaptively secure in the random oracle model.

1.3   Related Works

Table 1 shows a comparison of related works. Goh [11] first presented an encrypted search scheme using a Bloom filter. His method stores keywords of each document in one Bloom filter and searches all documents for a given keyword. Hence the size of the index is $O(N)$ and the search time is $O(n)$. Since Goh uses a Bloom filter, the size does not depend on a security parameter. Curtmola et al. [10] presented an encrypted index using an inverted index. They pointed out the drawback of the security models used in Goh [11] and introduced two new security models, a non-adaptive semantic security model and an adaptive semantic security model. They gave two secure SSE schemes, called SSE-1 and SSE-2. SSE-1 meets non-adaptive semantic security, but the index size is $O(\lambda N)$. The search time is proportional to the number of documents containing a keyword, that is, $O(n_w)$. SSE-2 meets adaptive semantic security in the standard model and runs in time proportional to the number of documents containing a keyword. However, the index size of SSE-2 becomes $O(\lambda n s_{max})$. In addition, SSE-2 generates n encrypted keywords (called a trapdoor) to search documents for any keyword. Goh's scheme only satisfies non-adaptive semantic security under Curtmola's security models. Chase et al. [4] proposed an adaptively secure SSE scheme in the standard model. However, the index size becomes much larger, that is, $O(nm)$ in the worst case. As seen in our experiment of Sect. 6, the scheme generates a huge index. Kamara et al. [21] presented a dynamic SSE (DSSE) scheme, which can add and delete a document. Kamara's DSSE scheme meets adaptive security and achieves an optimal index size in the random oracle. As with SSE-1, the size is proportional to a security parameter.

Since Kamara's paper, a number of DSSE schemes have been proposed and proved to be adaptively secure in the random oracle [6,14,20,28,30].

Table 1. Comparison of related works. m is the number of distinct keywords, n is the number of documents, n_w is the number of documents containing w, $N = \sum_{w \in \mathcal{K}(D)} n_w$, s_{max} is the maximum number over $\{|\mathcal{K}(d)| \mid d \in D\}$, k_I is the number of bits of an integer, which is usually 32 bits, and λ is a security parameter, which is typically 128 or more bits. RO means that the scheme uses a random oracle. The column "rounds" shows the number of rounds of communication and the column "commun" shows the complexity of communication (bits) except for encrypted documents finally sent by the server.

Scheme	Size of index	Search time	Security	Rounds	Commun	Dynamic
Goh [11]	$O(N)$	$O(n)$	non-adap	1	λ	No
Curtmola [10] (SSE-1)	$O(\lambda N)$	$O(n_w)$	non-adap	1	λ	No
Curtmola [10] (SSE-2)	$\Theta(\lambda n s_{max})$	$O(n_w)$	adap	1	λn	No
Chase [4]	$O(nm)$	$O(n_w)$	adap	1	2λ	No
Kamara^{RO} [21]	$O(\lambda N)$	$O(n_w)$	adap	1	3λ	Yes
This work	$O(m \log m + N)$	$O(\log m + n_w)$	adap	2	$\lambda + k_I(n_w + 3)$	No

The schemes [6, 28, 30] achieve an optimal index size, but the size is proportional to a security parameter like Kamara's scheme. For example, Chash et al. [6] and Stefanov et al. [28] proposed DSSE schemes with an optimal index size, but the size of the indexes becomes $O(\lambda N)$ because it stores encrypted keywords. They also show an implementation of the schemes without a random oracle. In this case, the complexity of communication gets much larger by removing a random oracle. In addition, their schemes need several rounds of communication and the client must generate many trapdoors. Yavuz [31] proposed an adaptively secure DSSE scheme in the standard model. Yavuz's scheme leaks less information than other schemes, but the index size of their scheme becomes $O(nm)$, which is much larger than N, and the search time is $O(n)$. Thus the existing schemes with an optimal index size has an index in proportion to a security parameter. In Sect. 6, we will evaluate the proposed scheme using a real dataset and compare our scheme with SSE-2 and Chase's scheme. Then the index size of SSE-2 and Chase's scheme can be much larger. In particular, the size of Chase's scheme can be huge for our dataset.

We design an adaptively secure SSE scheme with an optimal index size in the standard model by adding one more round of communication. As seen in Table 1, other SSE schemes use only one round of communication, but our scheme requires two rounds of communication. By this added communication, however, we achieve an adaptively secure SSE scheme with an almost optimal index size in the standard model. Hirano [15] recently proposed a general-purpose method to build a DSSE scheme from an SSE scheme. By applying their method to our scheme, we can modify our SSE scheme into a DSSE scheme which can add new documents.

2 Preliminaries

We consider the following search problem. Let $D = \{d_0, \ldots, d_{n-1}\}$ be the set of n documents and let $n = |D|$, where $|D|$ denote the number of elements in D. Each document d_i ($0 \leq i \leq n-1$) is assigned a unique number i called a *document ID* and we write $ID(d_i) = i$. For each document d_i, let $\mathcal{K}(d_i)$ be the set of keywords taken out from d_i. Furthermore let us define $\mathcal{K}(D) = \cup_{d \in D}\mathcal{K}(d)$ and $m = |\mathcal{K}(D)|$. Thus m denotes the total number of distinct keywords used in our scheme. For any $w \in \mathcal{K}(D)$, let $D(w)$ be the set of IDs of documents containing w. We let $n_w = |D(w)|$ and $N = \sum_{w \in \mathcal{K}(D)} n_w$. The search problem is to find all documents d_i containing w for a given keyword w. By $x||y$ we denote the concatenation of two strings x and y and by $[i, j]$ ($i \leq j$) we denote the set $\{i, i+1, \ldots, j\}$ of integers.

In this paper, we address the search problem using a symmetric encryption scheme. A symmetric encryption scheme is a set of three polynomial time algorithms $SKE = (\mathsf{KeyGen}, \mathsf{Enc}, \mathsf{Dec})$, where $\mathsf{KeyGen}(1^\lambda)$ takes as an input a security parameter λ and randomly outputs a secret key sk; $\mathsf{Enc}(sk, d)$ takes as inputs a secret key sk and a text d and returns a ciphertext c; $\mathsf{Dec}(sk, c)$ takes as inputs a key sk and a ciphertext c and returns d if sk is the key which is used to produce c. As seen in [17], we consider that a symmetric encryption scheme satisfies chosen-plaintext attack security (CPA-security). For simplicity, by $E_{sk}(\cdot)$ we denote an encryption function $\mathsf{Enc}(sk, \cdot)$ with a secret key sk. In addition to a symmetric encryption scheme, we make use of pseudo-random functions, which are polynomial-time functions that cannot be distinguished from random functions. Let $F : \{0,1\}^\lambda \times \{0,1\}^* \to \{0,1\}^\lambda$ be a pseudo-random function. Then we will write $F_K(x)$ instead of $F(K, x)$.

The search system consists of two parties, a client and a server. We assume that the server is honest but curious. A client is an owner of documents and stores them in encrypted form in a server. Given an encrypted keyword (which is called *a trapdoor*) from a client, a server performs a search using an encrypted index. In the following, we show an outline of an encrypted search scheme.

(1) A client randomly generates a secret key $SK = (sk_1, sk_2, sk_3)$ using a security parameter λ.

(2) Construction of an encrypted index. The client encrypts all documents using his secret key sk_1. In addition, the client makes an encrypted index from keywords using a pseudo-random function F_{sk_2} and the secret key sk_2. After that, the client sends encrypted documents and the encrypted index to the server.

(3) Search phase. For a keyword w, the client makes a trapdoor $T_w = F_{sk_2}(w\|0)$ using a pseudo-random function F_{sk_2}, and then sends them to the server. Receiving T_w, the server searches for documents d containing w using T_w and the encrypted index. After that, the server returns $\{c_i\}_{i \in D(w)}$ to the client. The client decrypts each c_i using the secret key sk_1 and gets original documents d_i.

3   Bloom Filter

A Bloom filter is a data structure proposed by Bloom [2], which consists of a bit sequence and is used for efficiently checking whether $x \in S$ for a set S and an element x. Let BF be a Bloom filter of m bits and let $W = \{w_1, \ldots, w_l\}$ be a set of l words, $h_1, \ldots, h_k$ be hash functions mapping words to $[0, m-1]$. Then, for $w_i \in W$, we set all bits at positions $h_1(w_i), \ldots, h_k(w_i)$ in BF to 1. For any word w, whether $w \in W$ or not is checked as follows. First we compute $h_1(w), \ldots, h_k(w)$, and then check bits at positions $h_1(w), \ldots, h_k(w)$ in BF. If all bits are 1, then we decide $w \in W$; otherwise $w \notin W$. The drawback of a Bloom filter is to make an error called *a false positive*, that is, there is a possibility such that for $v \notin W$, all bits at positions $h_1(v), \ldots, h_k(v)$ are 1. In this case, we get a wrong answer such that $v \in W$. Note that for $w \in W$ we always get a correct answer. Broder and Mitzenmacher [3] analyze the relationship among the size m_1 of a Bloom filter, the number m_2 of elements, the probability of false positive, and the number of hash functions. They showed that given m_1 and M_2, the probability of false positive becomes minimum when $k = (m_1/m_2)\ln 2$. In this case the false positive rate is approximately $(0.6185)^{m_1/m_2}$. We give an example of a Bloom

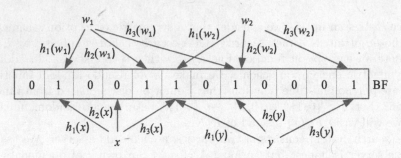

Fig. 1. An example of a Bloom filter.

filter and a counting Bloom filter in Fig. 1 for $W = \{w_1, w_2\}$. Let $W = \{w_1, w_2\}$ and let h_1, h_2, h_3 be hash functions. Then, to make a Bloom filter for W, we set bits at position $h_1(w_1), h_2(w_1), h_3(w_1)$ to 1 for w_1 and set bits at positions $h_1(w_2), h_2(w_2), h_3(w_2)$ to 1 for w_2. On the other hand, for a counting Bloom filter, $h_3(w_1)$ and $h_2(w_2)$ indicate the same position; therefore the position has a value 2. For example, for a word x, since the bit at the position $h_2(x)$ is 0, we see $x \notin W$. For a word y, since all bits of positions $h_1(w_2), h_2(w_2), h_3(w_2)$ are 1, we decide $y \in W$.

4   Proposed Scheme

The proposed SSE scheme consists of six polynomial-time algorithms $\textbf{SSE} = (\textsf{KeyGen}, \textsf{Enc}, \textsf{Dec}, \textsf{BuildIndex}, \textsf{Trapdr}, \textsf{Search})$ such that

- $\textsf{KeyGen}(1^\lambda)$ is a probabilistic algorithm which takes as an input a security parameter λ and returns secret keys $SK = (sk_1, sk_2, sk_3)$.
- $\textsf{Enc}(sk_1, d)$ is a probabilistic algorithm which takes as inputs a secret key sk_1 and a document d and returns an encrypted document $c = E_{sk_1}(d)$.
- $\textsf{Dec}(sk_1, c)$ is a deterministic algorithm which takes as inputs a secret key sk_1 and an encrypted document c and returns the decrypted document.
- $\textsf{BuildIndex}(SK, D, \varepsilon)$ is a probabilistic algorithm which takes as inputs secret keys, a set of documents, and a parameter ε and returns an encrypted index $\Pi = (\textbf{Kmap}, \textbf{Nmap}, \textbf{Dmap})$.
- $\textsf{Trapdr}(sk_2, w)$ is a deterministic algorithm which takes as inputs a secret key sk_2 and a keyword w and returns a trapdoor T_w. We define $\textsf{Trapdr}(sk_2, w) = F_{sk_2}(w\|0)$ $(= T_w)$ where $F_{sk_2}(\cdot)$ is a pseudo-random function.
- $\textsf{Search}(T_w)$ is a deterministic algorithm which takes an input as a trapdoor T_w and returns document IDs containing w.

We use a CPA-secure symmetric encryption scheme for $(\textsf{KeyGen}, \textsf{Enc}, \textsf{Dec})$. In the following, we will show an algorithm $\textsf{BuildIndex}$ and a search algorithm $\textsf{Search}$.

4.1   Constructing an Encrypted Index

An encrypted index $\Pi = (\mathbf{Kmap}, \mathbf{Nmap}, \mathbf{Dmap})$ consists of $\mathbf{Kmap}$, $\mathbf{Nmap}$, and $\mathbf{Dmap}$. $\mathbf{Kmap}$ is constructed by Bloom filters and is used to compute a keyword ID from a trapdoor. $\mathbf{Nmap}$ is an array of m entries, each of which has start position of a list of $D(w)$ and n_w for each keyword w. $\mathbf{Dmap}$ is an array of N entries which have a pair of a document ID and a next address. The encrypted index is built by BuildIndex given in Algorithm 1. The operator $\oplus$ denotes an exclusive-OR. Let π be a random permutation over $[1, N-1]$.

Construction of Kmap. We assign a unique number (called *an identifier*) from $[0, m-1]$ to each keyword of $\mathcal{K}(D)$. This time, we randomly take a unique number from $[0, m-1]$ for each keyword of $\mathcal{K}(D)$. The identifier of a keyword w is denoted by $ID(w)$. Let $\mathcal{K}(D) = \{w_0, \ldots, w_{m-1}\}$, where for $0 \leq i \leq m-1$, $ID(w_i) = i$. $\mathbf{Kmap}$ is an array consisting of Bloom filters, that is, $\mathbf{Kmap}[i]$ has a Bloom filter. Then, for any $w \in \mathcal{K}(D)$, $\mathbf{Kmap}$ realizes a mapping the trapdoor T_w of w to $ID(w)$. Now we present how to build $\mathbf{Kmap}$ in the following. We use a binary tree to manage the set $\mathcal{K}(D)$.

Let us define $h = \lceil \log_2 m \rceil$. We use 2 as the base of logarithm in this paper. First of all, for any level $0 \leq lev \leq h$, we define a decomposition $\mathcal{K}(D)^{lev}$ of $\mathcal{K}(D)$ at level lev. The decomposition $\mathcal{K}(D)^{lev}$ is the set $\{W_0^{lev}, W_{2^\alpha}^{lev}, W_{2 \cdot 2^\alpha}^{lev}, \ldots, W_{(2^{lev}-1) \cdot 2^\alpha}^{lev}\}$ such that for any $i = 0, 2^\alpha, 2 \cdot 2^\alpha, 3 \cdot 2^\alpha, \ldots, (2^{lev}-1) \cdot 2^\alpha$, $W_i^{lev} = \{w_i, w_{i+1}, \ldots, w_{i+2^\alpha-1}\}$, where $\alpha = h - lev$. For level lev, each W_i^{lev} is a subset of $\mathcal{K}(D)$ and consists of at most 2^{lev} elements. Note that for any lev, W_i^{lev} is numbered with an interval of 2^α starting from 0. Furthermore W_i^{lev} is a subset of $\mathcal{K}(D)$ consists of at most 2^α elements. And the number i is called the node ID of W_i^{lev}.

The decomposition forms a binary tree by regarding subsets as nodes. For W_i^{lev} ($1 \leq lev \leq h$), we define W_i^{lev+1} and $W_{i+2^{h-(lev+1)}}^{lev+1}$ to be children of W_i^{lev}. Then the decomposition constructs a binary tree such that each W_i^{lev} is a node and the root is W_0^0. We call this tree *a keyword tree* and i *the ID* of node W_i^{lev}. For a keyword tree, each leaf W_i^h is a set $\{w_i\}$ consisting of just one keyword w_i. We build one Bloom filter BF_{lev} for $0 \leq lev \leq h$ and store it in $\mathbf{Kmap}[lev]$. Since we use a Bloom filter of size $\varepsilon \cdot m$ bits for each level, the size of $\mathbf{Kmap}$ is $(h+1) \cdot \varepsilon \cdot m$ bits.

Example 1. We give an example of a keyword tree. Let $\mathcal{K}(D) = \{w_0, w_1, w_2, w_3, w_4\}$. Then, Fig. 2 depicts the keyword tree for $\mathcal{K}(D)$. In Fig. 2, BF_0, BF_1, BF_2, and BF_3 are a Bloom filter for each level, and each node consists of the following sets, respectively.

$W_0^0 = \{w_0, w_1, w_2, w_3, w_4\}$,
$W_0^1 = \{w_0, w_1, w_2, w_3\}$, $W_4^1 = \{w_4\}$,
$W_0^2 = \{w_0, w_1\}$, $W_2^2 = \{w_2, w_3\}$, $W_4^2 = \{w_4\}$, $W_6^2 = \emptyset$,
$W_0^3 = \{w_0\}$, $W_1^3 = \{w_1\}$, $W_2^3 = \{w_2\}$, $W_3^3 = \{w_3\}$, $W_4^3 = \{w_4\}$, $W_5^3 = \emptyset$,
$\quad W_6^3 = \emptyset$, $W_7^3 = \emptyset$.

Algorithm 1. BuildIndex($(sk_1, sk_2, sk_3), D, \varepsilon$)

```
1: n ← |D|, m ← |K(D)|, h ← ⌈log m⌉
2: initialize Kmap, Nmap, and Dmap
3: start ← 1
4: for lev = 0 to h do
5:     initialize a Bloom filter BF_lev of ε · m bits with 0
6:     for all W_i^lev ∈ K(D)^lev = {W_0^lev, W_{2^α}^lev, W_{2·2^α}^lev, ..., W_{(2^lev−1)·2^α}^lev} do
7:         for all w ∈ W_i^lev do
8:             X ← F_{sk_2}(w||0),
9:             p_1 ← F_X(i||1), ..., p_k ← F_X(i||k), and set all bits of positions p_1, ..., p_k of BF_lev
10:            if lev = h then
11:                let set n_w = |D(w)| and D(w) = {id_0, ..., id_{n_w−1}}
12:                Nmap[i] ← (F_X(0), (n_w||π(start)) ⊕ F_{sk_2}(w||1))
13:                for j = 0 to n_w − 2 do
14:                    Dmap[π(start + j)] ← (id_j||π(start + j + 1)) ⊕ F_{sk_3}(w||j)
15:                end for
16:                Dmap[π(start + n_w − 1)] ← (id_{n_w−1}||0) ⊕ F_{sk_3}(w||n_w − 1)
17:                start ← start + n_w
18:            end if
19:        end for
20:    end for
21:    store BF_lev to Kmap[lev]
22: end for
23: output Π = (Kmap, Nmap, Dmap)
```

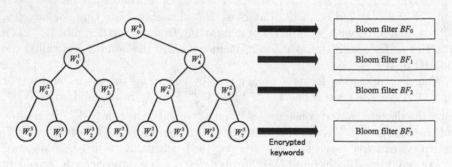

Fig. 2. Keyword tree and the corresponding Bloom filters (**Kmap**).

Construction of Nmap. For any keyword $w \in \mathcal{K}(D)$, let $n_w = |D(w)|$. Then we store $(F_X(0), (n_w||\pi(start)) \oplus F_{sk_2}(w||1))$ in **Nmap**$[ID(w)]$, where $X = F_{sk_2}(w||0)$. As we store document IDs in $D(w)$ in **Dmap**, the number $start$ is randomly chosen as the position of the first element in $D(w)$. Furthermore, we can exclude a false positive which may occurs on **Kmap** by checking if the first element of **Nmap**$[ID(w)]$ is equal to $F_X(0)$. The size of **Nmap** is $O(m)$. If we use a 4-byte integer, then the size becomes $12m$ bytes because $(F_X(0), (n_w||\pi(start)) \oplus F_{sk_2}(w||1))$ can be represented by 3 integers.

Construction of Dmap. Recall that $N = \sum_{w \in \mathcal{K}(D)} n_w$. **Dmap**$[i]$ is an array of N entries which are pairs of a document ID and a next address. Then, for

Algorithm 2. Search(T_w, Π)

Require: T_w:a trapdoor, Π Fan encrypted index
1:   $cid \leftarrow NULL$
2:   $ListOfID \leftarrow \emptyset$, $stackIDLevel \leftarrow \emptyset$
3:   push $(0,0)$ onto $stackIDLevel$ // $stackIdLevel$ is a stack holding $(nodeID, level)$
4:   **while** $stackIDLevel \neq \emptyset$ **do**
5:      pop (id, lev) from $stackIDLevel$
6:      **if** $cid \neq id$ **then**
7:         $p_1 \leftarrow F_{T_w}(id||1), \ldots, p_k \leftarrow F_{T_w}(id||k)$
8:         $cid \leftarrow id$
9:      **end if**
10:    **if** all bits of positions $p_1, \ldots, p_k$ in **Kmap**[lev] are 1 **then**
11:      **if** $lev = h$ **then**
12:         add id to $ListOfID$
13:      **else**
14:         $lev \leftarrow lev + 1$
15:         push $(id + 2^{h-lev}, lev)$ onto $stackIDLevel$
16:         push (id, lev) onto $stackIDLevel$
17:      **end if**
18:    **end if**
19: **end while**
20: **for all** $id \in ListOfID$ **do**
21:    $(X, Y) \leftarrow$ **Nmap**[id]
22:    **if** $F_{T_w}(0) = X$ **then**
23:      **return** Y
24:    **end if**
25: **end for**

---

any keyword w, document IDs in $D(w)$ are stored in **Dmap** as follows. Let $D(w) = \{id_0, \ldots, id_{n_w-1}\}$. Let $start$ be the start position stored in **Nmap** for w. For any $0 \leq j \leq n_w - 2$, we store $(id_j||\pi(start + j + 1)) \oplus F_{sk_3}(w||j)$ in **Dmap**[$\pi(start+j)$] and $(id_{n_w-1}||0) \oplus F_{sk_3}(w||n_w-1)$ in **Dmap**[$\pi(start+n_w-1)$]. Note that 0 is used as a sentinel value. Since **Dmap** becomes a simple array of $2N$ integers, the size of **Dmap** is $8N$ bytes if we use a 4-byte integer.

The total size of the index. From the above discussion, it is obvious that the total size of **Kmap**, **Nmap**, and **Dmap** becomes $O(m \log m + N)$.

4.2 Search Protocol

The following shows a search protocol which requires two rounds of communication between the client and the server.

1. The client makes a trapdoor $T_w = \mathsf{Trapdr}(sk_2, w)$ for a keyword w and sends it to the server.
2. Receiving T_w, the server performs Search on T_w and gets $(n_w||\pi(start)) \oplus F_{sk_2}(w||1)$. After that, the server sends it to the client.

Algorithm 3. Search2($first, T_0, \ldots, T_{n_w-1}$)

1: $addr \leftarrow first$, $j \leftarrow 0$, and $DocID \leftarrow \emptyset$
2: **while** $addr \neq 0$ **do**
3:     $(id, next) \leftarrow \mathbf{Dmap}[addr] \oplus T_j$
4:     $addr \leftarrow next$ and $j++$
5:     add id to $DocID$
6: **end while**
7: **return** $DocID$

---

3. The client decrypts $(n_w \| \pi(start)) \oplus F_{sk_2}(w\|1)$ and gets $(n_w, \pi(start))$. The client computes $T_0 = F_{sk_3}(w\|0), \ldots, T_{n_w-1} = F_{sk_3}(w\|n_w - 1)$, and sends $(\pi(start), T_0, \ldots, T_{n_w-1})$ to the server.
4. Receiving $(\pi(start), T_0, \ldots, T_{n_w-1})$, the server gets the set $DocID$ of document IDs by performing Search2$(\pi(start), T_0, \ldots, T_{n_w-1})$ given in Algorithm 3. After that, the server encrypted documents c_{id} ($id \in DocID$) to the client.
5. The client gets the original documents d_{id} by decrypting c_{id}.

We explain how Search behaves. Given a trapdoor T_w, Search computes $ID(w)$ using **Kmap**. Search visits each node of a keyword tree from the root in preorder using a stack $stackIDLevel$. For any node, if all bits related to T_w in **Kmap**[lev] are 1, then Search checks two children of the node. If not, then all descendants of the node are excluded for checking. For any leaf, if all bits related to T_w in **Kmap**[h] are 1, then Search adds id to $ListOfID$. Since a keyword ID is unique, the number of IDs in $ListOfID$ is exactly one if false positive does not occur. Search picks (X, Y) from **Nmap**[id] for all $id \in ListOfID$ and checks if $F_{T_w}(0) = X$. If id is a correct position for w, then $F_{T_w}(0) = X$ always holds; else the equation does not. Thus we can eliminate false positive if any. Finally Search returns $Y = (n_w \| \pi(start)) \oplus F_{sk_2}(w\|1)$.

Let us analyze the running time of Search. It takes $O(\log m)$ to compute a keyword ID using **Kmap**, and $O(1)$ to pick data from **Nmap**. The time to pick document IDs from **Dmap** is $O(n_w)$. Indeed, this job can be performed very fast because we can get the ID by simply computing an exclusive-OR of two integers. Thus the search time on the server is $O(\log m + n_w)$, which is almost optimal.

Let us analyze the communication complexity. Our scheme requires two rounds of communication. At the first round, the client sends a trapdoor to the server and the sever returns data $(n_w \| \pi(start)) \oplus F_{sk_2}(w\|1)$. Therefore the amount of communication data becomes λ bits plus $2k_I$ bits, where k_I is the number of bits of integer, typically 32 bits. At the second round, the client sends $\pi(start)$ and n_w trapdoors $T_0, \ldots, T_{n_w-1}$ to the server. Since these can be all represented as an integer, the amount of bits sent to the server becomes $k_I + k_I n_w$ bits. Finally the server sends encrypted documents to the client. Let us compare our scheme with other SSE schemes. We examine communication other than the encrypted documents sent by the server because it is common in all SSE schemes. Following the above discussion, the communication complexity of our scheme becomes $\lambda + k_I(n_w + 3)$ bits. Since Curtmola's SSE-2 [10] must sends n trapdoors to the server, the communication complexity becomes $\lambda \cdot n$ bits. Chase's scheme [4] sends one trapdoor; therefore it is λ bits. Therefore our scheme requires two rounds of communication, but the complexity of communication is less than Curtmola's SSE-2 and more than Chase's SSE.

5 Security Analysis

We show that the propose scheme satisfies adaptive security in the standard model. Other schemes with an optimal index size is proved to be adaptively secure in the random oracle model. We eliminate a restriction of a random oracle by increasing a round of communication by one. We define a negligible function for a security definition.

Definition 1. *A function f from natural numbers to positive real numbers is negligible in a security parameter λ if for every positive polynomial $p(\cdot)$ and sufficiently large λ, $f(\lambda) < 1/p(\lambda)$.*

To analyze the security, we consider two games, a real game $\mathbf{REAL}_{\mathcal{A}}$ and a simulation game $\mathbf{SIM}_{\mathcal{A},\mathcal{S}}$. As you see below, $\mathbf{REAL}_{\mathcal{A}}$ performs the proposed scheme, while $\mathbf{SIM}_{\mathcal{A},\mathcal{S}}$ simulates the proposed scheme using only information that an adversary (that is, a server) can get from leakage functions $\mathcal{L}_1$ and $\mathcal{L}_2$. The leakage functions $\mathcal{L}_1$ and $\mathcal{L}_2$ are defined as follows.

- $\mathcal{L}_1(\Pi, \mathbf{c})$: This is information leaked from the initial setting. Given an encrypted index Π and a collection $\mathbf{c} = \{c_0, \ldots, c_{n-1}\}$ of encrypted documents, $\mathcal{L}_1$ outputs the length of each document $|d_0|, \cdots, |d_{n-1}|$ and the document IDs $ID(d_0), \ldots, ID(d_{n-1})$. In addition, it outputs the size of Bloom filters, n, m, and N.
- $\mathcal{L}_2(\Pi, T_w, \mathbf{c}, t)$: This is information leaked by a search at time t. Given a trapdoor T_w at time t, $\mathcal{L}_2$ outputs the identifier $ID(w)$ of a keyword w, n_w, the access pattern consisting of $D(w)$ and the search pattern which is the set of trapdoors used previously by a search.

Now, we define two games to define adaptive semantic security. These games are played by three player, an adversary $\mathcal{A}$, a challenger $\mathcal{C}$, and a simulator $\mathcal{S}$.

$\mathbf{REAL}_{\mathcal{A}}(\lambda)$

- The adversary $\mathcal{A}$ chooses $D = \{d_0, \ldots, d_{n-1}\}$ and sends them $\mathcal{C}$
- $\mathcal{C}$ generates secret keys $SK = (sk_1, sk_2, sk_3)$ using $KeyGen(1^\lambda)$ and build an encrypted index $\Pi = (\mathbf{Kmap}, \mathbf{Nmap}, \mathbf{Dmap})$ using BuildIndex(SK, D, ε). After that, $\mathcal{C}$ sends $(\Pi, E_{sk_1}(D))$ to $\mathcal{A}$.
- repeat the following polynomially many times.
  1. $\mathcal{A}$ sends a keyword w to $\mathcal{C}$.
  2. $\mathcal{C}$ makes a trapdoor T_w for w using Trapdr(sk_2, w) and sends it to $\mathcal{A}$.
  3. $\mathcal{A}$ gets $(n_w || \pi(start)) \oplus F_{sk_2}(w || 1)$ using Search(T_w, Π) and sends it to $\mathcal{C}$.

4. $\mathcal{C}$ decrypts $(n_w||\pi(start)) \oplus F_{sk_2}(w||1)$ and gets $(n_w, \pi(start))$. After that, $\mathcal{C}$ computes $T_0 = F_{sk_3}(w||0), \ldots, T_{n_w-1} = F_{sk_3}(w||n_w - 1)$, and sends $(\pi(start), T_0, \ldots, T_{n_w-1})$ to $\mathcal{A}$.
5. $\mathcal{A}$ picks document IDs from **Dmap**.
- Finally, $\mathcal{A}$ outputs a bit $b \in \{0, 1\}$.

$\mathbf{SIM}_{\mathcal{A},\mathcal{S}}(\lambda)$

- The adversary $\mathcal{A}$ chooses $D = \{d_0, \ldots, d_{n-1}\}$ and sends them to $\mathcal{C}$. $\mathcal{C}$ sends information obtained from $\mathcal{L}_1$ to the simulator $\mathcal{S}$.
- $\mathcal{S}$ builds an encrypted index $\Pi^* = (\mathbf{Kmap}^*, \mathbf{Nmap}^*, \mathbf{Dmap}^*)$ and sends them to $\mathcal{C}$. $\mathcal{C}$ sends them to $\mathcal{A}$.
- Repeat the following polynomially many times.
  1. $\mathcal{A}$ sends a keyword w to $\mathcal{C}$ and $\mathcal{C}$ sends information obtained from $\mathcal{L}_2$ to $\mathcal{S}$.
  2. $\mathcal{S}$ generates a trapdoor T_w^*, and then sends it to $\mathcal{C}$. $\mathcal{C}$ sends it to $\mathcal{A}$.
  3. $\mathcal{A}$ gets Y using $\mathsf{Search}(T_w^*, \Pi^*)$ and sends it to $\mathcal{C}$. $\mathcal{C}$ sends it to $\mathcal{S}$.
  4. $\mathcal{S}$ generates $(p^*, T_0^*, \ldots, T_{n_w-1}^*)$ and sends them to $\mathcal{C}$. $\mathcal{C}$ sends them to $\mathcal{A}$.
  5. $\mathcal{A}$ picks document IDs from **Dmap**.
- Finally, $\mathcal{A}$ outputs a bit $b' \in \{0, 1\}$.

Definition 2. *An SSE scheme is $(\mathcal{L}_1, \mathcal{L}_2)$-secure against adaptive attacks if for any probabilistic polynomial time algorithm $\mathcal{A}$, there is a simulator $\mathcal{S}$ such that*
$$|Pr[\mathcal{A} \text{ outputs } 1 \text{ in } \mathbf{REAL}_{\mathcal{A}}(\lambda)] - Pr[\mathcal{A} \text{ outputs } 1 \text{ in } \mathbf{SIM}_{\mathcal{A},\mathcal{S}}(\lambda)]|$$
is negligible.

Theorem 1. *The proposed SSE scheme is $(\mathcal{L}_1, \mathcal{L}_2)$-secure against adaptive attacks.*

Proof. We will show that we can construct a simulator $\mathcal{S}$ step by step such that the adversary cannot distinguish **REAL** and **SIM**. $\mathcal{S}$ uses a random permutation π^* to simulate π.

1. Simulating encrypted documents for the document set D. Simulator $\mathcal{S}$ generates a random strings c_i^* of $|d_i|$ bits for a document d_i $(0 \le i \le n - 1)$. Since a symmetric encryption scheme is CPA-secure, c_i and c_i^* is indistinguishable.
2. Simulating **Kmap**. From the leakage information $\mathcal{L}_1$, the simulator $\mathcal{S}$ knows the number m of keywords. First $\mathcal{S}$ generates m random strings $T_{w_0}^*, \ldots, T_{w_{m-1}}^*$ of λ bits. Each $T_{w_i}^*$ $(0 \le i \le m - 1)$ is used as a trapdoor for a keyword w_i. $\mathcal{S}$ builds **Kmap*** in a similar way to $\mathsf{BuildIndex}$.
3. Simulating **Nmap**. To simulate **Nmap**, $\mathcal{S}$ builds **Nmap*** in the following way. For i $(0 \le i \le m - 1)$, $\mathcal{S}$ generates a random string r_i of $3k_I$ bits and stores r_i in **Nmap**$^*[i]$, where k_I is the number of bits of an integer. In **REAL**, **Nmap**$[i]$ has $(F_X(0), (n_w||\pi(start)) \oplus F_{sk_2}(w||1))$, which is represented by $3k_I$ bits. Since F_{sk_2} is a pseudo-random function, **Nmap** and **Nmap*** are indistinguishable.

4. Simulating **Dmap**. To simulate **Dmap**, S builds $\textbf{Dmap}^*$ in the following way. S knows the size N of **Dmap** from leakage information by $\mathcal{L}_1$. For all j ($1 \le j \le N - 1$), S generates two random strings r_j^1 and r_j^2 of k_I bits, and then stores $(r_j^1 || r_j^2)$ in $\textbf{Dmap}^*[j]$. In **REAL**, **Dmap** has $(id || next) \oplus F_{sk_3}(w || cnt)$ for some $cnt \in [0, N - 1]$. Thus, $\textbf{Dmap}^*$ and **Dmap** are indistinguishable due to pseudo-randomness of F_{sk_3}.

5. Simulating a search at time t. Let w be a keyword for which the adversary $\mathcal{A}$ searches at time t and let Q_{t-1} be the set of keywords which are searched for at some time $t' < t$. S knows an access pattern $D(w)$ from leakage information by $\mathcal{L}_2$. First, S checks if w is a keyword searched for before. If so, then S uses the same random string T_w^* as one used previously for w; otherwise S chooses T_w^* from $\{T_{w_0}^*, \ldots, T_{w_{m-1}}^*\}$ such that T_w^* is not used yet and uses it as a trapdoor of w. This time, S memorizes that T_w^* is used for w. S sends T_w^* to C. After performing Serach, $\mathcal{A}$ sends a random string r to C and C sends it to S. Then S acts as follows. Let $D(w) = \{id_0, \ldots, id_{n_w-1}\}$ and let $start_w = 1 + \sum_{w \in Q_{t-1}} n_w$.

   (a) S computes $p_j^* = \pi^*(start_w + j)$ for all $0 \le j \le n_w - 1$.

   (b) For $0 \le j \le n_w - 2$, S generates T_j^* such that $(id_j || \pi^*(start_w + j + 1) = \textbf{Dmap}[\pi^*(\ start_w + j)] \oplus T_j^*$. In addition, S generates $T_{n_w-1}^*$ such that $(id_{n_w-1} || 0) = \textbf{Dmap}[\pi^*(\ start_w + n_w - 1)] \oplus T_{n_w-1}^*$.

   (c) S sends $(p_0^*, T_0^*, \ldots, T_{n_w-1}^*)$ to C.

   Now let us show that $\mathcal{A}$ cannot distinguish **REAL** and **SIM** in a search protocol. Given T_w^* with $ID(w) = j$, $\mathcal{A}$ picks $(F_{T_w^*}(0), r_j)$ from $\textbf{Nmap}^*[j]$. If $\mathcal{A}$ accesses an element (X, Y) other than $\textbf{Nmap}^*[j]$ by a false positive, then X does not match with $F_{T_w^*}(0)$. Therefore $\mathcal{A}$ cannot distinguish **REAL** and **SIM** at this moment because two games behave in a similar way. Next $\mathcal{A}$ sends r_j to C and C sends it to S. Receiving r_j, S generates $(p_0^*, T_0^*, \ldots, T_{n_w-1}^*)$ as follows. If w is a keyword asked before, then S picks the same one as used before. If w is a new keyword, then S generates them using the procedure stated above. This time, we must note that T_j^* can be viewed as a string chosen randomly because each entry of $\textbf{Dmap}^*$ is set to a random string. Finally S gives $(p_0^*, T_0^*, \ldots, T_{n_w-1}^*)$ to $\mathcal{A}$. It follows from the property of $(p_0^*, T_0^*, \ldots, T_{n_w-1}^*)$ that $\mathcal{A}$ can get $D(w)$ using $\textbf{Dmap}^*$. In **REAL**, each ID of $D(w)$ is stored according a random permutation π and $T_0, \ldots, T_{n_w-1}$ are generated using a pseudo-random function. Therefore $\mathcal{A}$ cannot distinguish **REAL** and **SIM** with all but negligible probability.

Thus difference between the probabilities that the adversary $\mathcal{A}$ outputs 1 in **REAL** and in **SIM** is negligible. Hence the theorem holds.

6   Experimental Results

We present an experimental evaluation of the proposed method on a real world dataset: the Enron Email Dataset [9]. We use a total of 517431 mails and extract at most 500 keywords per a mail to built an encrypted index. Note that we regard a mail as a document. The total size of keywords extracted is 348MB and the number of distinct keywords is 307830. We implemented all algorithms in JAVA language on a Win10 (64 bits) machine with Intel Core i7-6700 Processor 3.4 GHz and a memory of 16 GB. We use AES of key-length 128 bits as a symmetric encryption scheme to encrypt a keyword, and HMAC-SHA256 as a pseudo-random function and a hash function to register an encrypted keyword in a Bloom filter.

The size of the index. In implementation of **Kmap**, we set $\varepsilon = 10$ and $h = 20$. First of all, let us evaluate the size of the index, **Kmap**, **Nmap**, and **Dmap**. **Kmap** consists of 20 Bloom filters each of which has a size of 3078300 bits. Therefore the size of **Kmap** becomes $307830 \times 10 \times 20 = 61566000$ bits. This is about 7.7 MB. **Nmap** is an array of 307830 entries, each of which has a form of $(F_X(0), (n_w||\pi(start)) \oplus F_{sk_2}(w||1))$. We use 4 bytes for $F_X(0)$. Furthermore we can express $(n_w||\pi(start)) \oplus F_{sk_2}(w||1)$ in 8 bytes because n_w and $\pi(start)$ are integers. Hence, we use 12 bytes per one entry. Therefore the size of **Nmap** becomes $307830 \times 12 = 3693960$ bytes, that is, about 3.7MB. **Dmap** is an array of 37776352 pairs of two integers. Since we use a 4-byte integer, the size of **Dmap** becomes $37776352 \times 8 = 302210816$ bytes, that is, about 303 MB. Consequently, the total size of the index is about 315 MB, which is almost same the size of the original data. This is because our index depends on only the number of bits of an integer type but not on a security parameter.

Curtmola's SSE-2 needs to store $517431 \times 500 = 258715500$ keyword-document pairs because every document ID needs to appear in the index the same number of times and there is a document ID with 500 keywords. A keyword-document pair is represented as a pair of a trapdoor of the keyword and document ID. In our setting, the trapdoor is 128-bit (32 bytes) and a document ID is represented by a 4-byte integer. Therefore the total size of the index becomes 258715500×36 bytes, that is, about 9.3 GB. Chase's scheme [4] further needs a larger index because it stores a pair (a keyword, a list of document IDs) in the index and a list of document IDs is padded so that all lists are of the same length. In our dataset, there is a keyword such that it appears in all documents. Therefore the length of all lists is 517431 and the size of the index becomes at least 517431×307830 bytes, that is, 150 GB. Thus Chase's scheme requires an extremely large index if there a high frequency keyword; that is, a keyword appears in many documents. How about SSE schemes with an $O(\lambda N)$ index size. We have $\lambda = 128$ bits (16 bytes) and $N = 37776352$ for our experiment. Therefore such SSE schemes need at least 604MB because $16 \times 37776352 = 604441632$ bytes.

The search time. Next, we examine the search time of all keywords. Note that we implement the search protocol as one program on a computer, so the time related to communication is not considered in a search time stated in the following. The time taken by the client is included. We measure the time taken to gets document IDs from a given keyword w for all keywords. The experiment results are shown in Fig. 3 and Table 2. In this experiment, a false positive did not occur in **Kmap**. This is because we made the size of Bloom filters large

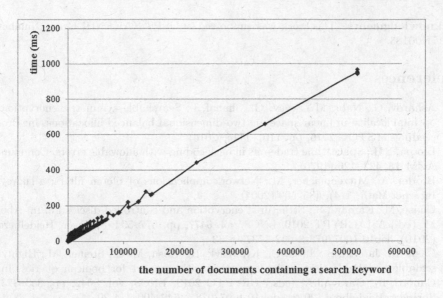

Fig. 3. Relationship between search time (ms) and the number of documents containing a search keyword.

Table 2. Summary of search time. The column "# of matching documents" denotes the number of documents containing a search keyword.

# of matching documents	Search time (ms)
1 to 100	0.17 to 0.23
101 to 1000	0.23 to 2.0
1001 to 10000	2.0 to 18
10001 to 100000	18 to 221
100001 to 517431	221 to 946
517431	946

sufficiently (that is, $\varepsilon = 10$. Note that the larger ε, the smaller a false positive.). Figure 3 shows the relation between the search time and n_w which is the number of documents containing w. Table 2 shows the outline of the search time. As shown in Fig. 3, the search time is proportional to n_w. Since the theoretical analysis tells us that the search time is $O(\log m + n_w)$, we can see that the search time almost depends on searching **Dmap**. In the worst case that $n_w = 517431$ (this means that w appears in all documents), it takes 946[ms]. This time includes the time for the client to generate 517431 trapdoors T_i for searching **Dmap**. We note that SSE-2 must always generate 517431 trapdoors for any keyword even though n_w is small. On the other hand, our scheme generates only n_w trapdoors.

Acknowledgments. This work was supported by JSPS KAKENHI Grant Number JP17K00183.

References

1. Asharov, G., Naor, M., Segev, G., Shahaf, I.: Searchable symmetric encryption: optimal locality in linear space via two-dimensional balanced allocations. In: Proceedings of STOC 2016, pp. 1101–1114 (2016)
2. Bloom, B.H.: Space/time trade-offs in hash coding with allowable errors. Commun. ACM **13**, 422–426 (1970)
3. Broder, A., Mitzenmacher, M.: Network applications of bloom filters: a survey. Internet Math. **1**(4), 485–509 (2004)
4. Chase, M., Kamara, S.: Structured encryption and controlled disclosure. In: Abe, M. (ed.) ASIACRYPT 2010. LNCS, vol. 6477, pp. 577–594. Springer, Heidelberg (2010). doi:10.1007/978-3-642-17373-8_33
5. Cash, D., Jarecki, S., Jutla, C., Krawczyk, H., Roşu, M.-C., Steiner, M.: Highly-scalable searchable symmetric encryption with support for boolean queries. In: Canetti, R., Garay, J.A. (eds.) CRYPTO 2013. LNCS, vol. 8042, pp. 353–373. Springer, Heidelberg (2013). doi:10.1007/978-3-642-40041-4_20
6. Cash, D., Jaeger, J., Jarecki, S., Jutla, C., Krawczyk, H., Rosu, M.-C., Steiner, M.: Dynamic searchable encryption in very-large database: data structures and implementation. In: Proceedings of NDSS 2014 (2014)
7. Chang, Y.-C., Mitzenmacher, M.: Privacy preserving keyword searches on remote encrypted data. In: Ioannidis, J., Keromytis, A., Yung, M. (eds.) ACNS 2005. LNCS, vol. 3531, pp. 442–455. Springer, Heidelberg (2005). doi:10.1007/11496137_30
8. Cao, N., Wang, C., Li, M., Ren, K., Lou, W.: Privacy-preserving multi-keyword ranked search over encrypted cloud data. In: Proceedings of INFOCOM 2011, pp. 829–837 (2011)
9. Cohen, W.W.: Enron email dataset. http://www.cs.cmu.edu/~enron/
10. Curtmola, R., Garay, J., Kamara, S., Ostrovsky, R.: Searchable symmetric encryption: improved definitions and efficient constructions. J. Comput. Secur. **19**(5), 895–934 (2011)
11. Goh, E.-J.: Secure indexes. Stanford Univ. Technical report. IACR ePrint Cryptography Archive (2003). http://eprint.iacr.org/2003/216
12. Golle, P., Staddon, J., Waters, B.: Secure conjunctive keyword search over encrypted data. In: Jakobsson, M., Yung, M., Zhou, J. (eds.) ACNS 2004. LNCS, vol. 3089, pp. 31–45. Springer, Heidelberg (2004). doi:10.1007/978-3-540-24852-1_3
13. Hacigümüş, H., Hore, B., Iyer, B., Mehrotra, S.: Search on encrypted data. In: Yu, T., Jajodia, S. (eds.) Secure Data Management in Decentralized Systems, vol. 33, pp. 383–425. Springer, Boston (2007). doi:10.1007/978-0-387-27696-0_12
14. Hahn, F., Kerschbaum, F.: Searchable encryption with secure and efficient updates. In: Proceedings of ACM CCS 2014, pp. 310–320 (2014)
15. Hirano, T., Hattori, M., Kawai, Y., Matsuda, N., Iwamoto, M., Ohta, K., Sakai, Y., Munaka, T.: Simple, secure, and efficient searchable symmetric encryption with multiple encrypted indexes. In: Ogawa, K., Yoshioka, K. (eds.) IWSEC 2016. LNCS, vol. 9836, pp. 91–110. Springer, Cham (2016). doi:10.1007/978-3-319-44524-3_6
16. Jho, N.-S., Hong, D.: Symmetric searchable encryption with efficient conjunctive keyword search. KSII Trans. Internet Inf. Syst. **7**(5), 1328–1342 (2013)

17. Katz, J., Lindell, Y.: Introduction to Modern Cryptography, 2nd edn. CRC Press, Boca Raton (2015)
18. Kurosawa, K., Ohtaki, Y.: UC-secure searchable symmetric encryption. In: Keromytis, A.D. (ed.) FC 2012. LNCS, vol. 7397, pp. 285–298. Springer, Heidelberg (2012). doi:10.1007/978-3-642-32946-3_21
19. Kurosawa, K., Sasaki, K., Ohta, K., Yoneyama, K.: UC-secure dynamic searchable symmetric encryption scheme. In: Ogawa, K., Yoshioka, K. (eds.) IWSEC 2016. LNCS, vol. 9836, pp. 73–90. Springer, Cham (2016). doi:10.1007/978-3-319-44524-3_5
20. Kamara, S., Papamanthou, C.: Parallel and dynamic searchable symmetric encryption. In: Sadeghi, A.-R. (ed.) FC 2013. LNCS, vol. 7859, pp. 258–274. Springer, Heidelberg (2013). doi:10.1007/978-3-642-39884-1_22
21. Kamara, S., Papamanthou, C., Roeder, T.: Dynamic searchable symmetric encryption. In: Proceedings of ACM CCS 2012, pp. 965–976 (2012)
22. Liu, L., Gai, J.: Bloom filter based index for query over encrypted character strings in database. In: Proceedings of CSIE 2009, pp. 303–307 (2009)
23. Liu, Q., Wang, G., Wu, J.: An efficient privacy preserving keyword search scheme in cloud computing. In: Proceedings of CSE 2009, pp. 715–720 (2009)
24. Moataz, T., Shikfa, A.: Boolean symmetric searchable encryption. In: Proceedings of ACM ASIACCS 2013, pp. 265–276 (2013)
25. Naveed, M., Prabhakarn, M., Gunter, C.A.: Dynamic searchable encryption via blind storage. In: Proceedings of IEEE on Security and Privacy, pp. 639–654 (2014)
26. Popa, R.A., Redfield, C.M.S., Zeldovich, N., Balakrishnan, H.: CryptDB: processing queries on an encrypted database. Commun. ACM **55**(9), 103–111 (2012)
27. Suga, T., Nishide, T., Sakurai, K.: Secure keyword search using bloom filter with specified character positions. In: Takagi, T., Wang, G., Qin, Z., Jiang, S., Yu, Y. (eds.) ProvSec 2012. LNCS, vol. 7496, pp. 235–252. Springer, Heidelberg (2012). doi:10.1007/978-3-642-33272-2_15
28. Stefanov, E., Papamanthou, C., Shi, E.: Practical dynamic searchable encryption with small leakage. In: Proceedings of NDSS 2014 (2014)
29. Song, D.X., Wagner, D., Perrig, A.: Techniques for searchers on encrypted data. In: Proceedings of IEEE Symposium on Security and Privacy, pp. 44–55 (2000)
30. Xu, P., Liang, S., Wang, W., Susilo, W., Wu, Q., Jin, H.: Dynamic searchable symmetric encryption with physical deletion and small leakage. In: Pieprzyk, J., Suriadi, S. (eds.) ACISP 2017. LNCS, vol. 10342, pp. 207–226. Springer, Cham (2017). doi:10.1007/978-3-319-60055-0_11
31. Yavuz, A.A., Guajardo, J.: Dynamic searchable symmetric encryption with minimal leakage and efficient updates on commodity hardware. In: Dunkelman, O., Keliher, L. (eds.) SAC 2015. LNCS, vol. 9566, pp. 241–259. Springer, Cham (2016). doi:10.1007/978-3-319-31301-6_15
32. Wang, C., Cao, N., Li, J., Ren, K., Lou, W.: Secure ranked keyword search over encrypted cloud data. In: Proceedings of ICDCS 2010, pp. 253–262 (2010)

Anonymous Certification
for an e-Assessment Framework

Christophe Kiennert, Nesrine Kaaniche, Maryline Laurent,
Pierre-Olivier Rocher, and Joaquin Garcia-Alfaro[✉]

SAMOVAR, Télécom SudParis, CNRS, Université Paris-Saclay, Paris, France
`garcia_a@telecom-sudparis.eu`

Abstract. We present an anonymous certification scheme that provides
data minimization to allow the learners of an e-assessment platform
to reveal only required information to certificate authority providers.
Attribute-based signature schemes are considered as a promising cryptographic primitive for building privacy-preserving attribute credentials, also known as anonymous credentials. These mechanisms allow the
derivation of certified attributes by the issuing authority relying on non-
interactive protocols and enable end-users to authenticate with verifiers
in a pseudonymous manner, e.g., by providing only the minimum amount
of information to service providers.

Keywords: Attribute-based signatures · Attribute-based credentials ·
Privacy · Bilinear pairings · Anonymous certification · e-Assessment
applications

1 Introduction

E-Assessment is an innovative form for the evaluation of learners' knowledge and
skills in online education, as well as in blended-learning environments, where
part of the assessment activities is carried out online. As e-assessment involves
online communication channel between learners and educators, as well as data
transfer and storage, security measures are required to protect the environment
against system and network attacks. Issues concerning the security and privacy
of learners is a challenging topic. Such issues are discussed under the scope of
the TeSLA project (cf. http://www.tesla-project.eu/ for further information), a
EU-funded project that aims at providing learners with an innovative environment that allows them to take assessments remotely, thus avoiding mandatory
attendance constraints.

In [16], security of the TeSLA e-assessment system were analyzed and discussed. A security proposal for securing the TeSLA platform according to the
General Data Protection Regulation (GDPR) [11] was proposed. With respect

---

N. Kaaniche and M. Laurent—Member of the Chair Values and Policies of Personal
Information.

H. Lipmaa et al. (Eds.): NordSec 2017, LNCS 10674, pp. 70–85, 2017.
https://doi.org/10.1007/978-3-319-70290-2_5

to the protection of learners' data, and more specifically in terms of learners' certification techniques, it was highlighted the necessity of enhancing the framework with privacy-preserving attribute credentials, in order to allow learners to authenticate with verifiers in a pseudonymous manner. Indeed, an open educational system like TeSLA has to be properly secured with classical measures, such as authentication, data ciphering and integrity checks, in order to mitigate cyber-attacks that may lead to disastrous consequences, such as data leakage or identity theft.

To meet the GDPR recommendations, it is also necessary to ensure a reasonable level of privacy in the system. Security and privacy are very close domains, and yet important differences have to be highlighted, since it is possible to build a very secure system that fails to ensure any privacy properties. Security, from a technological standpoint, consists in guaranteeing specific requirements at different levels of the architecture, such as confidentiality, integrity or authentication. It mainly targets the exchange and storage of data, which in the case of TeSLA may contain some traces of learner's biometric data, the learner's assessment results, and other sensitive information. In contrast with security, privacy consists in preventing the exploitation of metadata to ensure that no personal information leakage will occur. However, it always remains mandatory to comply with legal constraints, which may prevent full anonymization of the communications. Therefore, the main objective of privacy, from a technological perspective, is to reveal the least possible information about the user's identity, and to prevent any undesired traceability, which is often complex to achieve.

In the context of TeSLA, several privacy technological filters have been included in the underlying design of the architecture. The randomized TeSLA identifier (TeSLA ID for short) associated to each learner is a proper example. This identifier is used each time the learner accesses TeSLA, hence ensuring pseudo-anonymity to every learner—full anonymity not being an option in TeSLA for legal reasons. Yet, a randomized identifier alone cannot protect the learners against more complex threats such as unwanted traceability. The system can still be able to link two different sessions of the same learner. A technical solution that could be integrated in the TeSLA architecture to handle such issues is the use of anonymous certification.

Anonymous certification allows users to prove they are authorized to access a resource without revealing more than they need about their identity. For example, users can be issued with certified attributes that may be required by the system verifier, such as *older than 18*, or *lives in France*. When the users want to prove that they own the right set of attributes, they perform a digital signature based on the required attributes, allowing the system verifier to check if a precise user is authorized, sometimes without even knowing precisely which attributes were used.

Such an approach could be integrated in several points of the TeSLA architecture where it is not necessary to identify the learner. For example, to access course material on the VLE, it should be enough to prove that the learner comes from an allowed university and is registered for this course. That way, it becomes impossible for the VLE (Virtual Learning Environment) to follow the studying activity of each learner, while still letting the learners access the course material. Similarly, when a student has taken an assessment, the student's work can be anonymously sent to anti-cheating tools (such as anti-plagiarism). With anonymous certification, each tool might receive a request for the same work without being able to know which learner wrote it, but also without being able to correlate the requests and decide whether they were issued by the same learner.

Therefore, anonymous certification might prove to be a solid and innovative asset to enhance privacy in TeSLA, and to prevent traceability of the learners whenever it is not required. This paper reports an anonymous certification scheme that addresses the aforementioned challenges. It allows the learners of an e-assessment platform to reveal only required information to certificate authority providers. It builds on attribute-based signature schemes and allows the derivation of certified attributes by issuing authorities. The resulting construction provides a non-interactive protocol that allows the e-assessment users to authenticate with verifiers by providing only the minimum amount of information to service providers.

Paper Organization—Section 2 surveys some related work. Section 3 provides a short description of the mathematical details of our proposed anonymous certification mechanism. Section 4 presents a description of the TeSLA architecture, provides a use case towards validating our anonymous certification mechanism. Section 5 briefly discusses some details of the ongoing implementation of the solution and details about the security levels of the proposal. Section 6 concludes the paper.

2   Related Work

Privacy-preserving authentication mechanisms, called also anonymous certification schemes, are based on advanced cryptographic primitives, such as anonymous credentials, minimal disclosure tokens, self-blindable credentials, group signatures, sanitizable signatures or attribute-based signatures [3,4,6,8,10,14,28].

In these schemes, users obtain certified credentials for their attributes from trusted issuing organizations and later derive, without further assistance from any issuing authority, presentation tokens that reveal only the required attribute information that might be verified by the verifier under the issuing organization's public key. Well-known examples include Brands scheme [4], mainly relying on blind signatures, and Camenisch-Lysyanskaya scheme, using group signatures [6], which have been implemented in Microsoft U-Prove and IBM Identity Mixer, respectively.

Attribute-based signature schemes (ABS for short) are considered as a promoting cryptographic primitive for building privacy-preserving attribute credentials [19]. To use ABS, a user shall possess a set of attributes and a secret signing key per attribute. The signing key must be provided by a trusted authority. The user can sign, e.g., a document, with respect to a predicate satisfied by the set of attributes. Several ABS schemes exist in the related literature, considering different design directions. This includes ABS solutions in which (i) the attribute value can be a binary-bit string [13,18,19,21,23] or general-purpose data structures [29]; (ii) ABS solutions satisfying access structures under threshold policies [13,18,23], monotonic policies [19,29] and non-monotonic policies [21]; and (iii) ABS solutions in which the private keys associated to the attributes are either issued by a single authority [19,23,29] or by a group of authorities [19,21].

Kaaniche and Laurent present in [14] a complete anonymous certification scheme, called $\mathcal{HABS}$, and constructed over the use of ABS. In addition to common requirements such as *privacy* and *unforgeability*, $\mathcal{HABS}$ is designed with three additional properties: (i) signature traceability, in order to grant some entities the ability of identifying the user originating an ABS signature; (ii) issuing organization unlinkability, to avoid that colluding ABS authorities link user requests sharing a single public key; and (iii) mitigation of replayed sessions, by imposing the use of random nonces and secure timestamps.

In [26,27], some of the requirements imposed by $\mathcal{HABS}$ are questioned by Vergnaud. The concrete realization of the $\mathcal{HABS}$ primitive is presented as unsatisfactory with regard to the unforgeability and privacy properties under the random oracle model. $\mathcal{PCS}$ [15], built over $\mathcal{HABS}$, addresses the limitations pointed out by [26,27] is used in this paper as the underlying construction deployed as an AC scheme of TeSLA.

The work by Aïmeur et al. in [1,2] discusses about the necessity of extended analysis of security and privacy techniques for e-learning systems. E-learning systems are presented by Aïmeur et al. as a composition of Internet-based protocols and tools, that require from well-established cryptographic techniques, in order to allow learners to perform on-line studies while preserving a minimum of privacy requirements. The authors survey in their work a list of security challenges to address, as well as some common threats to the privacy of the learners. A high-level overview of research examples in terms of attribute-based encryption and anonymous credentials is reported—without providing any explicit construction.

3 Anonymous Certification (AC) Construction

3.1 Background

In [9], Chaum introduced the notion of Anonymous Credentials (AC). Camenisch and Lysyanskaya fully formalized the concept in [6,7]. AC, also referred to as privacy-preserving attribute credentials, involve several entities and procedures. It fulfills some well-identified security and functional requirements. In the sequel, we first present some further details about the type of entities and procedures associated to traditional AC schemes. Then, we provide our specific AC construction.

3.1.1 Entities

An AC involves several entities. Figure 1 identifies several AC entities. Some entities, such as the *user*, the *verifier* and *issuer* are mandatory, while other

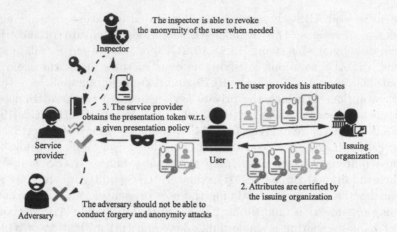

Fig. 1. Traditional AC entities

entities, such as the *revocation authority* and the *inspector* are optional [5]. These entities can be defined as follows:

- The *user* is the central entity, whose interest is to have privacy-preserving access to services, offered by service providers, known as *verifiers*. The user has first to collect credentials from various *issuing organizations*. Then, he selects the appropriate information from credentials, to present to the requesting verifier, under the presentation token.
- The verifier protects access to a resource or service that it offers by imposing restrictions on the credentials that users have to own and the information from these credentials that users must present to access the service. The verifier restrictions are referred to as *presentation policy*. The user generates from his credentials a *presentation token* that contains the required information and the supporting cryptographic evidence.
- The issuing organization issues credentials to users, while attesting the correctness of the information contained in the credential with respect to the user. Notice that before issuing a credential, the issuer may have to authenticate the user.
- The *revocation authority* has to revoke issued credentials and maintain the list of valid credentials in the system. So that, these credentials can no longer be used to derive presentation tokens. Both the user and the verifier have to obtain the most recent revocation information from the *revocation authority* to generate, respectively verify, presentation tokens.
- The *inspector* is a trusted entity, which has the technical capabilities to, when needed, remove the anonymity of a user.

3.1.2 Procedures

As depicted in Fig. 1, privacy-preserving ABC systems mainly rely on two main procedures (i.e., *issuance* and *presentation*).

The issuance of a credential is an interactive protocol, between the user and the issuing organization. At the end of this phase, the issuing organization provides a *signed* credential to the user, certifying the validity of the contained information. A user may have several credentials, each asserting some collection of attributes.

The presentation phase starts when a user requests access to the service provider's resources. Indeed, the verifier sends to the user the presentation policy, that describes which proofs must be sent, and which information from the credential(s) have to be revealed. The user then checks the combination of credentials that fulfill the policy in order to generate the response, referred to as *presentation token*, then sent to the verifier. Thus, a presentation token may reveal information about the user (reveal attribute values), but also prove certain facts about some other attributes (while hiding the values), such as proving that the birth date is earlier than a given day.

During a presentation procedure, the user may also need to prove not only that he possesses certain attribute values, but also that the credentials certifying those attributes have not been revoked.

3.1.3   Security and Functional Requirements

Privacy preserving authentication systems have to fulfill the following security requirements:

- *anonymity* – the user must remain anonymous during the authentication process.
- *unforgeability* – a party that does not belong to the set of authorized users should not be able to successfully run the protocol with the verifier.
- *unlinkability* – this property is important to preserve the privacy of users. Two sub-proprieties have to be identified: *issue-show unlinkability*, ensuring that any information gathered during the credential issuing cannot be used to later link the credential to its issuance while proceeding to its verification, and *multi-show unlinkability*, guarantying that multiple presentation sessions w.r.t. the same credential should not be linked.

Additionally, privacy preserving attribute-based credentials have to ensure several functional features, namely revocation, inspection and *selective disclosure*. The selective disclosure property refers to the ability provided to users, to present to the verifier partial information extracted or derived from their credentials.

3.2   Our Construction

In this section, we present our precise anonymous certification scheme, in order to extend the e-assessment framework reported in [16]. The solution is based on an existing attribute-based signature scheme previously presented in [15]. Our construction relies on the following list of algorithms:

- SETUP—It takes as input the security parameter ξ and returns the public parameters *params*. The public parameters are considered an auxiliary input to all the algorithms.

  Global Public Parameters params – the SETUP algorithm first generates an asymmetric bilinear group environment $(p, \mathbb{G}_1, \mathbb{G}_2, \mathbb{G}_T, \hat{e})$ where $\hat{e}$ is an asymmetric pairing function such as $\hat{e} : \mathbb{G}_1 \times \mathbb{G}_2 \to \mathbb{G}_T$.

  The random generators $g_1, h_1 = g_1{}^\alpha, \{\gamma_i\}_{i \in [1,U]} \in \mathbb{G}_1$ and $g_2, h_2 = g_2{}^\alpha \in \mathbb{G}_2$ are also generated, as well as $\alpha \in \mathbb{Z}_p$ where U denotes the maximum number of attributes supported by the span program.

  We note that each value γ_i is used to create the secret key corresponding to an attribute a_i.

  Let $\mathcal{H}$ be a cryptographic hash function. The global parameters of the system are denoted as follows:

  $$params = \{\mathbb{G}_1, \mathbb{G}_2, \mathbb{G}_T, \hat{e}, p, g_1, \{\gamma_i\}_{i \in [1,U]}, g_2, h_1, h_2, \mathcal{H}\}$$

- KEYGEN—It returns a pair of private and public keys for each participating entity (i.e., issuing organization and user). In other words, the user has a pair of keys (sk_u, pk_u) where sk_u is chosen at random from $\mathbb{Z}_p$ and $pk_u = h_1{}^{sk_u}$ is the related public key. The issuing organization also holds a pair of secret and public keys (sk_o, pk_o). The issuing organization secret key sk_o relies on the couple defined as $sk_o = (s_o, x_o)$, where s_o is chosen at random from $\mathbb{Z}_p$ and $x_o = g_1{}^{s_o}$. The public key of the issuing organization pk_o corresponds to the couple $(X_o, Y_o) = (\hat{e}(g_1, g_2)^{s_o}, h_2{}^{s_o})$.

- ISSUE—It is executed by the issuing organization. The goal is to issue the credential to the user with respect to a pre-shared set of attributes $\mathcal{S} \subset \mathbb{S}$, such that $\mathbb{S}$ represents the attribute universe, defined as: $\mathcal{S} = \{a_1, a_2, \cdots, a_N\}$, where N is the number of attributes such that $N < U$.

  The ISSUE algorithm takes as input the public key of the user pk_u, the set of attributes $\mathcal{S}$ and the private key of the issuing organization sk_o. It also picks an integer r at random and returns the credential C defined as:

  $$C = (C_1, C_2, \{C_{3,i}\}_{i \in [1,N]}) = (x_o \cdot [pk_u{}^{s_o \mathcal{H}(\mathcal{S})^{-1}}] \cdot h_1{}^r, g_2{}^r, \{\gamma_i{}^r\}_{i \in [1,N]})$$

  where $\mathcal{H}(\mathcal{S}) = \mathcal{H}(a_1)\mathcal{H}(a_2) \cdots \mathcal{H}(a_N)$ and $\gamma_i{}^r$ represents the secret key associated to the attribute a_i, where $i \in [1, N]$.

- OBTAIN—It is executed by the user. It takes as input the credential C, the secret key of the user sk_u, the public key of the issuing organization pk_o and

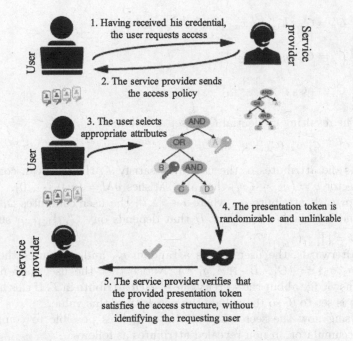

Fig. 2. ABS for the support of AC (Presentation Procedure)

the set of attributes $\mathcal{S}$. The algorithm returns 1 if Eq. 1 holds true; or 0, otherwise.

$$\hat{e}(C_1, g_2) \stackrel{?}{=} X_o \cdot \hat{e}(g_1^{sk_u \mathcal{H}(\mathcal{S})^{-1}}, Y_o) \cdot \hat{e}(h_1, C_2) \tag{1}$$

- SHOW ↔ VERIFY: this two-party algorithm is illustrated by Fig. 2. The different algorithms are defined as follows:
  - VERIFY: this algorithm is executed by the verifier upon receiving an authentication request from a user. In a first step, it outputs the presentation policy, including a randomized message $\mathtt{M} = g_1{}^m$, a predicate Υ and the set of attributes that have to be revealed. In the following, we note that:
    * m should be different for each authentication session to prevent replay attacks,
    * $\mathcal{S}_R$ denotes the set of attributes revealed to the verifier and $\mathcal{S}_H$ denotes the set of non-revealed attributes, such as $\mathcal{S} = \mathcal{S}_R \cup \mathcal{S}_H$,
    * Υ is represented by an LSSS access structure (M, ρ), where M is an $l \times k$ matrix, and ρ is an injective function that maps each row of the matrix M to an attribute.
  - SHOW: The SHOW algorithm takes as input the user secret key sk_u, the credential C associated to the attribute set $\mathcal{S}$ for pk_u, the message $\mathtt{M}$ and the predicate Υ. The showing process is as follows:
    1. The user first randomizes his credential in the following way: it selects uniformly at random an integer $r' \in \mathbb{Z}_p$ and sets:

$$\begin{cases} C_1' = C_1 \cdot {h_1}^{r'} = x_o \cdot [pk_u^{\,s_o \mathcal{H}(\mathcal{S})^{-1}}] \cdot {h_1}^{r+r'} \\[2mm] C_2' = C_2 \cdot {g_2}^{r'} = {g_2}^{r+r'} \\[2mm] C_{3,i}' = C_{3,i} \cdot {\gamma_i}^{r'} = {\gamma_i}^{r+r'} \end{cases}$$

The resulting credential C' is set as follows:

$$C' = (C_1', C_2', \{C_{3,i}'\}_{i\in[1,N]}) = (x_o \cdot [pk_u^{\,s_o \mathcal{H}(\mathcal{S})^{-1}}] \cdot {h_1}^{r+r'}, {g_2}^{r+r'}, \{{\gamma_i}^{r+r'}\}_{i\in[1,N]})$$

2. As the attributes of the user in $\mathcal{S}$ satisfy Υ, the user can compute a vector $\boldsymbol{v} = (v_1, \cdots, v_l)$ that also satisfies $\boldsymbol{v}M = (1, 0, \cdots, 0)$.
3. For each attribute a_i, where $i \in [1, l]$, the user computes $\omega_i = {C_2'}^{v_i}$ and calculates a quantity B that depends on $\{C_{3,i}'\}_{i\in[1,N]}$ such that $B = \prod_{i=1}^l (\gamma'_{\rho(i)})^{v_i}$.
4. Afterwards, the user selects a random r_m and computes the couple $(\sigma_1, \sigma_2) = (C_1' \cdot B \cdot \mathsf{M}^{r_m}, {g_1}^{r_m})$. Notice that the user may not have knowledge about the secret value of each attribute in Υ. If this happens, v_i is set to 0, so to exclude the necessity of this value.
5. Using now the secret key of the user, it is possible to compute an accumulator on non-revealed attributes as follows:

$$A = {Y_o}^{\frac{sk_u \mathcal{H}(\mathcal{S}_H)^{-1}}{r_m}}$$

The user returns the presentation token $\Sigma = (\Omega, \sigma_1, \sigma_2, C_2', A, \mathcal{S}_R)$, that includes the signature of the message M with respect to the predicate Υ, and where $\Omega = \{\omega_1, \cdots, \omega_l\}$ is the set of committed element values of the vector $\boldsymbol{v}$, based on the credential's item C_2'.

- VERIFY: In a second step, given the presentation token Σ, the public key of the issuing organization pk_o, the set of revealed attributes $\mathcal{S}_R$, the message m and the signing predicate Υ, the verifier first computes an accumulator A_R such as $A_R = \sigma_2^{\mathcal{H}(\mathcal{S}_R)^{-1}}$. Then, it picks uniformly at random $k-1$ integers $\mu_2, \cdots, \mu_k$ and calculates l integers $\tau_i \in \mathbb{Z}_p$ for $i \in \{1, \cdots, l\}$ such that $\tau_i = \sum_{j=1}^k \mu_j M_{i,j}$ where $M_{i,j}$ is an element of the matrix M. It accepts the presentation token as valid (i.e.; outputs 1) if and only if Eq. 2 holds:

$$\hat{e}(\sigma_1, g_2) \stackrel{?}{=} X_o \hat{e}(A_R, A)\hat{e}(h_1, C_2') \prod_{i=1}^l \hat{e}(\gamma_{\rho(i)} {h_1}^{\tau_i}, \omega_i)\hat{e}(\sigma_2, {g_2}^m) \qquad (2)$$

4   E-learning Use Case for $\mathcal{PCS}$

In this section, we describe how anonymous certification relying on attribute-based signatures may be integrated into an e-learning environment to enhance the learners' privacy. We first present the TeSLA architecture for e-learning and e-assessment, before detailing in which parts of the architecture anonymous certification may be implemented.

4.1   TeSLA Architecture

The TeSLA project aims at providing an e-learning environment that integrates secure e-assessment, in order to allow the learners to take assessments remotely while providing the necessary countermeasures to prevent cheating.

The TeSLA architecture is comprised of several components that may belong to two domains: the university domain and the TeSLA domain. Components that belong to the university domain must be present in the network of each university willing to make use of the TeSLA e-assessment framework, while components that belong to the TeSLA domain are completely independent of the university network. The two domains do not share data unless explicitly stated. The TeSLA domain contains the following components:

- The TeSLA E-assessment Portal (TEP), which acts as a service broker that gathers and forwards requests to the TeSLA components.
- The TeSLA Portal, that aims at gathering statistics regarding the e-assessment activities.
- Instruments that analyze the biometric samples and send their analysis results back to the client side.

The university domain contains the following components:

- A Virtual Learning Environment (VLE), which can be provided by a classic Learning Management System (LMS) such as Moodle[1].
- A plugin integrated to the VLE that acts as a client side interface with the TeSLA components.
- Various tools integrated to the VLE that send requests and data to the TeSLA components through the plugin. There are three categories of tools: the learner tool, the instructor tool, and external tools. The learner tool and instructor tool are respectively designed to take or setup an e-assessment. External tools are in charge of sampling the learner's biometric data and sending them to TeSLA instruments for evaluation, as part of the anti-cheating countermeasures.
- The TeSLA Identity Provider (TIP), which is in charge of generating an anonymized identity for each learner, called TeSLA ID, to be used in the communication with TeSLA components.

The TeSLA architecture is represented in Fig. 3. The communications between the components are secured by the TLS protocol [22], deployed on the whole architecture with mutual authentication, hence ensuring confidentiality and integrity of every data exchange. The underlying Public Key Infrastructure for TLS deployment and management is detailed in [16].

Taking an e-assignment in this architecture first requires to log in on the VLE that contains the client-side plugin. The learner can require the e-assignment using the learner tool available on the VLE as a third-party tool. The learner

---

[1] https://moodle.org/.

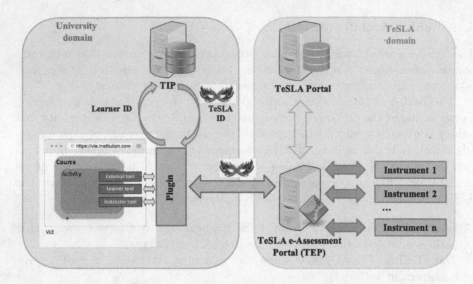

Fig. 3. Simplified TeSLA architecture representation

tool sends a request through the plugin to the TEP. The incoming request does not contain the name of the learner, but only the TeSLA ID, that the plugin requested from the TIP. Then, the TEP fetches the e-assignment in its database and sends it back to the VLE, where the learner will take the assignment while external tools sample biometric data that will be regularly sent to instruments for anti-cheating analysis.

4.2 Pseudonymity

A project for secure e-assessment such as TeSLA does not make it possible to implement full anonymity for the learners. Indeed, the very nature of the assessment makes it mandatory to store the association between the examinee number (e.g., the TeSLA ID) and the real name of the learner. Therefore, in such context, only partial anonymity, i.e. pseudonymity can be provided to learners during exchanges with the TeSLA components.

In this architecture, pseudonymity is ensured with a randomized TeSLA identifier named TeSLA ID, which becomes the learner's identity within the TeSLA domain. Therefore, no TeSLA component has ever access to the learner's true identity.

The TeSLA ID is generated by the TIP component as a random number computed according to version 4 of the UUID standard [17]. The matching between the learner's identity and the TeSLA ID is stored in the TIP database. The TIP database is placed at the university side and is not accessible from TeSLA. The TIP database shall be shared with all the VLEs. Since any interaction between the university domain and the TeSLA domain involve the plugin on one hand, and the TEP on the other hand, it is sufficient to make sure that any request sent to the TEP through the plugin is first redirected to the TIP to retrieve the learner's TeSLA ID and use it in place of the learner's identity. Notice that the TeSLA ID enables pseudonymity for all learners, who can take e-assessments without revealing their identity to the TeSLA system. However, it should be noted even though the learners are anonymized with respect to the e-assessment system, it is not enough to prevent the acquisition and correlation of personal data by the system. For example, the TeSLA ID does not ensure multi-session unlinkability, since the e-assessment system is obviously able to know when the same learner is logging in over two different sessions and gather data about this learner's actions, even without knowing his identity. Anonymous certification, as described in Sect. 3, is a solution that ensures many more privacy properties than a simple anonymized identifier. In the next subsection, we describe how the system can be integrated to an e-learning environment such as TeSLA.

4.3   Integrating Anonymous Certification to TeSLA

The purpose of anonymous certification is to perform anonymous access control, in order to certify that users are allowed to access a resource because they own some attributes required by the verifier. However, the verifier only knows that the users' attributes match the policy, without necessarily knowing which attributes they own exactly.

Therefore, anonymous certification cannot be used in a context where it is necessary to perform authentication in order to identify a specific user. Obviously, it can be adapted to such a situation by requiring the user identifier as an attribute that must be revealed, but it loses its interest by doing so. In the context of TeSLA, it means that anonymous certification cannot be used during e-assessment itself, since the e-assessment needs to be associated to the unique identifier of a learner.

However, anonymous certification can be naturally added to the VLE. Indeed, a LMS generally aims at informing learners about courses they registered at, and letting them access the course material. In both cases, the VLE does not need to identify the learner in a unique way, but only needs to prove that the learner is authorized. In this case, the following attributes could be defined and used to decide whether to authorize a learner:

- The university where the learner is enrolled
- The courses at which the learner registered

These attributes are enough to let every learner access to the VLE pages he is entitled to visit, without proceeding to an usual, nominative authentication (even using a pseudonym or an anonymized identifier). Thus, learners might be able to access the course document at any time without any possibility for the VLE to log and profile the learners' activity. This can be a significant advance for learners' privacy since learners may for example abhor to let the system know at which hours they are awake, and at which moment they accessed the course material.

Likewise, it is also possible to enhance the privacy of e-assignments' post processing. When an e-assignment is completed by a learner, it must first be sent to a number of external anti-cheating instruments, that will for example check if the assignment contains plagiarism. Instead of transmitting requests associated to the learner's TeSLA ID, the requests can be anonymized and authorized with anonymous certification. The attributes may be defined similarly as above. On top of preventing each instrument from profiling students based on their TeSLA ID, the unlinkability property of the anonymous certification scheme guarantees that two different instruments will not be able to know that the request was emitted from the same learner. This greatly limits the possibility for the instruments to correlate data, i.e., it enhances the learners' privacy.

5 Implementation and Security Details of $\mathcal{PCS}$

We briefly discuss in this section the ongoing implementation of the proposal reported in this paper, as well as some remarks about the security level of $\mathcal{PCS}$.

5.1 Implementation Details

Available at http://j.mp/PKIPCSgit as a multi-platform C++ software code, and mainly based on existing cryptographic libraries such as PBC [25] and MCL [24], the construction is available online to facilitate understanding, comparison and validation of the solution. Special attention has been paid to the nature of the elliptic curves required to validate the operations of the construction in Sect. 3. We recall that Anonymous Credentials (AC) are built on top of Attribute-Based Cryptography, which makes use of pairing-friendly elliptic curves, i.e., elliptic curves that satisfy certain conditions [12]. For instance, the degree of immersion of such curves. Some parts of the implementation and testing of the elliptic curve operations are based on either Ate or Tate pairing implementations (cf., for instance, the *Ate Pairing over Barreto-Naehrig Curves* implementation, available at https://github.com/herumi/ate-pairing). Extended versions of the Miller algorithm [20] from [12] are used to computing the pairings. Precise examples, and data computed to verify the security of the construction, are available at http://j.mp/PKIPCSgit as well.

5.2 Security Level Sketch of Our Proposal

We recall that brute-force attacks consist in checking all possible keys until the correct one is discovered (i.e., with a key of length k bits, there are 2^k possible keys). Thus, k denotes the security level in symmetric cryptography. In public-key cryptography, the security level of an algorithm is defined with respect to the hardness of solving a mathematical problem such as the Discrete Logarithm Problem (DLP). The time required to resolve the DL problem is much less important than trying the 2^k keys by a brute-force attack. For instance, a 1024-RSA key-length bits provides a 80 key-length equivalent key of a symmetric algorithm.

In order to generate security parameters for each security level of the $\mathcal{PCS}$ proposal, we shall investigate the structure of $\mathbb{G}_1$, $\mathbb{G}_2$ and $\mathbb{G}_T$. The attribute-based signature scheme depends on the pairing function $\hat{e} : \mathbb{G}_1 \times \mathbb{G}_2 \leftarrow \mathbb{G}_T$. Let $\mathcal{E}(\mathbb{F}_q)$ denote an elliptic curve [12] defined over the finite prime field $\mathbb{F}_q$ of order q. $\mathbb{G}_1$ is a finite additive subgroup of $\mathcal{E}(\mathbb{F}_q)$, $\mathbb{G}_T$ is a finite multiplicative subgroups of $\mathcal{E}(\mathbb{F}_{q^k})$ with order equal to p, such as k is the embedding degree of the curve $\mathcal{E}(\mathbb{F}_{q^k})$ relatively to p.

k defines the type of pairing function (type A, E, D, G) with respect to PBC library [25]. As such, the security level is related to the hardness of solving the DLP in the group $\mathbb{G}_T$. Let the order of G1 be the ECC key and the order of $\mathbb{G}_T$ be $p = q * k$. In order to generate the security parameters using PBC library, it is necessary to know the *rbit* order of $\mathbb{G}_1$ and the *qbit* order of $\mathbb{F}_q$. Table 1 shows the equivalent sizes of *rbits* and *qbits* for three considered security levels.

Table 1. Equivalent key sizes for some representative security levels (in bits)

Security level	Pairing type	$\mathbb{G}_T$ Size	$\mathbb{F}_q$ Size	$\mathbb{G}_1$ Size
80	A	1024	512	160
80	E	1024	1024	160
≥80	D	1050	175	167
≥80	G	1080	108	103
112	A	2028	1024	224
112	E	2048	2048	224
≥112	D	2082	347	332
≥112	G	3010	301	279
128	A	3072	1536	256
128	E	3072	3072	256
≥128	D	3132	522	514
≥128	G	5250	525	487

From Table 1, we notice that the computation duration of pairing functions, while considering different security levels, should be taken into consideration while implementing $\mathcal{PCS}$, since the size of $\mathbb{G}_T$, $\mathbb{F}_q$ and $\mathbb{G}_1$ groups size, mainly depend on the selected security level. For our $\mathcal{PCS}$ construction, the security level depends on the sensitivity level of handled e-learners data.

6 Conclusion

We have detailed an anonymous certification scheme for e-assessment systems. The proposed construction revisits an existing mechanism based on homomorphic attribute-based signatures, and offers a selective disclosure of features to enable anonymous certification of learners of an e-assessment system. A precise use case has been presented, and an ongoing implementation of the approach discussed. Perspectives of future work include extending the framework for additional use cases, as well as an exhaustive performance reporting of the full C++ implementation of the construction, will be released at http://j.mp/PKIPCSgit.

Acknowledgements. This work is supported by the H2020-ICT-2015/H2020-ICT-2015 TeSLA project *An Adaptive Trust-based e-assessment System for Learning*, Number 688520. The authors graciously acknowledge as well the support received from the Chair Values and Policies of Personal Information of the Institut Mines-Télécom.

References

1. Aïmeur, E., Hage, H.: Preserving learners' privacy. In: Nkambou, R., Bourdeau, J., Mizoguchi, R. (eds.) Advances in Intelligent Tutoring Systems. SCI, vol. 308, pp. 465–483. Springer, Heidelberg (2010). doi:10.1007/978-3-642-14363-2_23
2. Aïmeur, E., Hage, E., Onana, F.S.M.: Anonymous credentials for privacy-preserving e-learning. In: 2008 International MCETECH Conference on E-Technologies, pp. 70–80. IEEE (2008)
3. Belenkiy, M., Camenisch, J., Chase, M., Kohlweiss, M., Lysyanskaya, A., Shacham, H.: Delegatable anonymous credentials. Cryptology ePrint Archive, Report 2008/428 (2008)
4. Brands, S.A.: Rethinking Public Key Infrastructures and Digital Certificates: Building in Privacy. MIT Press, Cambridge (2000)
5. Camenisch, J., Krenn, S., Lehmann, A., Mikkelsen, G.L., Neven, G., Pederson, M.O.: Scientific comparison of ABC protocols: Part i - formal treatment of privacy-enhancing credential systems (2014)
6. Camenisch, J., Lysyanskaya, A.: An efficient system for non-transferable anonymous credentials with optional anonymity revocation. In: Pfitzmann, B. (ed.) EUROCRYPT 2001. LNCS, vol. 2045, pp. 93–118. Springer, Heidelberg (2001). doi:10.1007/3-540-44987-6_7
7. Camenisch, J., Mödersheim, S., Sommer, D.: A formal model of identity mixer. In: Kowalewski, S., Roveri, M. (eds.) FMICS 2010. LNCS, vol. 6371, pp. 198–214. Springer, Heidelberg (2010). doi:10.1007/978-3-642-15898-8_13
8. Canard, S., Lescuyer, S.: Protecting privacy by sanitizing personal data: a new approach to anonymous credentials. In: Proceedings of the 8th ACM SIGSAC Symposium on Information, Computer and Communications Security, ASIA CCS 2013. ACM, New York (2013)
9. Chaum, D.: Security without identification: transaction systems to make big brother obsolete. Commun. ACM **28**(10), 1030–1044 (1985)
10. Chaum, D., van Heyst, E.: Group signatures. In: Davies, D.W. (ed.) EUROCRYPT 1991. LNCS, vol. 547, pp. 257–265. Springer, Heidelberg (1991). doi:10.1007/3-540-46416-6_22
11. European Council: Proposal for a regulation of the European parliament and of the council on the protection of individuals with regard to the processing of personal data and on the free movement of such data. In General Data Protection Regulation (2016)
12. Hankerson, D., Menezes, A., Vanstone, A.: Guide to Elliptic Curve Cryptography. Springer Science & Business Media, New York (2006)
13. Herranz, J., Laguillaumie, F., Libert, B., Ràfols, C.: Short attribute-based signatures for threshold predicates. In: Dunkelman, O. (ed.) CT-RSA 2012. LNCS, vol. 7178, pp. 51–67. Springer, Heidelberg (2012). doi:10.1007/978-3-642-27954-6_4

14. Kaaniche, N., Laurent, M.: Attribute-based signatures for supporting anonymous certification. In: Askoxylakis, I., Ioannidis, S., Katsikas, S., Meadows, C. (eds.) ESORICS 2016. LNCS, vol. 9878, pp. 279–300. Springer, Cham (2016). doi:10.1007/978-3-319-45744-4_14

15. Kaaniche, N., Laurent, M., Rocher, P.-O., Kiennert, C., Garcia-Alfaro, J.: $\mathcal{PCS}$, a privacy-preserving certification scheme. In: Garcia-Alfaro, J., Navarro-Arribas, G., Hartenstein, H., Herrera-Joancomartí, J. (eds.) ESORICS/DPM/CBT-2017. LNCS, vol. 10436, pp. 239–256. Springer, Cham (2017). doi:10.1007/978-3-319-67816-0_14

16. Kiennert, C., Rocher, P.O., Ivanova, M., Rozeva, A., Durcheva, M., Garcia-Alfaro, J.: Security challenges in e-assessment and technical solutions. In: 8th International Workshop on Interactive Environments and Emerging Technologies for eLearning, 21st International Conference on Information Visualization, London, UK (2017)

17. Leach, P.J., Salz, R., Mealling, M.H.: A Universally Unique IDentifier (UUID) URN Namespace. RFC 4122, July 2005

18. Li, J., Au, M.H., Susilo, W., Xie, D., Ren, K.: Attribute-based signature and its applications. In: ASIACCS 2010 (2010)

19. Maji, H.K., Prabhakaran, M., Rosulek, M.: Attribute-based signatures. In: Kiayias, A. (ed.) CT-RSA 2011. LNCS, vol. 6558, pp. 376–392. Springer, Heidelberg (2011). doi:10.1007/978-3-642-19074-2_24

20. Miller, V.S.: Use of elliptic curves in cryptography. In: Williams, H.C. (ed.) CRYPTO 1985. LNCS, vol. 218, pp. 417–426. Springer, Heidelberg (1986). doi:10.1007/3-540-39799-X_31

21. Okamoto, T., Takashima, K.: Efficient attribute-based signatures for non-monotone predicates in the standard model. In: Catalano, D., Fazio, N., Gennaro, R., Nicolosi, A. (eds.) PKC 2011. LNCS, vol. 6571, pp. 35–52. Springer, Heidelberg (2011). doi:10.1007/978-3-642-19379-8_3

22. Rescorla, E., Dierks, T.: The Transport Layer Security (TLS) Protocol Version 1.2. RFC 5246, August 2008

23. Shahandashti, S.F., Safavi-Naini, R.: Threshold attribute-based signatures and their application to anonymous credential systems. In: Preneel, B. (ed.) AFRICACRYPT 2009. LNCS, vol. 5580, pp. 198–216. Springer, Heidelberg (2009). doi:10.1007/978-3-642-02384-2_13

24. Shigeo, M.: MCL - Generic and fast pairing-based cryptography library, version: release20170402. https://github.com/herumi/mcl

25. Stanford University: PBC - The Pairing-Based Cryptography Library, version: 0.5.14. https://crypto.stanford.edu/pbc/

26. Vergnaud, D.: Comment on "attribute-based signatures for supporting anonymous certification" by N. Kaaniche and M. Laurent (ESORICS 2016). IACR Cryptology ePrint Archive (2016)

27. Vergnaud, D.: Comment on attribute-based signatures for supporting anonymous certification by N. Kaaniche and M. Laurent (ESORICS 2016). Comput. J. 1–8, June 2017

28. Verheul, E.R.: Self-blindable credential certificates from the Weil pairing. In: Boyd, C. (ed.) ASIACRYPT 2001. LNCS, vol. 2248, pp. 533–551. Springer, Heidelberg (2001). doi:10.1007/3-540-45682-1_31

29. Zhang, Y., Feng, D.: Efficient attribute proofs in anonymous credential using attribute-based cryptography. In: Chim, T.W., Yuen, T.H. (eds.) ICICS 2012. LNCS, vol. 7618, pp. 408–415. Springer, Heidelberg (2012). doi:10.1007/978-3-642-34129-8_39

PARTS – Privacy-Aware Routing with Transportation Subgraphs

Christian Roth$^{(\boxtimes)}$, Lukas Hartmann$^{(\boxtimes)}$, and Doğan Kesdoğan

Chair of IT Security Management, University of Regensburg, Regensburg, Germany
{christian.roth,lukas.hartmann,dogan.kesdogan}@ur.de

Abstract. To ensure privacy for route planning applications and other location based services (LBS), the service provider must be prevented from tracking a user's path during navigation on the application level. However, the navigation functionality must be preserved. We introduce the algorithm *PARTS* to split route requests into route parts which will be submitted to an LBS in an unlinkable way. Equipped with the usage of dummy requests and time shifting, our approach can achieve better privacy. We will show that our algorithm protects privacy in the presence of a realistic adversary model while maintaining the service quality.

Keywords: Routing · Location privacy · Anonymity

1 Introduction

In most areas of life, people are using smart devices and thereby generating data with personally identifiable information. Smart services collect personal data and base their quality of service (QoS) on this information. For this reason, data mining is on a broad research agenda, being able to extract precise user information from massive databases. With this data, the user experience and convenience of a service can be improved. For example, Apple's operating system iOS 9 studies a user's behavior to suggest the probable next locations the user wants to drive to [4]. However, this can be misused to control a user and massively violate his privacy. This was demonstrated by Facebook by experimentally changing the emotional perception of the user through news feed consumption [5].

Consequently, the user has the choice either not to use the smart service and lose the QoS improvements or use the smart device and resign their privacy. Since opting out of a service is less attractive for most users, the service provider almost always can collect highly sensitive user data without any significant consequences. On the first look, there is no possibility to resolve this dilemma. But nowadays smart phones are powerful enough to apply intelligent techniques to combine sensitive data in offline mode with additional online (real-time) information. With this approach, it will be much harder for an adversary, like data miners, to collect sensitive personal data.

To demonstrate our approach, we chose the well-known and popular field of route planning. Route planning services need location information from the user,

H. Lipmaa et al. (Eds.): NordSec 2017, LNCS 10674, pp. 86–101, 2017.
https://doi.org/10.1007/978-3-319-70290-2_6

which can be linked and analysed so that micro- and macroscopic profiles can be easily generated. It is shown that the start and target of a route are enough to deanonymise an otherwise anonymous user with a high probability [3]. Thus, the location information is very sensitive and protection algorithms have to be carefully designed. In addition, traditional obfuscation techniques from other scenarios dealing with location privacy cannot be applied because the route must be calculated for the exact given locations to provide the best user experience.

It would be theoretically possible to exclusively use the smart phone in offline mode by downloading all routing information from an area. But this means relinquishing any helpful third party information, i.e. any useful additional services. Consequently, to soften the trade-off between privacy and utility in the case of routing, we introduce privacy-aware routing with route parts using several interim destinations on the path to the overall target location.

1.1 Contribution

To the best of our knowledge, this is the first paper to investigate straightforward mechanisms to protect a user's route. We implement a basic set of these mechanisms (see Sect. 4) in a routing algorithm which divides the route into several parts to hide location information. However, we can still guarantee navigation to an exact location from a specific start while also including real time data from the untrusted cloud. We believe that these are real crucial arguments to create a widely adopted application. In this paper, we follow the classical approach of the triple bottom line of security "algorithm, adversary and evaluation" and contribute with:

1. Our algorithm *PARTS* based on dynamical, unlinkable route parts using straightforward protection mechanisms.
2. A realistic adversary model in which a honest-but-curious location based service (LBS) provider wants to reconstruct full routes.
3. A detailed evaluation of our algorithm by means of a self generated data set based on our theoretical user model, where we derive the best combination of privacy protection methods.

Due to limited space, we plan to present a real smartphone application implementing our algorithm in future work. Therefore, we do not analyze extended performance figures such as battery impact and user experience.

1.2 Structure

The remainder of the paper is organised as follows: After a review of related work in Sect. 2, the general system model is introduced in Sect. 3. In Sect. 4 we present techniques for privacy-aware routing, introducing our *PARTS* algorithm based on route parts. Section 5 contains the description of the adversary model who wants to reconstruct the navigated route. The results of our evaluation are depicted in Sect. 6 where we analyse PARTS' overhead, its performance and ability to protect a user's privacy. We discuss and summarise the results in Sects. 7 and 8 respectively.

2   Related Work

Anonymity of location information is a relevant topic in academic research. The best discussed concept is the idea of *k-Anonymity* introduced by Sweeney [10]. A subject is anonymous within a set of k subjects reporting the same cloaked geographical region instead of the real locations of the subjects. Implementations of this concept tend to minimise the area of the cloaking region making it easy for adversaries with background knowledge to infer sensitive data [9]. An extension of this model called *l-Diversity* was introduced by Machanavajjhala et al. [6]. Here, the cloaking area should contain at least l different locations.

Wang and Liu [11] showed that anonymity and diversity of locations can be achieved by realising k-Anonymity with a graph based scalable model. However, this approach requires a certain number of active users. They state that different road segments per user can be used to realise l-diversity for disclosed locations.

Palanisamy and Liu [8] introduce their approach of *MobiMix* in which mix zones are placed on the road network to gain anonymity of locations. The users' location anonymity is ensured by unlinkable pseudonyms when entering or exiting a mix zone and within such a zone by non-traceability.

Michalevsky et al. [7] introduce *PowerSpy*, an application which is able to infer a user's location and his driving route by using the power consumption of his smart phone, without any permissions to access GPS, WiFi, or other location data on the phone. Their basic idea is to measure the power consumption of different routes in advance and train a machine learning algorithm.

While using semantic information to obtain a contextual service (e.g. recommendation of restaurants nearby) or to share information about a visited venue, Agir et al. [1] present a solution where the semantic dimension of a location can be protected by generalising the semantic tag and locations can be obfuscated.

To counteract location breaches by inference attacks or the reveal of semantic behaviour by membership inclusion attacks, Bindschaedler and Shokri [2] introduce synthesised plausible location traces which are separated into semantic and geographic features. This separation is required, because people with similar lifestyles share common semantic traces but differ in geographic patterns.

3   System Model

We assume that users utilise route planning devices that combine offline and online data to find the best possible route for a trip. The real start and target location should be hidden from a location based service (LBS). Therefore, a trusted application on the user's (mobile) device is used which employs anonymisation techniques and/or dummy requests. This application uses static offline data and sends the anonymised request to the LBS. Here, the route request is processed and online real-time information like traffic jams are taken into account. For the sake of simplicity, we assume that an LBS has access to this information, even though it gets harder for an LBS to collect such data when *PARTS* is widely applied. The app on the user's device deanonymises the calculated route and presents the user with the routing information from the actual

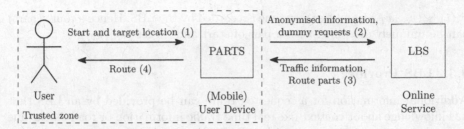

Fig. 1. Our *PARTS* framework works offline on the user's device and sends anonymised information to the LBS to protect the user's real start and target location (v_S, v_T). The LBS processes the requests and returns traffic information and route parts which are aggregated by our algorithm to the actual route.

start to target location. The user sticks to the route recommended by the local device, as depicted in Fig. 1. Here, the trusted application on the user's device is our *PARTS* framework (c.f. Sect. 4) which uses route parts as anonymisation technique.

In this setting, we consider an honest-but-curious LBS that is interested in deriving the start and target point of a user's route.

3.1 Road Network

We model the road network in the well-known way of a directed graph $G = (V, E)$. The road transitions form the set of vertices $v \in V$ and are connected via road segments represented by the edges $e \in E$. Since G is directed, a route segment from v_1 to v_2 has two corresponding edges $v_1 v_2$ and $v_2 v_1$, whereas one-way streets are represented by a single edge. The degree of a vertex $\deg(v)$ corresponds to the number of different roads one could reach from this point. Thus, road intersections are vertices v with $\deg(v) \geq 3$, whereas vertices with $\deg(v) = 1$ are equivalent to the end of a dead-end street. In our setting, points which connect road segments but are neither dead-ends nor intersections are of little relevance. To simplify the algorithm, we use the minor G^* of G in which the degree 2 vertices have been deleted by contracting one of the adjacent edges.

3.2 Users

Simplifying the presentation, we focus on route planning from intersection to intersection. Therefore, each location can be identified uniquely with a vertex v in the graph G representing the road network.

The set of users in our model is described with $U = (u_1, u_2, \ldots, u_m)$. There can be time intervals where no user moves and different users can travel to the same location at any time. Users submit their route requests with their current location v_S and their target v_T to the *PARTS* algorithm in a trusted zone, as depicted in Fig. 1. Our algorithm processes the user's request and submits several (anonymised) route requests to the LBS. Each request results in an event

$r_u(t) = (v_{S_i}, v_{T_i})$ from user u at time t recorded by the LBS. Hence a route (part) can be uniquely described with a pair of start and target (v_{S_i}, v_{T_i}).

3.3 LBS Provider

Additional information for a requested route can be provided by an LBS that has knowledge about context like real time traffic information or road blocks due to road constructions. Our algorithm *PARTS* uses this information to combine it with a locally generated route solely based on geographical information. By providing v_{S_i} and v_{T_i} as a result of a user request, an LBS collects personal data such as the request time. Furthermore, it is worthwhile to mention that the LBS has a similar amount of geographical information as our algorithm.

3.4 Adversary

The adversary in this paper is typically an LBS or an external observer who has access to all requests of all users, i.e. all received requests of an LBS. The adversary's main goal is to link the requests again and hence learn a route's overall start v_S and target vertex v_T. The adversary runs his attack *a posteriori*. Since he cannot be sure which requests are performed by a specific user, he collects a whole event log A for a specific amount of time and carries out his attack. It is reasonable for him to assume that A contains events from multiple users. However, it is possible that some of these events differ in time but share the same tuple of start and end vertices (v_S, v_T).

For his attack, the adversary takes into account geographical and temporal information described in Sect. 5. With limitation, the attack can also be performed by third-party services such as an Internet service provider (ISP) who has access to a user's request log but lacks knowledge about, for example, geographical information.

For our setup, we assume that the adversary knows all system parameters, i.e. he knows which privacy protection methods (PPMs) are applied and what specific parameters are used. This results in an even stronger adversary to stress our algorithm. Together with the PPMs and the event log, he starts his attack to deanonymise a route R, respectively a user u.

4 Strategies for Privacy-Enhanced Routing

Users request routing instructions from an LBS providing their geographical data v_S and v_T. To enforce their privacy goals, users apply PPMs, like our following *PARTS* framework, to hide their start and target location. To achieve this, *PARTS* can deploy different countermeasures which will be described in the following section.

In this work we focus on the route planning algorithm which resides in the application layer and tries to prevent information leakage from there. Hiding traffic data (such as IP addresses) is not the focus of our algorithm.

However, this data can be used to link different route requests including dummy requests to a specific user. In this case, our algorithm does not provide any anonymity at all. We therefore assume that route requests will be submitted using an anonymous channel like Tor in order to hide information from lower communication layers.

The PPMs mentioned in this section may be combined to provide a stronger protection against an honest-but-curious LBS. We will evaluate different settings and combinations of the following PPMs in Sect. 6.

4.1 Route Parts

PARTS splits every route R into multiple route parts R_i with start nodes v_{S_i} and target nodes v_{T_i}. The iterative combination of all R_i shall be the complete route R, thus $v_{S_{i+1}} = v_{T_i}$ holds. For each route part, a separate route request is necessary to obtain semantic information like traffic information. This request results in an event $r_u(t) = (v_{S_i}, v_{T_i})$ at the LBS. The LBS gains information that a user u wants to drive from v_{S_i} to v_{T_i} without knowing which of the start nodes $v_{S_1}, v_{S_2}, \ldots$ is the overall start v_S (resp. which v_{T_i} is the overall target v_T).

For each route part R_i, a subgraph G_i^* of G^* is constructed. The graph G_i^* shall contain all vertices having at most a specific distance to v_{S_i}. We call this adjustable distance *Hops* within our framework. Therefore, $dist_{G^*}(v_{S_i}, v) \leq Hops$ holds for all vertices v in G_i^*. The vertices v with $dist_{G^*}(v_{S_i}, v) = Hops$ are called *boundary vertices* $\partial V(G_i^*)$ and are candidates for being the target node v_{T_i}. The subgraph G_i^* contains at least one vertex with degree 3 by construction of G^* and therefore at least one intersection resulting in at least two candidates for v_{T_i}. It is obvious that a higher Hops count results in more target node candidates. To prohibit recursive routes, already used vertices are stored in a blacklist V_{BL} for the i-th iteration and will be ignored.

The node $v^* \in V_i^{cand} = \{v \in \partial V(G_i^*) \mid v \notin V_{BL}\}$ with the shortest straight line distance[1] $|v^* - v_T|$ to v_T is selected and used as target node v_{T_i} for the i-th iteration. A route request for v_{S_i} to v_{T_i} is submitted to the LBS and the calculated route forms the next route part R_i. For an example of a subgraph, see Fig. 2. For the next iteration, the algorithm sets $v_{S_{i+1}} = v_{T_i}$ and calculates a new subgraph G_{i+1}^*. This process stops, until the overall target node v_T is contained in one of the subgraphs G_i^*. In this case, the last route part R_i will be the route from v_{S_i} to v_T.

4.2 Dummy Traffic

We extend the PARTS algorithm from the previous section to include dummy requests to an LBS which may be needed in order to provide privacy. Using the previous setup, a route $R = \bigcup_{i=0}^n R_i$ consisting of n parts results in n events $r_u(t_i)$ at an LBS. It is trivial for an LBS to derive the complete route R by

---

[1] The straight line distance is used since it is locally computable and does not require any additional requests to a third party, thus not leaking any information.

Fig. 2. The subgraph G_3^* is depicted in black, whereas the subgraphs G_1^* and G_2^* are grey. The already calculated routes R_1 and R_2 are red. The vertex v^* has the lowest straight-line distance $|v^* - v_T|$ to the overall target v_T and will be used as the next step v_{T_3}. (Color figure online)

chronologically combining R_i. To oppose this, we introduce dummy requests in our system which should make it more challenging for an LBS to gain s and t.

Instead of submitting only one route request in the i-th iteration for the node v^* with the shortest straight-line distance to v_T, our algorithm produces one route request from v_{S_i} to v for some of the target vertex candidates $v \in V_i^{cand}$. Still, v^* is used as target vertex v_{T_i} and only the associated route is used as the route part R_i. In particular, the upper bound for dummy requests is $|V_i^{cand}| - 1$ since one request is always a legitimate request (v_{S_i}, v_{T_i}). The *PARTS* algorithm extended with the maximum number of dummy requests is depicted as Algorithm 1.

4.3   Time Shift Requests

As stated in the previous subsection, an event always contains time information (e.g. request time of a route by a user) which is outside of a user's sphere. Therefore, it is easy for an LBS to use this information to reconstruct the complete route using all part requests R_i just by chronologically sorting the request log. The attack is even possible if dummy traffic is used if an adversary uses statistical attacks. By time shifting the subsequent route requests, the attack can be hindered. Different methods of time shifting are possible:

- All route requests are executed after a specific but static amount of time, not allowing any insight into the length of a route part.
- All route requests occur at a random time resulting in no correlation between the length of a route and the request time.
- Route requests happen in batches, i.e. a specific number of requests are sent to the LBS at the same time to maintain unlinkability between requests.

Algorithm 1. PARTS with dummy requests

Input: Start vertex v_S, target vertex v_T
Result: Route R from v_S to v_T

1 $v_{S_1} \leftarrow v_S$;
2 $V_{BL} \leftarrow \{v_S\}$;
3 **while** $v_{T_i} \neq v_T$ **do**
4 | $i \leftarrow i + 1$;
5 | construct G_i^* with $V(G_i^*) = \{v \in G^* \mid dist_{G^*}(v_{S_i}, v) \leq Hops\}$;
6 | $V_i^{cand} \leftarrow \{v \in \partial V(G_i^*) \mid v \notin V_{BL}\}$;
7 | **if** $v_T \in V(G_i^*)$ **then**
8 | | $V_i^{cand} \leftarrow V_i^{cand} \cup \{v_T\}$;
9 | **foreach** $v \in V_i^{cand}$ **do**
10 | | submit route request (v_{S_i}, v) to LBS;
11 | $v_{T_i} \leftarrow argmin_{v \in V_i^{cand}}\{|v - v_T|\}$;
12 | $V_{BL} \leftarrow V_{BL} \cup \{v_{T_i}\}$;
13 | $R_i \leftarrow$ route from v_{S_i} to v_{T_i} calculated by LBS;
14 **return** $R = (R_1, R_2, \ldots)$

---

5   Adversary's Inference Model

In this section, we will present our inference model for the adversary which he uses to deanonymise specific users from the event log of route requests. We will explain how an adversary can exploit the route requests to reconstruct the full route from route parts. Based on this adversary model, we evaluate the privacy provided by our route planning algorithm in Sect. 6.

5.1   Background Knowledge

We assume that the adversary's inference model is based on the knowledge that users move along geographically valid routes using a valid behavior pattern. For example, users respect traffic regulations such as speed limits.

In our scenario, an adversary has extensive knowledge about the geographical structure of the road network represented. However, an adversary may utilise his own geographical database which can partially differ from the database on which our *PARTS* algorithm works (see Sect. 6.1). In general, we assume that such deviations are not harmful to perform the attack illustrated in Sect. 5.2. Thus, he is able to derive the same V_i^{cand} as our algorithm because all system parameters are known, as defined in Sect. 3.4. As a consequence, the adversary can, for example, compute the average travel time $t(R_i)$ for route segments.

5.2   Empirically Improved Guessing

The adversary plans to reconstruct complete routes with the given route requests submitted by different users. Since the adversary is an LBS, he has access to the full event log A. He selects one of the requests for which he has a high interest in recovering the corresponding route. Starting with this route request, the LBS calculates a likelihood tree Γ including possible past and future route parts.

Let $r_0 = (v_{S_0}, v_{T_0}) \in A$ be the route request for which the adversary is interested in reconstructing the full route. The adversary collects all route requests $A_{r_0}^{\rightarrow}$ whose start vertex is the target vertex v_{T_0} and which have been submitted to the LBS after r_0. For $r_1 = (v_{S_1}, v_{T_1}) \in A_{r_0}^{\rightarrow}$, we can define the function $f(r_1 \mid r_0)$ which models the likelihood that r_1 was the subsequent route request. Here, both temporal correlation and behavioural plausibility will be taken into account. With $t(R)$ being the travel time of the ideal route R from v_{S_0} to $v_{T_0} = v_{S_1}$ calculated by the LBS and t_0 and t_1 being the submission times of r_0 and r_1, respectively, the temporal correlation is denoted by $f_t(r_1 \mid r_0) = \exp\left(-c \cdot \frac{|(t_1 - t_0) - t(R)|}{t(R)}\right)$, where c is a scaling factor with $c \in [0, 1]$. If the request r_1 was submitted close to the travel time calculated by the LBS, it is more likely that r_1 will be the subsequent route request and $f_t(r_1 \mid r_0)$ increases.

For the behavioural plausibility $f_b(r_1)$, we count how many requests in the event log A have the same points as r_1 and divide this number by the total number of requests. Additionally, we introduce a factor $\lambda \in [0, 1]$ indicating the weight of the temporal correlation f_t in the likelihood f. Hence, we obtain

$$f(r_1 \mid r_0) = \lambda \cdot f_t(r_1 \mid r_0) + (1 - \lambda) \cdot f_b(r_1)$$

$$= \lambda \cdot \exp\left(-c \cdot \frac{|(t_1 - t_0) - t(R)|}{t(R)}\right) + (1 - \lambda) \cdot \frac{|\{r \in A \mid r = (v_{S_1}, v_{T_1})\}|}{|A|}$$

Since both values $f_t(r_1 \mid r_0)$ and $f_b(r_1)$ are in the range $[0, 1]$, we also have $f(r_1 \mid r_0) \in [0, 1]$. Here, a value close to 1 indicates a high likelihood that the route r_1 is the subsequent route part to r_0.

The adversary forms sets $A_{r_1}^{\rightarrow}$ for all $r_1 \in A_{r_0}^{\rightarrow}$ and evaluates the function $f(r_2 \mid r_1)$ for $r_2 \in A_{r_1}^{\rightarrow}$. He repeatedly continues with this approach and constructs a likelihood tree Γ with root r_0 and $r_1 \in A_{r_0}^{\rightarrow}$ being the first level nodes, etc. An edge (r, r') in this tree is weighted with the likelihood function $f(r' \mid r)$. To construct the prior route parts w.r.t. r_0, the adversary uses a similar approach by forming sets $A_{r_j}^{\leftarrow}$ containing route parts whose target vertex equals the start vertex of r_j. If the value of f falls below a predefined threshold value ε, the process will not be continued within the related subtree of Γ. Finally, the adversary chooses the path in Γ whose multiplied likelihood value along this path is the highest and uses the corresponding route requests as the reconstructed route R'.

5.3   Privacy Measurement

To evaluate the privacy of our algorithm, we compare the real route $R = (R_1, \ldots, R_k)$ from v_S to v_T from a user with the guess $R' = (R'_1, \ldots, R'_l)$ from $v_{S'}$ to $v_{T'}$ calculated by the adversary. His guess R' is the path in the tree Γ introduced in Sect. 5.2 with the highest likelihood. The following metrics will be applied, each measuring a different privacy aspect.

Distance to Start/Target Vertex. One of the goals of our algorithm was the protection of the start and target location of a route request. For this reason, we measure the straight-line distance $dist(R, R') = |v_S - v_{S'}| + |v_T - v_{T'}|$ between the start vertices and target vertices from R and R'.

Fit of Reconstructed Route. To protect movement patterns, reconstructing a route from route parts should not be easy for the adversary. Therefore, we measure $fit_{rou}(R, R')$, the percentage of the fit between guessed route and original one. Here, we count the number of correctly guessed route parts normalised over the total number of route parts in the original route.

Fit of Continuous Segments. In order to measure the attained linkability protection of our algorithm, we calculate the fit of continuous segments $fit_{seg}(R, R')$. We count the number of route segments successfully linked together by an adversary without any error in-between and normalise this figure over the total number of route segments. For instance, if R has $k = 6$ segments and an adversary was able to link segments (R_1, R_2, R_3) and (R_5, R_6) but missed to link R_4, we have $fit_{seg}(R, R') = 0.5$.

6 Evaluation

This section explains how we generated data according to our system model and applied the adversary's inference model to evaluate the quality of our algorithmic approach based on route parts, presented in Sect. 4.

6.1 Dataset and Simulator

PARTS is based on data from OpenStreetMap (OSM). In our setting we used a small portion of the whole dataset, more specifically a subregion of Bavaria, Germany. This region was converted to correspond to our graph setting.

Several users were created using a custom simulator. They move along the region in a predefined (but somewhat random) way[2]. Since our simulation should model users realistically, users follow different moving patterns resulting in route requests. All users submit route requests for activities which happen multiple times, like trips to a supermarket, whereas 80% of the users follow individual regular routines. Additionally, 30% of the users make random trips. On average, there are three moving patterns per user. We simulated a whole month which led to 398 moving patterns (every person therefore travels between 25–50 times a month) resulting in 36,167 (part) route requests. These are summed up values for the different combinations of our PPMs in place including direct requests.

---

[2] Since our users are exclusively moving within a city, we use the simplified assumption that travel speed is constant per road segment (homogeneous flow). Hereby, we use values from 37.5 kph to 62.5 kph. The adversary only knows that people will respect traffic regulations, therefore he uses a constant value of 50 kph for the whole route.

6.2   Experimental Setup

After generating the data as explained above, we applied the adversary's inference model to reconstruct the complete route from route segments (c.f. Sect. 5). To prove the power of our inference model, we showed that the usage of temporal correlation improves the chance of reconstructing a route correctly.

We successively applied more PPMs to get an insight into the adversary's ability to reconstruct the whole route of a user from the event log. We also tested different values for the parameters in the inference model to obtain the strongest adversary. More precise, we evaluated $c \in \{0.01, 0.1, 1\}$ for the scaling factor in the temporal correlation function f_t and $\lambda \in \{0, 0.5, 1\}$ for the weighting of f_t in the likelihood function f (see Sect. 5.2). Since the combination of $c = 0.01$ and $\lambda = 0.5$ results in the strongest adversary, we chose these values. Furthermore, we set $\varepsilon = 0.3$ for the threshold under which further route segments will not be considered in the likelihood tree Γ. If dummy traffic is used, we will send two dummy route requests per real request. A timeslot occurs every nine minutes.

6.3   Overhead of Segmented Routes

In this paragraph, we will investigate how the parameter $Hops$, as described in Sect. 4.1, will influence the route quality of our routing algorithm, i.e. the distance overhead of a route built with route parts compared to the ideal route.

Even though a user should have the possibility to select the ideal $Hops$ size, not every option makes sense regarding user experience and privacy – the main factors for the QoS. It is obvious that a higher $Hops$ size results in larger route segments and thus may disclose more private information. However, it often results in a better user experience because routes constructed of fewer segments lead to routes more similar to thoroughly constructed and therefore ideal routes.

In order to find a good trade-off between privacy and user experience we analysed routes with different numbers of intersections. Figure 3 shows the ratio to the optimal route for routes with 10 to 25 intersections (100% is the optimum). The different colors indicate the different values for the $Hops$ parameter. $MIXED$ uses a random value of $\{1, 3, 6, 12\}$ as the $Hops$ parameter for each iteration.

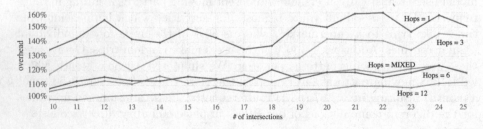

Fig. 3. Distance overhead for different values for the parameter $Hops$ with 250 iterations per number of intersections. (Color figure online)

Figure 3 illustrates that the smaller the *Hops* parameter gets the higher the distance overhead of a route is. Furthermore, $MIXED$ yields almost the same results as $Hops = 6$ does. In our simulation, on average all settings for the privacy enhanced routing resulted in an overhead to the optimal route. Obviously, there is no overhead if the complete route is shorter than the used *Hops* size since the complete route equals the first and only route segment (resulting in no privacy).

6.4 Privacy Related Results

We applied the privacy measures presented in Sect. 5.3 to our simulated data, namely the number of times in which an adversary was able to reconstruct the route (fit_{rou} and fit_{seg}) and the distance between the reconstructed route's start/end point and the real ones ($dist$). We chose *Hops* to be 6, 12 and $MIXED$.

In the remainder of this section and in the subsequent figures, we use the following abbreviations for the applied PPMs of our $PARTS$ algorithm: $DIRECT$ = route request without any PPMs applied, P = Route Parts, D = Dummy requests, Tb = Timeshift in batch mode, Tr = Timeshift in random mode, and Ts = Timeshift in timeslot mode.

fit_{rou} **and** fit_{seg}. Figure 4a shows the box plot for the metric fit_{rou} for different combinations of PPMs and $Hops \in \{6, 12, MIXED\}$ sizes, whereas Fig. 4b illustrates the results for fit_{seg}.

First, fit_{rou} and fit_{seg} have a value of 100% for the combinations $DIRECT$, $P + Tb$ and $P + Tb + D$, i.e. an adversary is able to reconstruct the full route in these cases without having a single outlier. Thus, there is no protection applying any combination which uses time shifting in batch mode. This seems reasonable since batch mode apparently eliminates all benefits of using route parts.

Regarding the *Hops* parameter, it can be stated that $MIXED$ performs best, followed by $Hops = 6$ and $Hops = 12$, independently from the used PPMs. This is especially pleasant, considering the little distance overhead added to a route when $Hops = MIXED$ is used (c.f. Fig. 3). Furthermore, $Hops = 6$ creates more route parts compared to $Hops = 12$. Therefore, the adversary has to do more work to relink all parts, providing more privacy.

We also analysed how combinations of PPMs affect the user's privacy. It can be stated, that P increases the privacy with every *Hops* parameter. Between the timeshifting modes, timeslot (Ts) performs best, followed by random (Tr) and trailed by the ineffective batch mode (Tb). Interestingly, the performance in $MIXED$ seems to lightly suffer from using dummy traffic (D) across the board.

Last, fig_{rou} and fit_{seg} show the same tendencies, although the spread of values for fit_{seg} is generally larger. This seems reasonable, since reconstructing a route in the correct order tends to be harder than finding used route segments.

$dist$. Figure 4c shows the box plot for the distance metric for the different combinations of PPMs and *Hops* size. Since an adversary is able to reconstruct the full route for $DIRECT$, $P + Tb$ and $P + D + Tb$, it is obvious that $dist$ is zero and that there is no difference between v_S and v_T and the guessed locations.

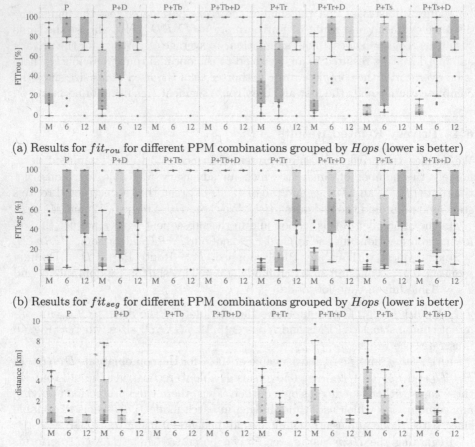

(a) Results for fit_{rou} for different PPM combinations grouped by $Hops$ (lower is better)

(b) Results for fit_{seg} for different PPM combinations grouped by $Hops$ (lower is better)

(c) $dist$ in km between the real and guessed start and target vertices (higher is better)

Fig. 4. The figures show the protection performance of our algorithm against the described inference model w.r.t our metrics (M is short for $MIXED$).

Overall, $dist$ follows the same trend as fig_{rou} and fit_{seg}: $MIXED$ yielded the highest distance across the board, i.e. an adversary guesses very distant points when reconstructing start and target. On the other hand, $dist$ is higher for $P + D$ than for P in contrast to fig_{rou} and fit_{seg} values.

It can further be seen that every combination of Ts with $Hops = MIXED$ shows the best results, even though D decreases $dist$.

6.5   Performance Analysis

We also analysed the performance of our algorithm w.r.t. runtime and data size of a request. Because of its easy-to-use API and the overall quality of service, Google Maps was chosen as the LBS for our performance measurement. All traffic to Google Maps is end-to-end encrypted, hence we used a man-in-the-middle

Table 1. Overview of 100 route requests to the Google API.

		Mean	Min	Max
Time [ms]	Roundtrip	185	162	500
Size [byte]	Request	296	296	296
	Response	2,765	997	5,263

proxy to gain access to the data. For the sake of reproducibility, we did not simulate connection resets or data loss which can occur in mobile networks.

Runtime. Table 1 shows the duration in milliseconds for a query to Google Maps. The mean duration is 185 ms per query (min 162 ms, max 500 ms). The bandwidth used is about 20 KB/s, a value easily achievable on mobile networks.

Executing requests in parallel does not seem to impact the duration. This is important since it proves that sending dummy requests does not significantly influence the performance and user experience. Furthermore, the time Google Maps needs to find a route is not significantly affected by the length of a route segment. In addition, we were unable to measure any difference regarding the time between real requests and dummy requests. This is obvious because both kinds of requests are using the same API calls. We thus assume that an LBS cannot distinguish real and dummy requests on a time basis.

Data Size. In a next step, we analysed the data size of requests for the different combinations of PPMs. It is obvious that $DIRECT$ uses the least amount of traffic. However, using route parts has almost the same data size. This seems reasonable since a direct route has a similar amount of routing instructions like the same route constructed from combining different route parts (i.e. number of navigation instructions). There is a slight amount of overhead since every Google Maps API call provides additional metadata. Table 1 shows that each request has the same size since it only contains start and target vertices encoded as coordinates. However, the resulting response differs in size due to a different number of instructions.

It is interesting that the growth of data is linear, i.e. every additional request is roughly the same in size. Arguing that an average route request needs about 3 KB, a route has 4 segments and 2 more dummy requests are sent, $PARTS$ requires acceptable 24 KB per privacy enhanced route request in addition.

We skipped the analysis of performance figures such as CPU and RAM usage because our proof of concept implementation is running on a desktop client. For future work, it is planned to create an Android application to measure performance figures on a real mobile device as well.

7 Discussion

One can observe that by using route parts, the overall privacy increases. However, the route part approach yields more privacy, as more requests of different users overlap. The required level of simultaneous requests can also be achieved by generating dummy traffic. One can see that the combination of route parts with dummy traffic always performs better than solely using route parts.

Another finding regarding dummy traffic is that it is sometimes easy for an adversary to filter. This fact may be connected to the number of requests at the same time and should improve if more users use the system and have overlapping route requests. Ideally, one dummy request from a user could be a real request from another user. It may also help to choose dummy target locations and run $PARTS$ in parallel for both the real and dummy locations instead of creating artificial requests per iteration.

In addition, a constant $Hops$ parameter adds a static component to $PARTS$ detectable by an adversary. Thus, it performs worse than $MIXED$ mode. This setting uses different $Hops$ parameters for each iteration and therefore weakens the adversary's ability to use behaviour knowledge by diluting his data pool.

A surprising result is that by sending route requests in batch mode, there is no privacy at all, regardless if dummy traffic is included or not. Hence, it is easy for an adversary to filter dummy requests and combine the different route segments just by comparing start and target vertices of each request since this indicates a chronological order. The attack is independent from the number of users in the system because it is very unrealistic that two requests from different users occur at the exact same time.

8 Conclusion

We presented the routing algorithm $PARTS$ which protects a user's movement pattern by splitting each route request into several route segments without revealing the real start and target locations v_S and v_T of the overall route. In this way it is possible to combine local offline knowledge, such as a geographical layout, with global online real-time data, like traffic information.

A simulation further emphasised the need for such an algorithm and revealed that it is trivial for an adversary to derive a movement pattern for a user. Our simulation has shown that route parts can provide additional privacy but need to be combined with further PPMs to achieve their full potential. Therefore we extended our algorithm to use dummy traffic and time shifting. It was shown that the application of time shifting with timeslots was very powerful to protect a user's privacy in almost every case. The usage of $MIXED$ mode for the $Hops$ parameter provides a good overall user experience, since it offers a very high privacy level and produces a reasonable distance overhead compared to the ideal route. In general, the $PARTS$ algorithm has no significant influence to the user experience in terms of performance and data consumption.

For future work, we plan to include semantic background information in our scenario. On the one hand, we want to strengthen our adversary with these capabilities. On the other hand, we want to improve the way dummy traffic is constructed since our experiments have identified that randomly generated dummy traffic is not that powerful. Furthermore, we plan to evaluate $PARTS$

against real-world datasets such as Microsoft Geolife [12] to prove its feasibility on a daily usage. In addition, we want to implement the algorithm as a mobile application to further elaborate its usability.

References

1. Ağır, B., Huguenin, K., Hengartner, U., Hubaux, J.P.: On the privacy implications of location semantics. In: Proceedings on Privacy Enhancing Technologies 2016, vol. 4, p. 1 (2016)
2. Bindschaedler, V., Shokri, R.: Synthesizing plausible privacy-preserving location traces. In: 2016 IEEE Symposium on Security and Privacy (SP), pp. 546–563. IEEE (2016)
3. Golle, P., Partridge, K.: On the anonymity of home/work location pairs. In: Tokuda, H., Beigl, M., Friday, A., Brush, A.J.B., Tobe, Y. (eds.) Pervasive 2009. LNCS, vol. 5538, pp. 390–397. Springer, Heidelberg (2009). doi:10.1007/978-3-642-01516-8_26
4. Hughes, N.: Inside ios 9: Apple's maps app. gets smarter with automatic directions based on user habits. Apple Insider (2015)
5. Kramer, A.D.I., Guillory, J.E., Hancock, J.T.: Experimental evidence of massive-scale emotional contagion through social networks. Proc. Natl. Acad. Sci. 111(24), 8788–8790 (2014)
6. Machanavajjhala, A., Kifer, D., Gehrke, J., Venkitasubramaniam, M.: L-diversity: privacy beyond k-anonymity. ACM Trans. Knowl. Discovery Data (TKDD) 1(1), 3 (2007)
7. Michalevsky, Y., Schulman, A., Veerapandian, G.A., Boneh, D., Nakibly, G.: PowerSpy: location tracking using mobile device power analysis. In: 24th USENIX Security Symposium (USENIX Security 2015), pp. 785–800 (2015)
8. Palanisamy, B., Liu, L.: MobiMix: protecting location privacy with mix-zones over road networks. In: IEEE 27th International Conference on Data Engineering (ICDE 2011), pp. 494–505. IEEE, Piscataway (2011)
9. Shokri, R., Troncoso, C., Diaz, C., Freudiger, J., Hubaux, J.P.: Unraveling an old cloak: k-anonymity for location privacy. In: Al-Shaer, E., Frikken, K. (eds.) Proceedings of the 9th Annual ACM Workshop on Privacy in the Electronic Society, p. 115. ACM, New York (2010)
10. Sweeney, L.: k-anonymity: a model for protecting privacy. Int. J. Uncertainty Fuzziness Knowl. Based Syst. 10(05), 557–570 (2002)
11. Wang, T., Liu, L.: Privacy-aware mobile services over road networks. Proc. VLDB Endowment 2(1), 1042–1053 (2009)
12. Zheng, Y., Xie, X., Ma, W.Y.: Geolife: a collaborative social networking service among user, location and trajectory. IEEE Data Eng. Bull. 33, 32–39 (2010)

Security and Privacy in Machine Learning

Bayesian Network Models in Cyber Security: A Systematic Review

Sabarathinam Chockalingam[(✉)], Wolter Pieters, André Teixeira, and Pieter van Gelder

Faculty of Technology, Policy and Management, Delft University of Technology, Delft, The Netherlands
{S.Chockalingam,W.Pieters,Andre.Teixeira,P.H.A.J.M.vanGelder}@tudelft.nl

Abstract. Bayesian Networks (BNs) are an increasingly popular modelling technique in cyber security especially due to their capability to overcome data limitations. This is also exemplified by the growth of BN models development in cyber security. However, a comprehensive comparison and analysis of these models is missing. In this paper, we conduct a systematic review of the scientific literature and identify 17 standard BN models in cyber security. We analyse these models based on 8 different criteria and identify important patterns in the use of these models. A key outcome is that standard BNs are noticeably used for problems especially associated with malicious insiders. This study points out the core range of problems that were tackled using standard BN models in cyber security, and illuminates key research gaps.

Keywords: Bayesian attack graph · Bayesian network · Cyber security · Information security · Insider threat

1 Introduction

The lack of data, especially historical data on cyber security breaches, incidents, and threats, hinders the development of realistic models in cyber security [1,2]. However, standard (or classical) Bayesian Networks (BNs) possess the potential to address this challenge. In particular, the capability to combine different sources of knowledge would help to overcome the scarcity of historical data in cyber security modeling.

Standard BNs belong to the family of probabilistic graphical models [3]. A standard BN consists of two components: qualitative, and quantitative [4]. The qualitative part is a Directed Acyclic Graph (DAG) consisting of nodes and edges. Specifically, each node represents a random variable, whereas the edges between the nodes represent the conditional dependencies among the corresponding random variables. The quantitative part takes the form of conditional probabilities, which quantify the dependencies between connected nodes in the DAG by specifying a conditional probability distribution for each node. A toy example of a standard BN model, representing the probabilistic relationships

© Springer International Publishing AG 2017
H. Lipmaa et al. (Eds.): NordSec 2017, LNCS 10674, pp. 105–122, 2017.
https://doi.org/10.1007/978-3-319-70290-2_7

between cyber-attacks ("Denial of Service Attack" and "Malware Attack") and symptoms ("Internet Connection" and "Pop-ups"), is shown in Fig. 1. Given symptom(s), the BN can be used to compute the posterior probabilities of various cyber-attacks as shown in Fig. 1. In this case, the user sets evidence for the "Pop-ups" node as "True", and "Internet Connection" node as "Normal" in the BN model based on his/her observations. Based on these evidences, the BN computes the posterior probabilities of the other nodes "Denial of Service Attack" and "Malware Attack" using Bayes rule. The BN model shown in Fig. 1 determines that the presence of pop-ups and normal internet connection are more likely due to a Denial of Service attack rather than to a Malware attack.

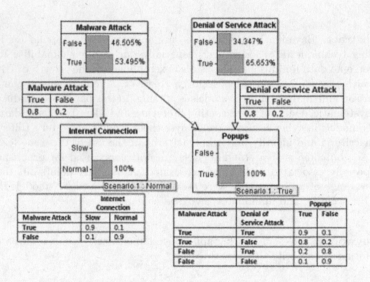

Fig. 1. Standard BN model - example

The major advantages of standard BNs include: the ability to combine different sources of knowledge, the capacity to handle small and incomplete datasets, and the availability of a broad range of validation approaches apart from data-driven validation approaches [5,6]. Some notable real-world applications of standard BNs include medical diagnosis [7] and fault diagnosis [8]. In addition, the advantages lead to the predominant use of standard BNs in domains where there is a limited availability of data, notably in Ecosystem Services (ESS) [5], water resource management [9], and security [10]. Similarly, we have seen the use of standard BNs in cyber (or information) security in recent years [11–29]. However, an overarching comparison and analysis of standard BN models in cyber security which could help to identify important usage patterns is currently lacking. Kordy et al. give a broader overview of modeling approaches based on DAGs, and thus only briefly mention BNs [10]. In contrast to Kordy et al., we specifically focus on BN models with the aim of performing comparison and analysis of these models to identify important usage patterns and key research gaps. This review would benefit the practical application of BN models in cyber security by providing important usage patterns and key research gaps. Therefore, this research aims to fill this gap by addressing the research question: "What are the important patterns in the use of standard Bayesian Network (BN) models in cyber security?". The research objectives are:

- **RO 1.** To identify standard BN models in cyber security literature.
- **RO 2.** To identify the important patterns in the use of standard BN models in cyber security based on the analysis of identified models.

In this paper, we focus on comparison and analysis of standard BN models [11–29] which also include Bayesian Attack Graphs (BAGs) [11–13] as they possess more comparable features. This would help to identify consistent patterns in the use of standard BN models in cyber security. However, the approaches in cyber security modeling that extend BN such as Bayesian Decision Network (BDN) [30], Causal event graph [31], Dynamic BN [32–34], Extended influence diagram [35,36], and Multi-entity BN [37,38] are beyond the scope of this paper as they are incomparable especially based on their structure development. For instance, decision and utility nodes are specific to BDN/Influence Diagram which would allow decision making under uncertainty. In contrast, these types of nodes are not applicable to standard BN.

The scope of this comparison and analysis is the structured development, application and validation of the existing standard BN models in cyber security. The comparison and analysis of identified models is performed using the characteristics that were chosen based on related literature and domain-specific objectives as described in Sect. 2. The key contributions of this work are: important patterns in the use of standard BN models in cyber security, and key research gaps in the use of standard BN models in cyber security.

The remainder of this paper is structured as follows. Section 2 describes the review methodology. In Sect. 3, we perform the comparison and analysis of identified BN models using the characteristics that we chose, followed by a discussion on the key findings in Sect. 4. Finally, we highlight important patterns in the use of standard BN models in cyber security followed by future work directions in Sect. 5.

2 Review Methodology

We perform the systematic literature review based on the guidelines provided by Okoli et al. [39]. The methodology which we used to select the standard BN models in cyber security literature and the appropriate characteristics to perform the comparison and analysis of the selected BN models is described below.

The selection of standard BN models in cyber security literature consists of two stages:

- Searches were performed on ACM Digital Library, DBLP, Google Scholar, IEEE Xplore Digital Library, Scopus, and Web of Science – All Databases.

Search-strings were constructed from keywords "Bayesian", "Bayesian Belief Network", "Bayesian Network", "BBN", "BN", "Cyber*", "Information*", and "Security". The wildcard "*" was used for "Cyber" and "Information" to match all words around these two keywords.

- Models were selected from the search results according to the listed criteria:
- The model should employ standard BN.
- The model should address problem(s) associated with cyber (or information) security.
- The literature should have basic information about both DAGs and Conditional Probability Tables (CPTs). This criterion is important taking into account the scope of our comparison and analysis which is the structured development, application and validation of the existing standard BN models in cyber security.
- The literature should be in English language.

Once a standard BN model in cyber security was selected, the scientific literature that cited it was also traced.

The characteristics used to perform the analysis of the selected BN models were chosen based on related literature and domain-specific objectives as described in Sects. 2 and 3. Landuyt et al. presented 47 BN models in ESS published from 2000 to 2012 [5]. In addition, they analysed these models based on 9 characteristics. Similarly, Phan et al. presented 111 BN models in water resource management [9]. Moreover, they analysed these models based on 10 characteristics. We adopted the characteristics from Landuyt et al. and Phan et al. that are generic and relevant to the scope of our analysis, as shown in Table 1. Also, we adapted and used the characteristic: *Citation details* provided by Phan et al. to perform the analysis of BN models in cyber security as described in Sect. 3.

Table 1. Adopted characteristics from Landuyt et al. and Phan et al.

Characteristics used in our analysis	Adopted from Landuyt et al.	Adopted from Phan et al.
I. Citation details		✓
II. Data sources used to construct DAGs and populate CPTs	✓	✓
III. The number of nodes used in the model	✓	
IV. Type of threat actor		
V. Application and Application sector		
VI. Scope of variables		
VII. The approach(es) used to validate models	✓	✓
VIII. Model purpose and type of purpose		

3 Analysis of Standard Bayesian Network Models in Cyber Security

This section aims to address *RO 1. To identify standard BN models in cyber security literature*, and *RO 2. To identify the important patterns in the use of standard BN models in cyber security based on the analysis of identified models*. Based on the methodology described in Sect. 2, we identified 17 standard BN models in cyber security. The corresponding article titles are listed in Table 2. Furthermore, this section performs the analysis of identified BN models based on the following characteristics.

- Citation details
- Data sources used to construct DAGs and populate CPTs
- The number of nodes used in the model
- Type of threat actor
- Application and Application Sector
- Scope of variables
- The approach(es) used to validate models
- Model purpose and Type of purpose

3.1 Citation Details

We adapted and used the components of the characteristic *"Citation details"* provided by Phan et al. Specifically, we used an additional component citations in our definition of *"Citation details"* because this will help us to assess the research impact/quality of each BN model [40]. In Table 2, citations is the number of citations of the article according to Google Scholar Citation Index as on 15th September 2017. The number of articles covering standard BN model in cyber security varies between 0 and 3 per year. No noticeable increase in the number of papers over time is encountered. The largest number of citations (247) is acquired by Poolsappasit et al. [11] published in 2012. The second most cited paper, among analysed, with 136 citations, is Frigault et al. [12] which is published in 2008. Interestingly, BAG-based standard BN models [11–13] are extensively used compared to the other standard BN models [14–29] in cyber security based on the number of citations.

3.2 Data Sources Used to Construct DAGs and Populate CPTs

We used the characteristic *"Data sources used to construct DAGs and populate CPTs"* to identify the type of data sources utilised in the reviewed BN models. We employed the coding scheme provided by Phan et al. in Table 2 where "Expert Knowledge (K)" refers to domain expert(s) and/or article's author(s) knowledge, "Empirical Data (D)" refers to observational or experimental evidence or data, either available directly to the authors or derived from the literature [9]. From Table 2, we observe that 5 out of 17 BN models used only expert knowledge to construct DAGs, whereas 5 out of 17 BN models employed only empirical data to construct DAGs. 7 out 17 BN models made use of both expert knowledge and empirical data to construct DAGs. In particular, 10 out of 12 BN models which utilised empirical data to construct DAGs relied on the literature. In contrast, 2 out of 12 BN models which utilised empirical data to construct DAGs relied on the inputs from vulnerability scanner [11] and incidents data [18].

From Table 2, we infer that 11 out of 17 BN models utilised only expert knowledge to populate CPTs, whereas 3 out of 17 BN models used only empirical data to populate CPTs. On the other hand, there were 3 out of 17 BN models which employed both expert knowledge and empirical data to populate CPTs. Specifically, the sources of empirical data includes literature, incidents data, National Vulnerability Database (NVD), Open Source Vulnerability Database (OSVDB), and Exploithub to populate CPTs. Notably, the review of BN models in water resource management and ESS pointed out *model simulations* as another data source used to construct DAGs and populate CPTs [5,9]. *Model simulations* refers to outputs of other empirical, deterministic or stochastic models [5]. Interestingly, there was no standard BN model in cyber security that used *model simulation* as the data source to construct DAGs and populate CPTs.

3.3   The Number of Nodes Used in the Model

The number of nodes can be used to describe the model complexity [5]. A high number of nodes often lead to a lot of intermediary layers between the layer of input nodes and the layer of output nodes. This could weaken the relation between input and output nodes. Marcot et al. recommended to limit the number of node layers or sequential relationships to less than five to prevent this dilution of interactions [41].

Landuyt et al. indicate that BN models with nodes lower than 40 can safeguard the functionalities of BNs [5]. Based on our analysis, we conclude that the amount of nodes is relatively kept low in the identified BN models in cyber security as 16 out of 17 BN models have a node number lower than 40. On the other hand, the BN model developed by Shin et al. exceeds the node number 40 [19]. However, the BN model developed by Shin et al. is a combination of two networks. If it is not possible to keep the model structure shallow, Marcot et al. suggested to break up the model into two or more networks [41]. Shin et al. utilised this idea to prevent the dilution of interactions between the input and output nodes.

3.4   Type of Threat Actor

We used the characteristic *"Type of threat actor"* because this will allow us to understand whether the BN model in cyber security was developed with a focus on particular type of threat actor(s). We classified threat actors as insider versus outsider [42]. Furthermore, we also considered their intentions, which could be either malicious/deliberate or accidental [42]. Figure 2 shows the general distribution of the BN models reviewed according to the type of threat actors and their intent.

Table 2. List of Bayesian network models in cyber security (ordered by the number of citations)

Article title (year)	Citations	Data source (DAG)	Data source (CPT)	Application	Application sector
Dynamic security risk management using Bayesian attack graphs [11] (2012)	247	D	K	Risk Management	Non-specific
Measuring network security using Bayesian network-based attack graphs [12] (2008)	136	K	K	Risk Management	Non-specific
Network vulnerability assessment using Bayesian networks [13] (2005)	106	K	K	Risk Management	Non-specific
Reasoning about evidence using Bayesian networks [14] (2008)	39	K	K	Forensic Investigation	Law Enforcement
A Bayesian network model for predicting insider threats [15] (2013)	35	D, K	D, K	Threat Hunting (Insider Threat)	Non-specific
Identifying at-risk employees: modeling psychosocial precursors of potential insider threats [16,17] (2012, 2010)	31,24	D, K	K	Threat Hunting (Insider Threat)	Non-specific
Identifying compromised users in shared computing infrastructures: a data-driven Bayesian network approach [18] (2011)	23	D	D	Forensic Investigation	University
Development of cyber security risk model using Bayesian networks [19] (2015)	21	D, K	K	Risk Management	Nuclear
Studying interrelationships of safety and security for software assurance in cyber physical systems: approach based on Bayesian belief networks [20] (2013)	20	K	K	Risk Management	Petroleum (Oil)
Vulnerability categorization using Bayesian networks [21] (2009)	10	D	D	Vulnerability Management (Classification)	Software
Quantitative assessment of cyber security risk using Bayesian network-based model [22] (2009)	8	D	D, K	Risk Management	Non-specific
A Bayesian network model for likelihood estimations of acquirement of critical software vulnerabilities and exploits [23] (2015)	7	D, K	D, K	Governance	Software

(continued)

Table 2. (*continued*)

Article title (year)	Citations	Data source (DAG)	Data source (CPT)	Application	Application sector
Analysis of the digital evidence presented in the Yahoo! case [24] (2009)	2	K	K	Forensic Investigation	Law Enforcement
Modeling information system availability by using Bayesian belief network approach [25] (2016)	1	D, K	K	Risk Management	Non-specific
A Bayesian network model for predicting data breaches [26] (2016)	0	D, K	K	Risk Management	Health Care
Information security risk assessment of smartphones using Bayesian networks [27, 28] (2016, 2015)	0,0	D, K	K	Risk Management	Smartphone (In Finland)
Bayesian network modelling for analysis of data breach in a bank [29] (2011)	0	D	D	Risk Management	Banking

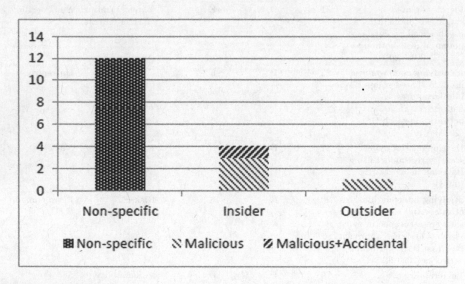

Fig. 2. Characterization of threat actors in the BN models reviewed

From Fig. 2, we infer that 4 out of 17 BN models are used only for problems associated with insiders [15, 16, 26, 29]. In particular, we observe that 4 out of these 4 BN models are appropriate for malicious insiders [15, 16, 26, 29], and only 1 out of these 4 BN models is relevant for accidental insiders in addition to malicious insiders [26]. Holm et al. developed a BN model with a focus on malicious outsider (professional penetration tester) [23].

Importantly, there was no integrated BN model that considers problem(s) associated with both insider and outsider type of threat actors, and their interactions. This type of BN models would help to combat especially social engineering attacks, and outsider collusion attacks [43]. Finally, there were 12 out of 17 BN models which did not focus on any specific type of threat actor [11–14,18–22,24,25,28]. For instance, the BN model developed by Pecchia et al. is used to identify compromised users in shared computing infrastructures based on alerts [18]. This model did not focus on any specific type of threat actor, but rather focused on alerts which could be appropriate to any type of threat actor. Therefore, we categorized it as 'non-specific'.

3.5   Application and Application Sector

We used the characteristic *"Application"* to understand the type of applications that partially or completely benefit from these BN models. We used the Chief Information Security Officer mind map as the basis to classify the reviewed BN models based on their application [44]. In addition, we used the characteristic *"Application Sector"* to identify the type of application sectors in which these BN models were demonstrated. From Table 2, we infer that 10 out of 17 BN models in cyber security completely or partially benefit Risk management. In addition, Forensic investigation, Governance, Threat hunting, and Vulnerability management were the other applications which completely or partially benefit from these BN models. From Table 2, we observe that the application sectors were quite diverse. However, 15 out of 17 BN models focused on the cyber security of Information Technology (IT) environment. In contrast, 2 out of 17 BN models focused on the cyber security of Industrial Control Systems (ICS) environment [19,20].

3.6   Scope of Variables

We used the characteristic *"Scope of Variables"* to identify the entities to which the variables used in the reviewed BN models are related. In addition, we classify the variables used in the reviewed BN models based on the key elements of cyber security. Cyber security is a combination of three key elements: People, Process and Technology [45].

From Table 3, we observe that the variables used in the BN models that focus on the cyber security of ICS environment did not consider the 'people' element of cyber security [19,20]. Importantly, the variables used in these BN models are mainly related to the technological components of ICS ('Technology' focussed) [19,20]. In addition, we infer that the variables used in 2 out of 4 BN models employed for the problems associated with insiders consider the three key elements of cyber security [26,29] which are application-specific, whereas the variables used in 2 out of 4 BN models employed for the problems associated with insiders take into account only the 'people' element of cyber security [15,16] which might be applicable to different organizations.

Table 3. Scope of variables used in the BN models reviewed

Authors	Variables - entities	Variables - key element(s) of cyber security
Poolsappasit et al. [11]	Mail server, DNS server, SQL server, NAT Gateway server, Web server, Administrator machine, Local desktops	Technology
Frigault, Wang [12]	N/A	N/A
Liu, Man [13]	Network hosts	Technology
Kwan et al. [14]	Seized computer	Technology
Axelrad et al. [15]	Employee	People
Grietzer et al. [16,17]	Employee	People
Pecchia et al. [18]	User profile, Shared computing infrastructure	People, Technology
Shin et al. [19]	Organization (Management) checklist, Reactor Protector System (RPS) components	Process, Technology
Kornecki et al. [20]	Components of ICS used to control oil pipeline flow	Technology
Wang, Guo [21]	Software	Technology
Mo et al. [22]	Organization (Management), Attack pathway	Process, Technology
Holm et al. [23]	Software	Technology
Kwan et al. [24]	Suspect, Seized computer, Yahoo! email account, Internet service provider	People, Technology
Ibrahimovic, Bajgoric [25]	Organization (Management)	Process
Wilde [26]	Employee, Organization (Management), Mobile Device	People, Process, Technology
Herland et al. [27,28]	Smartphone	Technology
Apukhtin [29]	Employee, Organization (Management), Security controls	People, Process, Technology

3.7   The Approach(es) Used to Validate Models

We used the characteristic *"The approach(es) used to validate models"* to identify the type of validation approaches used in the reviewed BN models. Based
on our analysis, we observe that real-world case study [14,24], cross-validation
[15,18], goodness of fit [16], Monte-Carlo simulation [25], expert evaluation
[26,27], and sensitivity analysis [26,29] were the approaches used to validate
the reviewed BN models. Importantly, there was no validation performed in

8 out 17 BN models [11–13, 19–23]. Finally, there was only one BN model which utilized several approaches such as sensitivity analysis, and expert evaluation to perform the validation [26]. However, the reviewed BN models validated different aspects depending on their objectives. For instance, Wilde [26] validated the usefulness of their model in practice, whereas Herland et al. [27, 28] validated the accuracy and completeness of the qualitative BN model.

3.8 Model Purpose and Type of Purpose

We used the characteristic *"Model Purpose"* to point out the problems that were tackled using BN models in cyber security. In addition, we used the characteristic *"Type of Purpose"* to identify the corresponding category of model purpose. Table 4 highlights the authors of the BN model, the corresponding purpose of the BN model, and the corresponding type based on the model purpose.

From Table 4, we observe that the reviewed BN models in cyber security were mainly used for two types of purposes based on their model purpose: I. Diagnostic: To reason from effects to causes, and II. Predictive: To reason from causes to effects. Importantly, 13 out of 17 BN models in cyber security were used for predictive purposes.

Table 4. BN model purpose and type of purpose

Authors	Model purpose	Type of purpose
Poolsappasit et al. [11]	To quantify the chances of network compromise at various levels	Predictive
Frigault, Wang [12]	To determine the likelihood of attaining the goal state by exploiting vulnerabilities in a network	Predictive
Liu, Man [13]	To perform quantitative vulnerability assessment of a network of hosts	Predictive
Kwan et al. [14]	To reason about digital evidence in the BitTorrent case	Diagnostic
Axelrad et al. [15]	To predict degree of interest in a potentially malicious insider	Predictive
Greitzer et al. [16, 17]	To predict the psychosocial risk level of an individual	Predictive
Pecchia et al. [18]	To detect compromised users in shared computing infrastructures	Diagnostic
Shin et al. [19]	To evaluate the cyber security risk of the reactor protection system	Predictive
Kornecki et al. [20]	To jointly assess safety and security of a SCADA system used to control oil pipeline flow	Predictive

(*continued*)

Table 4. (*continued*)

Authors	Model purpose	Type of purpose
Wang, Guo [21]	To categorise software security vulnerabilities	Diagnostic
Mo et al. [22]	To evaluate the security readiness of organizations	Predictive
Holm et al. [23]	To estimate the likelihood that a penetration tester is able to obtain information about critical vulnerabilities and exploits for these vulnerabilities corresponding to a desired software and under different circumstances	Predictive
Kwan et al. [24]	To reason about digital evidence in the Yahoo! Case	Diagnostic
Ibrahimovic, Bajgoric [25]	To predict information system availability	Predictive
Wilde [26]	I. To predict the probability of a data breach caused by a group of insiders who lose employee- and employer-owned mobile devices or misuse the employer-owned mobile devices, II. To help health care organizations determine which additional measures they should take to protect themselves against data breaches caused by insiders	Predictive, Diagnostic
Herland et al. [27,28]	To assess information security risks related to smartphone use in Finland	Predictive
Apukhtin [29]	To predict the probability of a data breach in a bank caused by a malicious insider	Predictive

4   Discussion

In the previous section, we identified key usage patterns of BNs in cyber security. This section discusses potential reasons for the key findings and suggests future research directions.

There is an emphasis on problems associated with insiders compared to outsiders in the use of standard BN models in cyber security. In general, this emphasis could be due to the most significant threat posed by insiders. This was elucidated by IBM's cyber security intelligence index which concluded that 60% of all attacks were carried out by insiders [46]. In connection with the use of standard BNs, the availability of characteristics associated with insiders in the literature provided a good starting point to determine appropriate variables and their relationships which form an integral part of a standard BN. In addition, the variables and their relationships determined from the literature were fine-tuned and/or complemented with other suitable variables based on expert knowledge in a few instances. This is one of the major advantages of standard BNs described in Sect. 1 which is the ability to combine different sources of knowledge. This could be the rationale behind the predominant use of standard BNs for problems associated with the insiders.

Special importance is given to problems associated with malicious insiders compared to accidental insiders in the use of standard BN models in cyber security. In general, this could be due to the fact that malicious insiders are more natural than accidental insiders in security contexts, as malicious insiders have a clear intent of compromising security, while accidental insiders do not. Moreover, malicious insiders have been shown to be the cause of more incidents than accidental insiders, as it was demonstrated by IBM's cyber security intelligence index which concluded that 44.5% of attacks were carried out by malicious insiders, and accidental insiders were responsible for 15.5% of attacks [46]. In order to use standard BNs for problems associated with accidental insiders compared to malicious insiders, it is important to identify features associated with accidental insiders in the literature to determine appropriate variables and their relationships, which form an essential part of a standard BN. There are studies which identify features associated with accidental insiders in the literature [43,47]. Once the appropriate variables and their relationships are determined for problems associated with accidental insiders, this could always be updated based on expert knowledge. It would also be useful to explore variables and their relationships in the reviewed BN models that focus on problems associated with malicious insiders, as some of the indicators might also apply for problems associated with accidental insiders [43].

The focus on insiders may also explain why there is little research on applications in the ICS domain. The reviewed BN models that focus on problems associated with the insiders might not be suitable for ICS environments, especially for control rooms with an operator. This is prevalent in control rooms that are used to operate sluices in the Netherlands. Not accepting feedback, Anger management issues, Confrontational issues, Counterproductive behaviour towards individuals (CPB-I), Counterproductive behaviour towards the organization (CPB-O) were some of the variables used in the reviewed BN models [15,16]. Most of these variables might be measured/observed based on interactions of the particular individual with the co-workers. However, this would not be possible in the control rooms where there would be no co-worker. It would be interesting to explore in the future whether the variables and their relationships in the reviewed BN models focused on problems associated with the insiders are suitable for ICS environment, and also whether the size of the organization in which the BN model would be applied have an effect on these variables and their relationships. In general, the limited use of standard BN models in cyber security on problems associated with ICS environment could be due to the shortage of ICS security expertise [48] as majority of the reviewed BN models relied on expert knowledge especially to construct DAGs and populate CPTs.

There is no integrated BN model which takes into account the problem(s) associated with both insiders and outsiders, and their interactions. The German steel mill incident is a typical example of a cyber-attack which involves both accidental insiders and malicious outsiders, and their interactions [49]. As an initial step, the adversaries used both the targeted email and social engineering techniques to acquire credentials for the plant's office network. Later, once they acquired credentials for the plant's office network, they worked their way into the plant's control system network and caused damage to the blast furnace. Standard BNs would help to tackle problem(s) associated with both insiders and outsiders, and their interactions, for instance a standard BN model that could predict the probability of an individual being deceived by outsider(s) to cause a cyber-attack in an organization, given certain risk factors and symptoms. This BN model would especially help to identify vulnerable individuals in an organization against social engineering attacks, and effective measures which could reduce the likelihood of an individual deceived by outsiders to cause a cyber-attack in an organization.

It is evident that the initial attempts in the use of standard BN models in cyber security were using BAG-based standard BN models [11–13]. BAG-based standard BN model combines acyclic attack graph which acts as the DAG with computational procedures of BN. Attack graph is one of the extensively used approaches in security modeling which was introduced in 1998 [10,50]. The use of BAG-based standard BN models in the intial attempts could be due to practicality. It could be practical to build attack graphs first which had been extensively studied in this domain and use BN computational procedures for quantification during the early stages in the use of standard BN models in cyber security. Similarly, there were attempts in the safety domain which mapped fault tree to BN [51,52]. Importantly, BAG-based standard BN models model static systems. Therefore, they are not directly applicable to multi-step attacks.

Risk management, forensic investigation, governance, threat hunting, and vulnerability management were the applications of standard BNs in cyber security. However, it would also be useful to investigate the potential of standard BNs to benefit other applications. Chockalingam et al. highlighted the importance of integrating safety and security especially in the context of modern ICS [53]. BNs possess the potential to develop an integrated BN model that could diagnose the root cause of an abnormal behavior in the ICS especially whether the abnormal behavior is caused by an attack (security-related) or technical failure (safety-related) by taking into account certain risk factors and symptoms. This would allow the operator(s) to point out the best possible response strategy. For instance, the process of routing traffic through a scrubbing center would be a suitable response strategy for a Distributed Denial of Service (DDoS) attack whereas this may not be an appropriate response strategy for a sensor failure.

The sources of empirical data used to construct DAGs and populate CPTs include: literature, incidents data, NVD, OSVDB, and exploithub. It is important to identify other domain-specific empirical data sources which would help to develop realistic models in cyber security. For instance, Capture-The-Flag (CTF) events like SWaT security showdown (S3) [48] could be a potential data source to construct DAGs and populate CPTs. CTF events could generate datasets that are realistic in nature [54]. However, this could have been overlooked because the data generated in these events would be in most cases specific to that particular system, and the quality of data generated could depend on the participants.

5   Conclusions and Future Work

In this paper, we have identified 17 standard BN models in cyber security. Based on the analysis, we identified important patterns in the use of standard BN models in cyber security.

- The standard BN models in cyber security were significantly used for problems associated with malicious insiders.
- There is an emphasis on the use of standard BN models in cyber security for problems associated with IT environment compared to ICS environment. In addition, the standard BN models that focus on the cyber security of ICS environment did not consider the 'people' element of cyber security. This implies that there is no standard BN model which deal with problem associated with insiders in ICS environment.
- There is a lack of standard BN models usage for problems associated with insiders and outsiders, and their interactions.
- Expert knowledge, and empirical data predominantly from literature were the data sources utilised to construct DAGs and populate CPTs.
- The standard BN models in cyber security completely or partially benefited risk management, forensic investigation, governance, threat hunting, and vulnerability management.
- The approaches used to validate standard BN models in cyber security were real-world case study, cross-validation, goodness of fit, monte-carlo simulation, expert evaluation, and sensitivity analysis.

These patterns in the use of standard BN models in cyber security would help to make full use of standard BNs in cyber security in the future especially by pointing out the current trends, limitations and future research gaps.

In the future, it is important to investigate whether the BN models used for problems associated with insiders are applicable for ICS environments, especially for a control room with an operator. It would be useful to demonstrate the capacity of standard BNs to tackle problems associated with both insiders and outsiders, and their interactions like social engineering attacks, collusion attacks. It would be intriguing to investigate how to deal with multi-step attacks using standard BNs. The potential of alternative data sources like model simulations, CTF events to construct DAGs and populate CPTs in cyber security also needs to be explored, as well as the capability of standard BNs to completely or partially benefit the other applications in cyber security.

Acknowledgements. This research received funding from the Netherlands Organisation for Scientific Research (NWO) in the framework of the Cyber Security research program under the project *"Secure Our Safety: Building Cyber Security for Flood Management (SOS4Flood)"*.

References

1. WEF: Partnering for Cyber Resilience: Towards the Quantification of Cyber Threats (2015)
2. Yu, S., Wang, G., Zhou, W.: Modeling malicious activities in cyber space. IEEE Netw. **29**, 83–87 (2015)
3. Ben-Gal, I.: Bayesian Networks. Encyclopedia of Statistics in Quality and Reliability. Wiley, Hoboken (2008)
4. Darwiche, A.: Chapter 11 - Bayesian networks. In: Foundations of Artificial Intelligence, vol. 3, pp. 467–509 (2008). doi:10.1016/S1574-6526(07)03011-8
5. Landuyt, D., et al.: A review of Bayesian belief networks in ecosystem service modelling. Environ. Model. Softw. **46**, 1–11 (2013)
6. Uusitalo, L.: Advantages and challenges of Bayesian networks in environmental modelling. Ecol. Model. **203**, 312–318 (2007)
7. Nikovski, D.: Constructing Bayesian networks for medical diagnosis from incomplete and partially correct statistics. IEEE Trans. Knowl. Data Eng. **12**(4), 509–516 (2000)
8. Nakatsu, R.T.: Reasoning with Diagrams: Decision-Making and Problem-Solving with Diagrams. Wiley, Hoboken (2009)
9. Phan, T.D., et al.: Applications of Bayesian belief networks in water resource management: a systematic review. Environ. Model. Softw. **85**, 98–111 (2016)
10. Kordy, B., Piètre-Cambacédès, L., Schweitzer, P.: DAG-based attack and defense modeling: don't miss the forest for the attack trees. Comput. Sci. Rev. **13**, 1–38 (2014)
11. Poolsappasit, N., Dewri, R., Ray, I.: Dynamic security risk management using bayesian attack graphs. IEEE Trans. Dependable Secure Comput. **9**, 61–74 (2012)
12. Frigault, M., Wang, L.: Measuring network security using Bayesian network-based attack graphs. IEEE (2008)
13. Liu, Y., Man, H.: Network vulnerability assessment using Bayesian networks. In: Proceedings of the SPIE, pp. 61–71 (2005)
14. Kwan, M., Chow, K.-P., Law, F., Lai, P.: Reasoning about evidence using Bayesian networks. In: IFIP International Conference on Digital Forensics, pp. 275–289 (2008)
15. Axelrad, E.T., Sticha, P.J., Brdiczka, O., Shen, J.: A Bayesian network model for predicting insider threats. In: Security and Privacy Workshops, pp. 82–89 (2013)
16. Greitzer, F.L., et al.: Identifying at-risk employees: modeling psychosocial precursors of potential insider threats. In: Hawaii International Conference on System Science (HICSS), pp. 2392–2401 (2012)
17. Greitzer, F.L., et al.: Identifying at-risk employees: a behavioral model for predicting potential insider threats. Pacific Northwest National Laboratory (2010)
18. Pecchia, A., et al.: Identifying compromised users in shared computing infrastructures: a data-driven bayesian network approach. In: 2011 30th IEEE Symposium on Reliable Distributed Systems (SRDS), pp. 127–136. IEEE (2011)
19. Shin, J., Son, H., Heo, G.: Development of a cyber security risk model using Bayesian networks. Reliab. Eng. Syst. Saf. **134**, 208–217 (2015)

20. Kornecki, A.J., Subramanian, N., Zalewski, J.: Studying interrelationships of safety and security for software assurance in cyber-physical systems: approach based on bayesian belief networks. In: 2013 Federated Conference on Computer Science and Information Systems (FedCSIS), pp. 1393–1399. IEEE (2013)
21. Wang, J.A., Guo, M.: Vulnerability categorization using Bayesian networks. In: Proceedings of the Sixth Annual Workshop on Cyber Security and Information Intelligence Research, p. 29. ACM (2010)
22. Mo, S.Y.K., Beling, P.A., Crowther, K.G.: Quantitative assessment of cyber security risk using Bayesian network-based model. In: 2009 Systems and Information Engineering Design Symposium, SIEDS 2009, pp. 183–187. IEEE (2009)
23. Holm, H., Korman, M., Ekstedt, M.: A bayesian network model for likelihood estimations of acquirement of critical software vulnerabilities and exploits. Inf. Softw. Technol. **58**, 304–318 (2015)
24. Kwan, M., Chow, K.-P., Lai, P., Law, F., Tse, H.: Analysis of the digital evidence presented in the Yahoo! case. In: Peterson, G., Shenoi, S. (eds.) DigitalForensics 2009. IAICT, vol. 306, pp. 241–252. Springer, Heidelberg (2009). doi:10.1007/978-3-642-04155-6_18
25. Ibrahimović, S., Bajgorić, N.: Modeling information system availability by using Bayesian belief network approach. Interdisc. Description Complex Syst. **14**, 125–138 (2016)
26. Wilde, L.: A Bayesian Network Model for predicting data breaches caused by insiders of a health care organization. University of Twente (2016)
27. Herland, K., Hammainen, H., Kekolahti, P.: Information security risk assessment of smartphones using Bayesian networks. J. Cyber Secur. Mobility **4**, 65–85 (2016)
28. Herland, K.: Information security risk assessment of smartphones using Bayesian networks. Aalto University, Finland (2015)
29. Apukhtin, V.: Bayesian Network Modeling for Analysis of Data Breach in a Bank. University of Stavanger, Norway (2011)
30. Khosravi-Farmad, M., Rezaee, R., Harati, A., Bafghi, A.G.: Network security risk mitigation using Bayesian decision networks. In: 4th International eConference on Computer and Knowledge Engineering (ICCKE), pp. 267–272. IEEE (2014)
31. Pan, S., Morris, T.H., Adhikari, U., Madani, V.: Causal event graphs cyber-physical system intrusion detection system. In: Proceedings of the Eighth Annual Cyber Security and Information Intelligence Research Workshop, p. 40. ACM (2013)
32. Frigault, M., et al.: Measuring network security using dynamic Bayesian network. In: Proceedings of the 4th ACM Workshop on Quality of Protection, pp. 23–30 (2008)
33. Sarala, R., Kayalvizhi, M., Zayaraz, G.: Information security risk assessment under uncertainty using dynamic Bayesian networks. Int. J. Res. Eng. Technol. **3**, 304–309 (2014)
34. Tang, K., Zhou, M.-T., Wang, W.-Y.: Insider cyber threat situational awareness framwork using dynamic Bayesian networks. In: 2009 4th International Conference on Computer Science and Education, ICCSE 2009, pp. 1146–1150. IEEE (2009)
35. Sommestad, T., Ekstedt, M., Johnson, P.: Cyber security risks assessment with Bayesian defense graphs and architectural models. In: 2009 42nd Hawaii International Conference on System Sciences, HICSS 2009, pp. 1–10. IEEE (2009)
36. Ekstedt, M., Sommestad, T.: Enterprise architecture models for cyber security analysis. In: Power Systems Conference and Exposition, pp. 1–6. IEEE (2009)
37. Laskey, K., et al.: Detecting threatening behavior using Bayesian networks. In: Conference on Behavioral Representation in Modeling and Simulation, p. 33 (2006)

38. AlGhamdi, G., et al.: Modeling insider behavior using multi-entity Bayesian networks (2006)
39. Okoli, C., Schabram, K.: A guide to conducting a systematic literature review of information systems research. Sprouts: Working Papers on Information Systems, vol. 10 (2010)
40. Meho, L.I.: The rise and rise of citation analysis. Phys. World **20**, 32 (2007)
41. Marcot, B.G., Steventon, J.D., Sutherland, G.D., McCann, R.K.: Guidelines for developing and updating Bayesian belief networks applied to ecological modeling and conservation. Can. J. For. Res. **36**, 3063–3074 (2006)
42. Alberts, C., Dorofee, A.: OCTAVESM Threat Profiles
43. Bureau, F.I.P.: Unintentional Insider Threats: A Foundational Study (2013)
44. Rehman, R.: CISO MindMap (2017). http://rafeeqrehman.com/wp-content/uploads/2017/07/CISO_Job_MindMap_v9.png
45. Andress, A.: Surviving Security: How to Integrate People, Process, and Technology. CRC Press, Boca Raton (2003)
46. Cyber Security Intelligence Index. IBM Security (2016)
47. Greitzer, F.L., et al.: Unintentional insider threat: contributing factors, observables, and mitigation strategies. In: 2014 47th Hawaii International Conference on System Sciences (HICSS), pp. 2025–2034. IEEE (2014)
48. Antonioli, D., et al.: Gamifying Education and Research on ICS Security: Design, Implementation and Results of S3. arXiv preprint arXiv:1702.03067 (2017)
49. Database, R.: German Steel Mill Cyber Attack (2017). http://www.risidata.com/database/detail/german-steel-mill-cyber-attack
50. Lippmann, R.P., Ingols, K.W.: An annotated review of past papers on attack graphs. Massachusetts Institute of Technology Lincoln Laboratory, Lexington (2005)
51. Bobbio, A., Portinale, L., Minichino, M., Ciancamerla, E.: Improving the analysis of dependable systems by mapping fault trees into Bayesian networks. Reliab. Eng. Syst. Saf. **71**, 249–260 (2001)
52. Khakzad, N., Khan, F., Amyotte, P.: Safety analysis in process facilities: comparison of fault tree and Bayesian network approaches. Reliab. Eng. Syst. Saf. **96**, 925–932 (2011)
53. Chockalingam, S., et al.: Integrated safety and security risk assessment methods: a survey of key characteristics and applications. In: International Conference on Critical Information Infrastructures Security (CRITIS), Paris (2016)
54. Salem, M.B., Hershkop, S., Stolfo, S.J.: A Survey of Insider Attack Detection Research. In: Stolfo, S.J., Bellovin, S.M., Keromytis, A.D., Hershkop, S., Smith, S.W., Sinclair, S. (eds.) Insider Attack and Cyber Security. Advances in Information Security, vol. 39. Springer, Boston (2008)

Improving and Measuring Learning Effectiveness at Cyber Defense Exercises

Kaie Maennel, Rain Ottis, and Olaf Maennel[✉]

Tallinn University of Technology, Tallinn, Estonia
{kaie.maennel,rain.ottis,olaf.maennel}@ttu.ee

Abstract. Cyber security exercises are believed to be the most effective training for the training audiences from top professional teams to individual students. However, evidence of learning outcomes is often anecdotal and not validated. This paper focuses on measuring learning outcomes of technical cyber defense exercises (CDXs) with Red and Blue teaming elements. We studied learning at Locked Shields, which is the largest unclassified defensive live-fire CDX in the world. This paper proposes a novel and simple methodology, called the "5-timestamp methodology", aiming at accommodating both effective feedback (including benchmarking) and learning measurement. The methodology focuses on collection of timestamps at specific points during a cyber incident and time interval analysis to assess team performance, and argues that changes in performance over time can be used to evidence learning. The timestamps can either be collected non-intrusively from raw network traces (such as pcaps, logs) or using traditional methods, such as interviews, observations and surveys. Our experience showed that traditional methods, such as self-reporting, fail at high-speed and complex exercises. The suggested method enhances feedback loop, allows identifying learning design flaws, and provides evidence of learning value for CDXs.

Keywords: Cyber defence exercise · Training and education · Learning outcomes · Measuring learning

1 Introduction

Cyber security exercises are quickly gaining popularity as a teaching method for cyber-readiness. Globally there are over 200 cyber security exercises and more than 50% have a performance objective focusing on learning [17]. The European Union Agency for Network and Information Security survey describes the state of art: "... after-action reports and 'lessons learned' documents have become increasingly at risk of becoming fantasy documents. There is an increased demand that lessons must have been successfully learned, and that noting such instances of lesson-drawing is all there is to it. Few, if any, controls are actually made to verify that they can even be called lessons by any sensible definition, or that anything has actually been learned" [17]. The evidence of learning outcomes

© Springer International Publishing AG 2017
H. Lipmaa et al. (Eds.): NordSec 2017, LNCS 10674, pp. 123–138, 2017.
https://doi.org/10.1007/978-3-319-70290-2_8

is limited and evaluation methodologies focus on the improvement of one exercise to the next [2]. On one side the literature describes enthusiasm of participants for the knowledge gained and lessons learned [11]. At other end of spectrum, Pusey et al. [19] claim that evidence is often anecdotal and little work has been done to validate learning outcomes.

This paper focuses on cyber defense exercises (CDXs) with the Red (RT) and Blue Team (BT) elements and looks at measuring learning effectiveness from an organizer's perspective. We use the NATO Cooperative Cyber Defense Centre of Excellence's (CCD COE) Locked Shields (LS) as testing platform. LS, that took place 26–28 April 2017 (LS17), is one of the largest and advanced team based live-fire RT/BT technical exercise with nearly 900 participants [14]. The exercise is a hybrid of competition, assessment and complex scenario-based learning event. The training audience comprises of the national BTs that in the exercise context take the role of the computer emergency response teams tasked to defend the pre-built virtualized networks of fictional organizations against the RT attacks. The other teams involved in the exercise are: Green Team (GT) responsible for game network and infrastructure development, White Team (WT) for game scenario development and execution control, and Yellow Team (YT) for monitoring and situational awareness [15]. One additional advantage of such CDXs is that permission to study individual and team performances and learning can be easily obtained by the organizers before the exercise starts.

2   Learning Measurement Dimensions in CDX's

CDXs in the current form are often not sufficiently instrumented for learning measurement and existing measurements focusing on scoring are not using learning related metrics. We recommend an overall CDX's learning measurement approach that brings together pre-exercise, execution and post-exercise phases and individual/team/organizational aspects. The measurement should include mixture of quantitative and qualitative methods. As a novel method we discuss the idea of the 5-timestamp methodology that focuses on unobtrusive data collection and comparable data analysis linked to learning objectives. This methodology is only a part of overall learning measurement (including traditional methods, such as surveys, interviews, etc.).

2.1   5-Timestamp Methodology

Learning in CDXs is affected by many variables, however the basic measurements, such as timing and accuracy metrics are still key elements that provide comparable trends in learning process and benchmarking for the teams. For example, Henshel et al. measurements in Cyber Shield 2015 showed that when teams took 20 or more minutes to identify an inject's NIST categorization, they were more accurate [9]. That means an overly time-constraint game-rule may prove to be an unrealistic expectation, which will not contribute to learning. Instead it forces teams to learn, share and store wrong behaviors and later

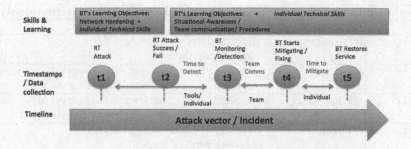

Fig. 1. 5-timestamp non-intrusive methodology

retrieve learned, but wrong behavioral models in real life situations [26]. Such metrics support development of appropriate exercise learning design.

Furthermore, measuring learning effectiveness and collecting data in order to provide feedback can be combined. The learning potential is not fully realized, if the BTs do not know what their weaknesses are, and how they progressed in the exercise. Scoring might give some indication of how teams compare, however, without knowing a baseline or standard in more detail, the overall score is worthless from learning viewpoint. For example, scoring may not take into account how much resistance the BTs put up and how efficient they were in responding.

As a part of solution we propose a non-intrusive methodology to collect and analyze timestamps from both the RT and BT actions from their digital footprint. The analysis of time intervals between the proposed timestamps enables to measure technical skills, but also soft skills (including leadership, team communications, decision making). The methodology is analysing data at a cyber incident/attack vector/target machine level, but provides metrics for different learning objectives (Fig. 1). For example, the assessment whether the BTs are effective and achieve incident handling related learning objectives, needs basic timing and accuracy metrics—how long does it take to respond to an attack, how long did teams take to respond to a significant threat vs. minor issues, what is correlation between the teams' detection time and quality of reporting. Further analysis can be carried out whether the most effective strategy from qualitative aspects was applied by the BTs, but having timing and accuracy metrics will provide input and focus to such qualitative analysis and feedback.

The analysis breaks an incident into phases to demonstrate strengths and bottlenecks in individual and team skills in each phase, and provides the basis for effective feedback. The model follows the incident timeline, and information can be collected non-intrusively (Fig. 2) from game-net/management network[1]. Even when t_1 and t_2 are intrusive for the RTs, data collection is non-intrusive

---

[1] The exercise runs on separate virtualised machines which are accessed remotely over the VPN [15]. The BTs can reside in their home location and connect via a dedicated management network to the game-environment. The game-net resides typically on a different interface and is where the attacks are happening.

Timestamp	Description	Non-intrusive Data	Intrusive (optional)
t_1	RT starts attacking	RT activity reporting	N/A
t_2	RT compromises	RT activity reporting and scoring data	N/A
t_3	BT detects	Possibly by access pattern	BT observation or self-reporting via inject
t_4	BT mitigates	management network (showing traffic activity)	BT observation or self-reporting via inject
t_5	BT restores	scoring, management network (end of session)	N/A

Fig. 2. Data sources for 5-timestamps, non-intrusive for the BTs (as the training audience). Intrusive methods can be used for cross-checking and validation.

for the BTs. For cross-checking, a sample using intrusive methods should be selected.

In order to fully understand this methodology, it should be noted that there are typically several target machines in a game-network that can be attacked repeatedly using the same attack methods. However, one of the advantages of a live-fire RT/BT exercise is defending against a "thinking" adversary, which implies that the same target can be attacked using different methods.

Collecting Timestamps Non-intrusively From PCAPs. The idea relies on the fact that the organizers are able to collect all raw network traffic (e.g., pcaps) not only from within the game-net, but also from the management network. From those traces it is possible to automatically detect the times of a BT activity for each target machine (e.g., when a BT member is working on a machine or not). This can be done by observing a ssh or remote desktop connection from the BT-network through management network. Even if the traffic is encrypted, and the BT member remains logged-in in the background, simply observing the traffic volume and packet inter-arrival times allows automatically detecting times at which someone is working on a specific game-net target. With traditional methods, this can also be achieved by asking the team member to keep a detailed log about timestamps.

The time intervals between timestamps provide basic learning metrics as shown in Fig. 3. In addition to measuring technical skills, these metrics also give insight to:

Team vs. individual—how long an individual and/or sub-team takes to resolve an issue, e.g., several members connecting to the same machine to work together.

Soft skills (leadership and decision making)—as the teams must make quick decisions (likely to have immediate and significant consequences), teams also learn decision-making. The OODA (Observation, Orientation, Decision, and Action) loop is a decision-making theory where time is the dominant parameter [23], and thus supports this framework using time intervals. The teams need to perform reliably and adapt their responses to mitigate adverse scenarios, and

Timestamps Interval	Description	Learning Objectives	Team vs. Individual
t_5–t_2	incident response time	Overall performance (organizer's objectives=scoring)	team
t_5–t_4	time to mitigate	Responding to attacks (technical skills)	individual, sub-team
t_4–t_3	time between mitigation and detection	Time management and prioritization; Teamwork: delegation, dividing and assigning roles, leadership; Handling cyber incident	team
t_3–t_2	time between compromise and detection	Monitoring networks, detecting of attacks	individual, sub-team
t_2–t_1	time to compromise	Learning the network; System administration and prevention of attacks	individual, sub-team

Fig. 3. Learning metrics from 5-timestamps intervals

that can be measured by t_4–t_3, i.e., time needed for intra-team communication, prioritization, task allocation.

Benefits and Application in Learning Process. The 5-timestamp methodology provides several advantages. Firstly, during a post-exercise debrief, it helps to create a general mental map of the events. For example, letting participants search for events in the BTs of pcaps or logfiles to figure out what happened is not useful for learning. In a similar analogy, where security cameras have become more effective when combined with a motion sensor—logfiles become more easily "searchable" when combined with accurate timestamp annotations. Debriefing an attack from the high-level objectives together with accurate timestamps, facilitates finding the relevant information. As the participants have already been in the situation during exercise, they understand the RT objectives, and are able to "relive" the events. Useful feedback can only be given, if the exercise can be debriefed in a meaningful way, and accurate timestamps are a first critical step towards achieving this.

Secondly, the timestamps can be used in building a baseline for performance or effectiveness. When grouped by the attack methods (not the target systems), those values become comparable. These can be further analyzed in several ways: (1) as an average overall performance against defending against a certain type of attack, (2) viewed over time for the same target machine (e.g., looking at repeated attacks using the same attack method) whether anything has been learned during the exercise—or potentially, even between exercises (if similar team composition returns to an event in which the same attack vector is repeated), and (3) for understanding whether the BTs are able to transfer learned knowledge (e.g., is a BT able to detect and defend the same type of attack against a different target system provided they have learned it earlier).

Thirdly, analyzing the timestamps provides insight into the BTs' strategies. Do the BTs only focus on certain class or difficulty-level of attacks, and maybe miss some more important/unknown challenges? Do they invest time during the exercise to understand the systems? The metrics enable a way of getting some basic baseline and benchmarking for the organizers and participants.

It is important to note that the timestamps themselves only measure effectiveness. However, there is an implicit assumption that measuring changes in effectiveness over time (e.g., repeated comparable events, such as repeated attacks), shows changes in performance. This is an indicator for learning, a dynamic process, together with other qualitative data. The complete exercise data analysis and projections are left for future work, and the scope of this paper discussing the suitability of proposed methodology with the community.

Challenges and Limitations. It is also important to acknowledge the challenges and limitations. The learning measurement process needs to be pre-planned, agreed with the stakeholders, and form an integral part of a CDX organization and evaluation process. Selection of what to measure is a challenging task and depends on training objectives. What learning metrics are "must have", "nice to have" and "wasteful" metrics from learning perspective? Having comparative metrics from several CDXs, would enable developing comparable standardized set of learning metrics.

Data monitoring and collection may fail to capture timing metrics and team actions with perfect reliability. Also a challenge is to develop clearly defined measures that integrate both qualitative and quantitative inputs. Metrics for future evaluation should include appropriateness and quality of responses and actions. Some training goals (such as incident handling procedures) may prove difficult to measure due to teams following different operating procedures, standards, and practices. Separating learning impact from other behavior effects (i.e., learning might not be visible straight away or recognized by participants by themselves, or overestimated and not result in behavior change) will remain a challenging area to assess.

2.2   Data Collection and Sources

The data collected as part of CDXs may vary based on training objectives and software environment, but it should not be an additional burden to the organizers. As shown by the 5-timestamp methodology, often such data is already collected. The learning related data is obtained from several sources:

1. RT reporting—failed attacks, resistance time to the attacks, number of repeated attacks;
2. YT reporting—reporting about situational awareness;
3. Scores—scoring for availability, usability and injects (trends over time);
4. Traffic from game-net and management-net;
5. Surveys—pre-exercise and post-exercise survey with pre- and post knowledge assessment if possible;

6. Injects—can be used to qualitatively verify a data sample from overall dataset (see below) and collecting learning feedback during the exercise;
7. Information from the RT—ratings for resistance level, classification of attack type (this might also be semi-automated by using Cobalt Strike [13] or similar);
8. Observations of the BTs;
9. Communication channels—chat logs, GT management network traffic (volumes and trends);
10. Interviews with participants (and management)—assessing the immediate reaction to exercise and long term impact on the job.

Sample Selection for Qualitative Validation. Due the large volume of virtual machines, attacks, and activities, it is not be possible to confirm all incidents during an exercise qualitatively as it may distract the BTs from learning. However, for a sample of attacks qualitative feedback can obtained from the BTs in order to cross-check the metrics. Such sample should be designed into the exercise as inquiries to the BTs via injects and/or observations.

The sample selection depends on the exercise training objectives, however should cover differing aspects, such as complexity, method of attacking, ease of detecting and mitigating the attack. There is no widely accepted taxonomy that can be applied from learning perspective in CDXs context. In order to measure learning impact, a comparison between easy tasks (potentially nothing learned and knowledge is already existing) and complex tasks (more challenging, more potential to learn) is valuable. As the teams have differing skillset any such criteria classification is somewhat forced and arbitrary, however it provides a comparison and feedback on the appropriate difficulty levels and learning opportunities created by the organizer. We propose the following classification matrix in Fig. 4, when a selection of specific events for learning impact measurement is based on: (1) detection and analysis—some attacks and incidents have visible signs that can be easily detected, whereas others are almost impossible to detect. (2) mitigation and recovery—responding to an incident involves different skillset and actions to be taken containing the damage, eradicating the incident components, and restoring systems to normal operation, and remediating vulnerabilities to prevent similar future attacks.

Easy to Detect and Easy to Mitigate	Easy to Detect and Difficult to Mitigate
Difficult to Detect and Easy to Mitigate	Difficult to Detect and Difficult to Mitigate

Fig. 4. Sample selection matrix

In addition to those two criteria and as an incident is part of whole exercise (scenario, mission), the prioritization of attacks (strategy) needs to be considered.

3   LS17—Learning Measurement

LS provides full experience of managing a major cyber incident to the BTs. The exercise consists of different attacks and tasks based on a scenario over two days and the data set is over 2500 attacks [14]. The measurement plan needs to ensure that the intrusive data collection is not distracting the participants' focus from learning efforts.

LS17 learning measurement included a mixture of quantitative and qualitative methods with focus on gaining some experience using the 5-timestamp methodology, combined analysis of participants' feedback and metrics collected, and identification of plausible learning correlations for further work.

3.1   5-Timestamp Methodology Experience

We illustrate how the 5-timestamp methodology works using the example of LS17. We picked one high-profile RT objective—a Siemens system part of Industrial Control System (ICS) segment for all timestamps to be recorded. The timestamps were obtained from the BTs self-reporting (through Injects), RT attack reports, and scoring data for all teams. Furthermore, those RT members conducting the attack on those systems were asked to keep a detailed log of all events, as accurately as possible. Regarding the pcaps from management interfaces, there was a technical issue and very unfortunately, the GT was unable to record the traffic from management interfaces; leaving analysis of the inter-arrival times for future work in next exercises.

The purpose of this objective was to gain control of the airport fueling station and cause a fuel leak. The BTs had time to mitigate before "all fuel was spilled". Before the exercise the RT had prepared some attack vectors, but which vector would work or not depends on the BT defenses. Starting the fuel spill is a very "noisy" attack, which means even if the initial compromise remained undetected, the BT had some time respond.

Four teams were successfully attacked by the RT (i.e., all fuel was spilled). For two more the RT managed to compromise the systems and start spilling, however, those two BTs managed to mitigate the attack before all fuel was spilled. The remaining 13 teams defended their systems well (e.g. no spilling started).

While all teams were analyzed, for anonymity and clarity reasons only one timeline is presented here. Figure 5 shows detailed timeline of events recorded according to the 5-timestamp methodology for BT Z (Z anonymized).

Before the RT is allowed to attack in the exercise, the respective objective must be opened. For this specific team and objective, this was done at 06:59 UTC, which corresponds roughly to the time first phase of attacks was allowed to start. The objectives are not opened individually in the RT reporting system, but rather for all teams at the same time—and because a RT member might have to "entertain" several teams at a time, the opening of an objective and actual start of an attack might differ. In this example case, BT Z was only attacked at 07:35 UTC (about 1/2 h later), i.e., the timestamp reported from the detailed

Incident Timeline	Time	Description	Data source
t_1 RT starts an attack	06:59	Campaign officially opened	RT reporting system
$t_{1.1}$ RT starts 1st attack	07:35	Attack started	RT members
t_2 RT compromises	07:40	Spilling started	Scoring
t_2 RT compromises	07:43	Spilling started	RT members
$t_{2.1}$ BT mitigates	07:44	Spilling stopped	Scoring
$t_{2.1}$ BT mitigates	07:45	Spilling stopped	RT members
$t_{1.1}$ RT starts 2nd attack	07:58	Attack repeated	RT members
t_3 BT detects	09:00	Suspicious activity noted	BT Inject
$t_{2.1.3}$ RT reporting	09:18	Partial RT objectives scored	RT reporting system
t_4 BT starts mitigating	09:20	Timestamp or interval reported	BT Inject
t_2 RT compromises	09:23	Spilling started	Scoring
$t_{2.1}$ BT mitigates	09:30	Spilling stopped	Scoring
t_5 BT fights back	09:30	Timestamp or interval reported	BT Inject
t_5 BT resolves	09:40	Suspicious user removed	BT Inject

Fig. 5. Example of 5-timestamps reconstructed timeline for an incident

RT member logs. LS has a comprehensive and automated scoring system, which recorded at 07:40 UTC that the attack has been successful and spilling started. However, the RT members reported that spilling started at 07:43 UTC. This small time difference is an artifact of the self-reporting, and understandable, as all teams are very busy during the exercise. It also highlights that self-reporting timestamps should be avoided, if possible. This is not only for accuracy reasons, but also to reduce the work-load for various teams during the exercise. Similarly, the scoring system reported that the BT mitigated the attack at 07:44 UTC, while the RT member recorded a timestamp of 07:45 UTC. Such minor discrepancies were observed throughout. As this attack has only partially been successful, the RT does not give up and manages to gain foothold in the systems again at 07:58 UTC (reported by RT member log), but this time the RT does not manage to cause any fuel spilling. This is not recorded in the scoring (and should not be scored as the BT successfully defended), but it is an important factor that hints at resistance and team performance. Having such timestamps facilitate a reflective team debrief after the event.

However, when analyzing the BT self-reporting then the BT only reports detecting any suspicious activities for the very first time at 09:00 UTC. Clearly, some BT members must have mitigated the attack already before 07:44 UTC, so this points to an intra-team communication/reporting problem. Therefore asking the BTs to self-report accurate timestamps, while defending systems during a "crisis situation", is not going to work (neither observations). The team's internal reporting systems do not capture such information, or at least not accurately enough. It is therefore of vital importance to obtain such timestamps from the management network (e.g., by observing in the pcaps when a BT member logs into the target system, or in case they are already logged in when the activity of system changes by a changed inter-arrival frequency of packets on the management network).

Overall during the exercise, spilling attempts start 7 more times using at least two different attack vectors. The first time spilling was for 3′39″ (3 min and 39 s), the second spilling continued for 6′44″. The next day spilling durations were significantly reduced, in the end only taking 0′07″ (7 s) to mitigate—despite the fact that different attack vectors were used.

The main challenges encountered in the process and assumptions for data quality are:

1. RT scoring timestamps from the system need to be sufficiently accurate—when attacking multiple teams the objectives are started for all teams simultaneously and final scoring is often delayed, so scoring data is not accurate;
2. BTs self-reporting is not reliable and more accurate data collection method is required—this supports the argument that non-intrusive methods for collecting and analysing data from logs (network traffic, log, etc.) is helpful;
3. Traditional observations methods are not possible as in a technical exercise there is nothing to see.

Of course, this is a first attempt to understand the feasibility of proposed methodology. Before drawing any conclusions on learning more data and measurements needs to be obtained in future work, however, such initial tests appear to be promising.

3.2 Discussion and Findings from LS17 Learning Measurement

In addition to the 5-timestamps methodology experience, we also discuss selected findings relevant from other learning measurements, as the 5-timestamps methodology is one integral part of the overall measurement framework. Only aggregated statistics are presented due to the confidential nature and privacy of participants. We used pre-survey, injects, post-survey and interviews to collect the feedback from individuals and teams. Our overall response rate was 21% for individual pre-survey, and team based injects had 89% response rate. Due to the timing constraints, post-survey results have not been included.

Learning in Pre-exercise Phase. We collected information about the participants, their experiences and learning process in the pre-execution phase, team environment, learning expectations about the execution and evidence of long-term learning from previous exercise participation. Our findings confirm that pre-exercise phase is vital part of the overall exercise with 53% of respondents spending 10–50 h preparing. Majority of participants (73%) however report that they prepared individually (over half of the preparation time); whereas sub-teams preparations were taking place either half of the time (35%) or seldom (31%). Despite that the exercise is team-based, whole team preparations were mostly seldom 37% and 22% of participants claim they never attended whole team preparation sessions.

No clear distinction between learning knowledge/skill on technical vs. soft skills learned in pre-exercise phase is visible—on average in each learning area 40% minor and 13% significant improvement was reported. This links to the fact that training audience consist mainly of the professionals, who assess their knowledge and skills in majority at medium (43%) to high level (37%), related to the similar working experience level (both medium and high 39%). When comparing what the participants have learned in preparation phase and what they expect to learn during the exercise, there is no clear distinction that some training objectives (e.g., teamwork) are more relevant for execution than pre-execution phase or that technical skills are mainly obtained in the pre-exercise phase.

Feedback on the Exercise Design. Feedback was focused on the Industrial Control System (ICS) segment design that has been designed and seen as one of the most complex and technically challenging areas in the exercise. By comparison we can draw conclusions also on other parts of the exercise. The attempt was also made to assess, how teams perceive individual team members/sub teams and whole team learning outcome.

Based on pre-survey 52% of respondents felt they do not have ICS capabilities in their teams, despite of nearly all teams reporting dedicated ICS team member(s). Average self-believed resistance level in the ICS segment was surprisingly low compared to the RT members' assessment—44% believed that their resistance was at medium level, 33% at high level 22% strong. This links positively to an assumption that learning can happen when team acknowledges they lack some knowledge or skills and "sensing more than see" (OODA loop). It is also interesting to see how the teams perceive level of difficulty to defend against those attacks—41% find it easy to detect and easy to mitigate, 39% easy to detect, but difficult to mitigate, 12% difficult to detect, easy to mitigate and 8% find it difficult to detect and difficult to mitigate. In comparison to other attacks in the exercise 44% of teams assessed level of difficulty at same level. Priority for ICS attacks was consistently (78% of teams) at high or critical priority level, as expected in the scenario. 52% of the BT reported that they managed to track the root cause of malicious activity and 42% not (showing missed learning opportunity without proper feedback).

An attempt to evaluate how individual and team learning outcomes are perceived, shows that team learning has quite even distribution (25%) from slight to significant improvement, then individual learning was in majority (59%) assessed as significant. This is somewhat expected, due to the technical specialization but therefore needs further focus of learning transferability within a team.

Furthermore, a question collecting narratives about the teams' learning experience to uncover and understand the big picture was asked. Top 5 expressions that emerged were "successes in learning" ("learning curve"), "challenges in learning", "complexity/variety of system", "preparations" and "team learning". All these themes confirmed the focus of measurement objectives.

Long-Term Learning Impact. We looked at LS16 for long-term impact indicators, based on responses of returning participants, and enquiries with previous LS participants. When asked individually about skills learned and maintained 69%

responded that they recall a skill from participating earlier LS—average for technical training objective is 67% and soft skill related training objective 69%. Sadly survey results were limited in the participants' comments what exactly they learned.

58% agreed that their team has become more coherent, confident and collaborative. Similarly, 64% agreed that their team's knowledge has increased (as a result of individuals sharing). However, as majority (59%) of the teams have changed significantly (less than 50% old team members)—long-term impact need to be interpreted carefully.

Feedback from few participants who participated over five years back tends to indicate long term impact of CDXs on mindset (e.g., "to have an emergency procedure in place, as when you're in the middle of the event there is no time to think, just to act.", "... key is thinking and mindset and learning why something was done, not what.").

Despite of limited evidence, the survey and interview results support the learning value of the LS. Further work needs to be conducted to evaluate long-term impact for specific training objectives.

4   Related Work in Learning Measurement Context

Unfortunately, there are no widely accepted methodological evaluation methods published and scientifically proven measuring learning impact or assessing cyber security skills/competencies obtained through CDXs. Some general guidance, such as [10,18], describe how organizers should look at design and performance (training success) measurements. The related work includes articles published on learning (and other) measurements at cyber exercises, and also interdisciplinary papers, game based learning and team learning, as relevant.

Cyber Exercises. Dr. Ahmad [1] investigated how a cyber crisis exercise benefits participants' individual learning and how their experience in the exercises is transferred to their organization using the four-level Kirkpatrick training post-assessment model (reaction about the exercise, learning skills experienced during the exercise, behavior developed during the exercise, and result, i.e., how the benefits are transferred to their organization). This approach lacks team aspects of learning. U.S. Army Research Institute for the Behavioral and Social Sciences Research [22] measured game-based simulations by different questionnaires and complemented those interviews with probing questions.

Game Based Learning and Serious Games. Connolly et al. [3] proposes a model for the evaluation of games for learning that includes motivational variables such as interest and effort, as well as learners' preferences, perceptions and attitudes to games in addition to looking at learner performance. Outcomes relate to learning and skill acquisition but also affective and motivational aspects. Methods to evaluate learning outcomes include meta-analyses, randomized controlled trials, quasi-experimental designs, single case experimental designs—pre- and post test, and non experimental designs—surveys, correlation, qualitative [7]. The effect on learning (acquisition of skills or knowledge) was measured by calculating the difference between pre-test and post-test scores on the questionnaires or cognitive tests, and comparison to control group [5]. Game-based learning and serious games provide excellent environments for mixed-method data gathering (i.e., triangulation), including crowd sourcing, panel discussions, surveys and observations, in-game logging and tracking on hundreds of events and results, including distances, paths, play time and avoidable mistakes, etc. [12]. Not yet explored issues are seamless, or "stealth" data-gathering and assessment as well as performance based evaluation [7]. Stealth assessment (i.e., non-invasive, non-intrusive assessment) could potentially increase the learning efficacy given that much of the learning remains relatively "implicit" and "subjective" [12]. These issues are very relevant in the CDXs context, and the cornerstones for the proposed 5-timestamp methodology in this paper.

Team Learning. Measuring team learning is a complex task with many factors, such as learning impact has not been identified (i.e., simply there is no similar event in reality), change can be environmental (i.e., not due to learning) and learning could be dysfunctional (i.e., false connections made between actions and outcome) [26]. Most common methods used are combination of interviews, surveys, questionnaires, observations, and learning maps. Edmondson [4] used observation and interviews (based on "informant sampling approach") to study role of teams in learning and based on her study half of the teams engaged in reflective discussion about process that led to subsequent changes, and would constitute a team learning. Newman et al. [16] measured critical thinking during group learning using a questionnaire and the content analysis method, whereas Hay [8] used concept mapping on the topic before and after. Learning maps or curves at team and organization level were used by Uzumeri et al. and Chiva et al. [24,25]. Two valuable points to note are: (1) the learning is not necessarily consciously accessible, thus asking the team members (survey or interview) what they have learned may not uncover any changes, however there might be learned patterns that members were not consciously aware of [4,26]; (2) measuring long-term learning effect requires detailed and multiple real-time observations of the same group over time [26].

Other Measurements Conducted at CDXs. Some team performance and effectiveness metrics also relate to learning measurement. Study about Baltic Cyber Shields 2010 team effectiveness [6] used different interdisciplinary methods and concluded that a combination of technical performance measurements and behavioral assessment techniques are needed to assess team effectiveness, and cyber situation awareness is required for the defending teams, but equally for the observers and the game control. In Cyber Shield 2015, Henshel et al. [9] attempted predicting proficiency in the teams and to identify the best training and assessment methods by pre- and post-event survey and data collection during the event, and developed proficiency metrics, such as Time-to-Detect, Time-to-apProval, Time-to-End and Category Correct. Reed et al. [20] evaluated cyber defender situation awareness, and showed that the most pervasive form of competition-based exercise is comprised of jeopardy-style challenges, which compliment a fictional cyber-security event. Silva et al. [21] study considered factors of successful performance in Tracer FIRE exercise with emphasis on the use of software tools and bring out a relevant consideration that speed is often not the main consideration—participants who devoted more time to challenges tended to make more correct submissions (similar finding to [9]).

Findings for CDXs. From CDXs viewpoint, all these methods are applicable. However, similar challenges are faced as by researches so far—i.e., separating learning from other factors and that learning might not be necessarily visible. Also, for the incident response teams activities are conducted on the computers/network—thus observations of behavior (sitting quietly behind computer screen but at the same time mitigating a significant risk or attack) might not provide sufficient information. Observation method should be seen with a different kind of eyes—on the network and system-level and to learn observing at such technical levels.

The key takeaways for CDXs are that learning measurement needs to use mixed-method approach with qualitative and quantitative data, have wide scope, provide comparison ("benchmarking"), consider both individual and team aspects, and ideally be non-intrusive (not distracting participants from main goal of learning).

5   Conclusion

Learning is such a complex and intractable process that its study is difficult and contentious. However, methodological measurement is required to conclude whether an exercise design was appropriate and effective, and planned learning outcomes were achieved.

We presented an idea for non-intrusive data collection and measurement, i.e., the 5-timestamp methodology as an integral part of overall learning measurement framework. Future work should continue with performing the data analysis of an exercise to compile learning metrics and trends benchmark. Identification and analysis of the data trends, will provide solid baseline and demonstrate learning improvement achieved in CDXs. This will complement often anecdotal and positive feedback obtained via traditional methods (surveys, interviews) that participants have actually learned. As we demonstrated, incorporating non-intrusive, social and behavioral research methods into the cyber security field can give new insights and possibilities in effective training for cyber defense teams in the future.

We explored CDXs' learning measurement state of play and presented interdisciplinary literature review, incorporating relevant findings from team (group) and game based learning studies. The findings support the proposed novel non-intrusive 5-timestamp methodology for mainly timing and accuracy metrics for measuring technical skills improvements, but equally incorporating team aspects and soft skills. As part of the methodology proposal, we also considered some practicalities of data collection and proposed practical validation approaches with the qualitative measurements.

With work performed in this paper, we have attempted to provide practical steps how the organizers can evidence the learning value and lessons learned at CDXs, and at the same time improve the participants' and teams' learning experience by providing valuable feedback based on such measurement data.

Acknowledgments. This work would not have taken place without the NATO CCD COE open-minded and friendly organizing team of LS17, who allowed the authors to experiment on this large cyber exercise.

References

1. Ahmad, A.: A cyber exercise post assessment framework. Malaysia perspectives. Ph.D. thesis, University of Glasgow (2016)
2. Ahmad, A., Johnson, C., Storer, T.: A cyber exercise post assessment: adoption of the Kirkpatrick model. Adv. Inf. Sci. Serv. Sci. **7**(2), 1 (2015)
3. Connolly, T.M., Boyle, E.A., MacArthur, E., Hainey, T., Boyle, J.M.: A systematic literature review of empirical evidence on computer games and serious games. Comput. Educ. **59**(2), 661–686 (2012)
4. Edmondson, A.C.: The local and variegated nature of learning in organizations: a group-level perspective. Organ. Sci. **13**(2), 128–146 (2002)
5. Girard, C., Ecalle, J., Magnan, A.: Serious games as new educational tools: how effective are they? A meta-analysis of recent studies. J. Comput. Assist. Learn. **29**(3), 207–219 (2013)
6. Granasen, M., Andersson, D.: Measuring team effectiveness in cyber-defense exercises-a cross-disciplinary case study. Cogn. Technol. Work **18**(1), 121–143 (2016). Springer-Verlag, London
7. Hauge, J.B., Boyle, E., Mayer, I., Nadolski, R., Riedel, J.C., Moreno-Ger, P., Bellotti, F., Lim, T., Ritchie, J.: Study design and data gathering guide for serious games' evaluation. In: Connolly, T.M., Hainey, T., Boyle, E., Baxter, G., Moreno-Ger, P. (eds.) Psychology, Pedagogy, and Assessment in Serious Games, pp. 394–419 (2014)
8. Hay, D.B.: Using concept maps to measure deep, surface and non-learning outcomes. Stud. High. Educ. **32**(1), 39–57 (2007)
9. Henshel, D.S., Deckard, G.M., Lufkin, B., Buchler, N., Hoffman, B., Rajivan, P., Collman, S.: Predicting proficiency in cyber defense team exercises. In: 2016 IEEE Military Communications Conference, MILCOM 2016, pp. 776–781. IEEE (2016)
10. Kick, J.: Cyber exercise playbook. Technical report, DTIC Document (2014)
11. Mattson, J.A.: Cyber defense exercise: a service provider model. In: Futcher, L., Dodge, R. (eds.) Fifth World Conference on Information Security Education. IFIP – International Federation for Information Processing, vol. 237, pp. 81–86. Springer, Boston (2007)
12. Mayer, I., Bekebrede, G., Warmelink, H., Zhou, Q.: A brief methodology for researching and evaluating serious games and game-based learning. In: Psychology, Pedagogy, and Assessment in Serious Games, pp. 357–393. IGI Global (2014)
13. Mudge, R.: Cobalt Strike. https://www.cobaltstrike.com. Accessed 18 Sept 2017

14. NATO CCD COE: Locked Shields (2017). https://ccdcoe.org/locked-shields-2017. html. Accessed 18 Sept 2017
15. NATO CCD COE: Locked Shields 2016 After Action Report. NATO Cooperative Cyber Defence Centre of Excellence Publication (2016)
16. Newman, D.R., Webb, B., Cochrane, C.: A content analysis method to measure critical thinking in face-to-face and computer supported group learning. Interpersonal Comput. Technol. 3(2), 56–77 (1995)
17. Ogee, A., Gavrila, R., Trimintzios, P., Stavropoulos, V., Zacharis, A.: The 2015 Report on National and International Cyber Security Exercises. https://www.enisa.europa.eu/publications/ latest-report-on-national-and-international-cyber-security-exercises
18. Patriciu, V.-V., Furtuna, A.C.: Guide for designing cyber security exercises. In: Proceedings of the 8th WSEAS International Conference on E-Activities and Information Security and Privacy. World Scientific and Engineering Academy and Society (WSEAS), pp. 172–177 (2009)
19. Pusey, P., Gondree, M., Peterson, Z.: The outcomes of cybersecurity competitions and implications for underrepresented populations. IEEE Secur. Priv. 14(6), 90–95 (2016)
20. Reed, T., Nauer, K., Silva, A.: Instrumenting competition-based exercises to evaluate cyber defender situation awareness. In: Schmorrow, D.D., Fidopiastis, C.M. (eds.) AC 2013. LNCS, vol. 8027, pp. 80–89. Springer, Heidelberg (2013). 10.1007/ 978-3-642-39454-6_9
21. Silva, A., McClain, J., Reed, T., Anderson, B., Nauer, K., Abbott, R., Forsythe, C.: Factors impacting performance in competitive cyber exercises. In: Proceedings of the Interservice/Interagency Training, Simulation and Education Conference, Orlando, FL (2014)
22. Singer, M.J., Knerr, B.W.: Evaluation of a game-based simulation during distributed exercises. Army Research Institute for the Behavioral and Social Sciences, Orlando, FL (2010)
23. Stytz, M.R., Banks, S.B.: Addressing simulation issues posed by cyber warfare technologies. SCS M&S Mag., no. 3 (2010)
24. Svetlik, I., Stavrou-Costea, E., Chiva, R., Alegre, J., Lapiedra, R.: Measuring organisational learning capability among the workforce. Int. J. Manpower 28(3/4), 224–242 (2007)
25. Uzumeri, M., Nembhard, D.: A population of learners: a new way to measure organizational learning. J. Oper. Manage. 16(5), 515–528 (1998)
26. Wilson, J.M., Goodman, P.S., Cronin, M.A.: Group learning. Acad. Manag. Rev. 32(4), 1041–1059 (2007)

Privacy-Preserving Frequent Itemset Mining for Sparse and Dense Data

Peeter Laud[1(⊠)] and Alisa Pankova[1,2]

[1] Cybernetica AS, Tartu, Estonia
{peeter.laud,alisa.pankova}@cyber.ee
[2] Software Technologies and Applications Competence Centre (STACC),
Tartu, Estonia

Abstract. Frequent itemset mining is a data mining task that can in turn be used for other purposes such as associative rule mining. The data may be sensitive. There exist multiple privacy-preserving solutions for frequent itemset mining, which should consider the tradeoff between efficiency and privacy. Leaking some less sensitive information such as density of the datatable may improve the efficiency. In this paper, we consider secure multiparty computation setting, where the final output (the frequent itemsets) is public, and no other information should be inferred by the adversary that corrupts some of the computing parties. We devise privacy-preserving algorithms that have advantage when applied to very sparse and very dense matrices. We compare them to related work that has similar security requirements, estimating the efficiency of our new solution on a similar secure multiparty computation platform.

Keywords: Secure multiparty computation · Frequent itemset mining

1 Introduction

Frequent itemset mining (FIM) is a standard data mining task. Given a collection of sets, the goal is to find the subsets that are contained in sufficiently many of these sets. After finding out which elements are more likely to occur together, one may search for the reason for that co-occurrence, and whether the existence of one item implies the existence of the other one, extracting some more interesting knowledge such as association rules. Not only the task itself, but also its privacy-preserving variants have been well studied in related work.

In this paper, we consider the security model where the final output is public, and the adversary, corrupting some of the parties, should be unable to infer any other information in addition to these public outputs. Our goal is to see if we can gain more efficiency given some additional assumptions about the matrix density. This allows us to make use of FIM algorithms whose efficiency depends on data density, that would not give advantage in privacy-preserving settings otherwise. Since the standard algorithms for FIM are iterative, even if data has not been sparse on the first iteration, it may become very sparse or very dense on

© Springer International Publishing AG 2017
H. Lipmaa et al. (Eds.): NordSec 2017, LNCS 10674, pp. 139–155, 2017.
https://doi.org/10.1007/978-3-319-70290-2_9

later iterations. We propose algorithms that allow to combine dense and sparse columns in the same computation.

2 Preliminaries

Let the sets be called *transactions*, and their elements *items*. These names come from one possible use case of FIM, where the items are goods sold in the supermarket, and each transaction corresponds to contents of a shopping cart. The task of FIM is to find out which subsets of items occur together in sufficiently many transactions. In general, the shopping carts are nothing more than just sets T over some universal set U (e.g. all the goods sold in the shop), and the task is to find the subsets of items $I \subseteq U$ that are encountered in sufficiently many sets T. A subset I is considered frequent iff $|\{T \mid I \subseteq T\}| \geq t$, where $t \geq 1$ is some threshold that is given as a parameter.

2.1 Secure Multiparty Computation

In this paper, we solve FIM using secure multiparty computation. We adjust our algorithms to a specific platform Sharemind [3], which is based on secret sharing. The main security domain of Sharemind supports 3 *computing* parties (P_1, P_2, P_3), tolerating at most one passively corrupted party. The number of *input* parties who provide the inputs by secret-sharing them among computing parties is unbounded. No party should learn any private data besides the final output that we consider public. There is some less sensitive information that we agree to leak, such as the total number of items and transactions.

Sharemind uses mainly *additive* and *bitwise* secret sharing. The costs of their standard operations are different. In the *additive* sharing, the value is shared as $a = a_1 + a_2 + a_3$ over some ring (in Sharemind $\mathbb{Z}_{2^m}$), where a_i is the share that belongs to the party P_i, and linear combinations can be computed without communication. In the *bitwise* sharing, the value is shared bitwise as $a = a_1 \oplus a_2 \oplus a_3$, and bitwise linear combinations of bitvectors can be computed without communication. Also, bitwise sharing is more efficient for comparison operations.

This paper builds algorithms from standard operations of Sharemind that it uses as black boxes. The algorithms would work for any platform in which these basic operations are universally composable. However, the particular costs that we get in this paper are very related to particular Sharemind protocols. On some other platform, the new methods could give even more advantage, but they also could be too inefficient to make sparse representation reasonable.

2.2 Notation

Throughout this paper, we use the following quantities:

- capital letters X denote sets, and calligraphic font $\mathcal{X}$ denotes a set of sets;
- $\sigma(I)$ is the support of I, i.e. the set of transactions that contain itemset I;

- $\Delta(I_1, I_2) := \sigma(I_1) \setminus \sigma(I_2)$ is the difference of supports of I_1 and I_2;
- secret shared value (either additive or bitwise sharing) $\langle a \rangle$;
- additively shared value $[\![a]\!]$; $a = a_1 + \cdots + a_n$;
- bitwise shared value $\langle\!\langle a \rangle\!\rangle$; $a = a_1 \oplus \cdots \oplus a_n$;
- i-th element of a vector $\boldsymbol{a}$: $\boldsymbol{a}[i]$;
- element of a matrix $\mathbf{A}$ in the i-th row and j-th column: $\mathbf{A}[i,j]$;
- i-th row and j-th column of a matrix $\mathbf{A}$: $\mathbf{A}[i,:]$, $\mathbf{A}[:,j]$;
- vector concatenation: $\boldsymbol{x}\|\boldsymbol{y}$;

Protocol Cost. We measure the number of rounds as well as the total number of bits communicated through the network. Formally, we define a type $Cost = \mathbb{N} \times \mathbb{N}$, where the first component is the number of communicated bits, and the second component is the number of rounds. We define the operations $\otimes$: $Cost \times Cost \rightarrow Cost$ (parallel composition) and $\oplus$: $Cost \times Cost \rightarrow Cost$ (sequential composition) as follows:

- $(a,b) \otimes (c,d) = (a+c, \max(b,d))$;
- $(a,b) \oplus (c,d) = (a+c, b+d)$.

We will use the shorthand $(a,b)^{\otimes n}$ to denote $(a,b) \otimes \cdots \otimes (a,b)$, and $(a,b)^{\oplus n}$ to denote $(a,b) \oplus \cdots \oplus (a,b)$, where (a,b) occurs n times. Let the operation $\otimes$ have higher priority than $\oplus$.

For a protocol Prot, we write Prot_k^n to denote the cost of application of Prot to a vector of length n of k-bit values. The number n is omitted if the protocol is applied to a single input, and $\mathsf{Prot}_k^{n_1,\ldots,n_m}$ denotes applying the protocol to several inputs of different lengths.

2.3 General FIM Algorithms

There exist several variations of the standard (not privacy-preserving) FIM algorithms. We give some examples in this section. A similar property of these algorithms is that, on each iteration, all they compute a frequent itemset of size k, based on the frequent itemsets of size $k-1$ found so far. The basis of finding a k-set from $k-1$ subsets is *set intersection*, or *set difference*. We will turn special attention to these set operations when we use these algorithms in a privacy-preserving setting.

Apriori. This algorithm sequentially constructs all the frequent itemsets of size 1, then of size 2, until all the frequent sets are obtained in this way. Any infrequent itemsets are immediately discarded. The frequent sets of size k are constructed only for those sets whose all subsets of size $k-1$ have been frequent. The way in which these subsets are constructed depends on the particular algorithm instance. One possible implementation of this method is given in Algorithm 1.

Eclat. Similarly to Apriori, this algorithm constructs a set of size k from sets of size $k-1$. The main difference from Apriori is that Eclat uses depth-first search, considering on one step not all the possible subsets of size k, but rather

---

Algorithm 1. Apriori

Data: $\mathcal{M}$: frequent itemsets of size $k-1$ found so far
Result: Frequent itemsets of size at least k

1 $\mathcal{F} \leftarrow \emptyset$;
2 **foreach** $X_i \in \mathcal{M}$ **do**
3   **foreach** $X_j \in \mathcal{M}, j > i$ **do**
4          $R \leftarrow X_i \cup X_j$;
5   **if** $|\sigma(R)| \geq t$ **then**
6              $\mathcal{F} \leftarrow \mathcal{F} \cup \{R\}$;

7 **if** $\mathcal{F} \neq \emptyset$ **then**
8      $\mathcal{F}' \leftarrow$ Apriori($\mathcal{F}$) ;
9 **return** $\mathcal{F} \cup \mathcal{F}'$;

Algorithm 2. Eclat

Data: $\mathcal{P}$ all the frequent sets of size $k-1$ with a prefix P
Result: Frequent itemsets of size at least k with a prefix P

1 **foreach** $X_i \in \mathcal{P}$ **do**
2      $\mathcal{F}_i \leftarrow \emptyset$;
3   **foreach** $X_j \in \mathcal{P}, j > i$ **do**
4          $R \leftarrow X_i \cup X_j$;
5          $\sigma(R) \leftarrow \sigma(X_i) \cap \sigma(X_j)$;
6   **if** $|\sigma(R)| \geq t$ **then**
7              $\mathcal{F}_i \leftarrow \mathcal{F}_i \cup \{R\}$;

8   **if** $\mathcal{F}_i \neq \emptyset$ **then**
9          $\mathcal{F}_i' \leftarrow$ Eclat($\mathcal{F}_i$) ;
10 **return** $\bigcup_i \mathcal{F}_i'$;

---

constrains one step to the sets of size k with a common *prefix* P of length $k-1$ (sets of the form $P \cup \{x\}$ for $x \notin P$). Let $\sigma(P)$ be the support of P. For each item x, all possible frequent sets with prefix $P' := P \cup \{x\}$ can be constructed as $\sigma(P \cup \{x\}) \cap \sigma(P \cup \{y\})$ from the sets $P \cup \{y\}$ such that $y \neq x$. The new prefix is then processed recursively. The description of this method is given in Algorithm 2.

Diffset. If the matrix columns are very dense, then instead of keeping transactions that *contain* the given dataset, one could try to keep transactions that *do not contain* the given dataset. Actually, even something more interesting can be done. Another FIM algorithm Diffset [14] is similar to Eclat, but instead of keeping the set of transactions in each itemset, it keeps the *sizes of supports* of sets of size $k-1$, and the *differences* between a set of size k and its subsets of size $k-1$. In this way, even if the initial matrix is not dense, the set differences that this algorithm keeps may become very small on later iterations.

Algorithm 3. Diffset

Data: $\mathcal{P}$ all the frequent sets of size $k - 1$ with a prefix P
Result: Frequent itemsets of size at least k with a prefix P

1 **foreach** $X_i \in \mathcal{P}$ **do**
2      $\mathcal{F}_i \leftarrow \emptyset$;
3   **foreach** $X_j \in \mathcal{P}, j > i$ **do**
4          $R \leftarrow X_i \cup X_j$;
5          $\Delta(P,R) \leftarrow \Delta(P,X_j) \setminus \Delta(P,X_i)$;
6          $|\sigma(R)| \leftarrow |\sigma(P)| - |\Delta(P,R)|$;
7   **if** $|\sigma(R)| \geq t$ **then**
8             $\mathcal{F}_i \leftarrow \mathcal{F}_i \cup \{R\}$;
9   **if** $\mathcal{F}_i \neq \emptyset$ **then**
10        $\mathcal{F}_i' \leftarrow \mathsf{Diffset}(\mathcal{F}_i)$;

11 **return** $\bigcup_i \mathcal{F}_i'$;

Let the itemsets $P \cup \{x\}$ and $P \cup \{y\}$ be frequent. We want to know whether the itemset $P \cup \{x\} \cup \{y\}$ is frequent. Let $\Delta(P, P \cup \{x\})$ be the difference in supports of the itemsets P and $P \cup \{x\}$. We can compute the support as $\sigma(P \cup \{x\}) = \sigma(P) \setminus \Delta(P, P \cup \{x\})$, and the difference as $\Delta(P, P \cup \{x\} \cup \{y\}) = \Delta(P, P \cup \{x\}) \setminus \Delta(P, P \cup \{y\})$. The description of this method is given in Algorithm 3.

3 Privacy-Preserving FIM

Since FIM can in turn be used for various purposes such as associative rule mining, preserving privacy may be very important. For example, several shops may want to make some statistics of the contents of shopping carts without revealing what exactly has been sold. Privacy is especially important in cases where the shopping cart is associated with the customer.

Privacy-preserving versions of Apriori and Eclat have been implemented and optimized in [1,5,11]. There are also some solutions designed for specific initial data sharing, such as vertical or horizontal partitioning [9]. Implementing an algorithm such as FP-tree is not suitable for our security model since its structure leaks more information than the frequent itemsets themselves. In [11], the FP-tree is constructed *after* the frequent itemsets have been found (using Apriori-based algorithm), and the goal is to introduce noise into the public output.

Besides making the computation secure, it may be also important to consider how much the public output leaks by itself. Differential privacy guarantees that the adversary will not learn too much from the public output, providing statistical privacy for each individual record in the dataset. There exist systems such as PINQ [12], PDDP [4], RAPPOR [6], that allow to make statistical computations that achieve this property. In the context of FIM, differential privacy has been considered in [5,11,15]. Similar distortion-based approach is also used in [13].

In this work, we do not consider differential privacy. To achieve it, we could add noise to the initial data, so that the final output (that is, all the frequent itemsets up to certain size) would provide privacy for each individual record. We would need to define what exactly an individual record is (a column or a row), while in this paper the initial distribution of the records is not important.

In this work, we mainly extend the results of [1], where Eclat and Apriori algorithms (but not Diffset), as well as hybrid solutions, have been implemented. The algorithms of [1] are based on bit matrix representation. Their efficiency does not depend on matrix sparsity, and they could not use the advantages of Diffset. We give a solution that works better with sparse matrices. Our algorithms use some blackbox operations that in general depend on the underlying secure multiparty computation platform, and whose implementation is not the part of FIM algorithms. The cost estimations for our algorithms are based on the blackbox operation costs of Sharemind [3].

Bit Matrix Representation. First, we describe the existing implementations of [1] based on representing the data table as a secret shared bit matrix. Initially, the n items and m transactions are assigned unique indices $\{1, \ldots, n\}$ and $\{1, \ldots, m\}$ respectively. A matrix $\mathbf{B}$ is defined, such that $\mathbf{B}[i, j] = 1$ iff the i-th transaction contains the j-th item. On further iterations of FIM algorithms, the columns correspond not to single items, but to itemsets.

Although all matrix elements are bits, in order to determine whether an itemset is frequent, at some moment the sum of column elements has to be computed. Therefore, at least before the addition, the bits should be converted to at least $(\log m)$-bit values to avoid addition overflow, since the maximal value that the sum may take is m. In [1], the matrix elements are permanently stored as $\log m$-bit values to avoid conversion overheads. For very sparse sets, such an encoding may consume excessive space due to large amount of zeroes that will not be needed anyway. In Sect. 3.2, we discuss whether it is better to keep the bits in $\log m$ format, or to convert 1-bit values to $\log m$-bit values on demand.

Finding an intersection of two itemsets i and j and checking its cardinality is implemented as follows:

1. multiply pointwise two $\log m$-bit vectors of length m;
2. sum the obtained m products up;
3. compare the obtained $\log m$-bit number with a $\log m$-bit threshold t.

Sparse Set Representation. In this paper, we propose another way to represent transactions that contain the given itemset. This representation makes sense for sparse matrices, i.e. when each column of the matrix contains at most m' entries for $m' \ll m$. We will now use an $m' \times n$ matrix for data table representation. Each column will now contain not the characteristic bit vector, but the indices of transactions. The order of indices in a column does not matter. Encoding a number from $\{1, \ldots, m\}$ requires $\log m$ bits. If the table contains at most nm' nonzero entries, then $nm' \cdot \log m$ bits are sufficient to encode it. If the size of some column is $m_j < m'$, then some $m' - m_j$ of its entries are set to 0. Since 0 now has special meaning, we start indexation with 1. Alternatively, we could leave exactly m_j elements in the column, but that would leak too much additional information about the data.

3.1   Algorithms for Privacy Preserving FIM

Existing Building Blocks. We describe the building block algorithms that we use for our constructions. Some very basic operations that do not require additional description are given in Table 1. We now describe some more complicated algorithms. The summary of the costs of used building blocks is given in Table 2.

Radix Sort (Rsort) and Counting Sort (Csort). We use the algorithms from [2, Algorithm 3]. Let k be the number of bits needed to encode each value. The sorting protocol runs in k iterations. On each iteration, the elements are sorted according to one bit using counting sort (we denote it by Csort), starting from the least significant bit. In our algorithms, we will use CSort when we need to sort values just by one bit. Using numbers of [2], the cost of a single counting sort, applied to a vector of length n of k-bit values, is $\mathsf{Csort}_k^n = \mathsf{ShareConv}_k^n \oplus \mathsf{Mult}_k^n \oplus \mathsf{Shuffle}_{2k}^n \oplus \mathsf{Declassify}_k^n$ (the description of used subprotocols is given in Table 1). The cost of the entire radix sort is $\mathsf{Rsort}_k^n = (\mathsf{Csort}_k^n)^{\oplus k}$.

Quicksort (Qsort). We use the algorithm from [8, Protocol 1] to sort n elements of k bits each. First of all, the array is shuffled, and then ordinary quicksort algorithm is run, declassifying only the outcomes of comparisons to decide where the element should be placed. As far as all elements are distinct, this does not cause any privacy breach. A Sharemind version of this algorithm is described in [2], and its average complexity is $\mathsf{Shuffle}_k^n \oplus (\mathsf{LessThan}_k^n \oplus \mathsf{Declassify}_1^n)^{\oplus \log n}$. Because of the random shuffle, the worst case comes with negligible probability, and we may indeed expect the average cost in practice.

In some cases, we apply quicksort to arrays that have already been shuffled. In this case, we denote the sorting itself by PQsort (*plain quicksort*), and its cost is $\mathsf{PQsort}_k^n = (\mathsf{LessThan}_k^n \oplus \mathsf{Declassify}_1^n)^{\oplus \log n}$. A good property of this sort is, that the sorting permutation is public.

Table 1. Basic blackbox operations

Operation call	Returned value
$\mathsf{Sum}(\langle\!\langle x \rangle\!\rangle)$	$\langle\!\langle \sum_{j=1}^m x[j] \rangle\!\rangle$
$\mathsf{Mult}(\langle\!\langle x \rangle\!\rangle, \langle\!\langle y \rangle\!\rangle)$	$\langle\!\langle x \cdot y \rangle\!\rangle$
$\mathsf{OuterProd}(\langle\!\langle x \rangle\!\rangle, \langle\!\langle y \rangle\!\rangle)$	$\langle\!\langle \mathbf{Z} \rangle\!\rangle$, where $\mathbf{Z}[i,j] = x[i] \cdot y[j]$
$\mathsf{ShareConv}(\langle\!\langle x \rangle\!\rangle, k)$	$\langle\!\langle y \rangle\!\rangle$, where $x \in \mathbb{Z}_2$, $y \in \mathbb{Z}_{2^k}$, $x = y$
$\mathsf{Shuffle}(\langle\!\langle x \rangle\!\rangle)$	$\langle\!\langle y \rangle\!\rangle$, where y is a random reordering of x
$\mathsf{UnShuffle}(\langle\!\langle y \rangle\!\rangle)$	$\langle\!\langle x \rangle\!\rangle$, where x is a restored previously shuffled vector
$\mathsf{Equal}(\langle\!\langle x \rangle\!\rangle, \langle\!\langle y \rangle\!\rangle)$	$\langle\!\langle x == y \rangle\!\rangle \in \mathbb{Z}_2$
$\mathsf{LessThan}(\langle\!\langle x \rangle\!\rangle, \langle\!\langle y \rangle\!\rangle)$	$\langle\!\langle x \leq y \rangle\!\rangle \in \mathbb{Z}_2$
$\mathsf{Declassify}(\langle\!\langle x \rangle\!\rangle)$	x

Table 2. Building block operations (k-bit elements, $|x| = |b| = n$)

Protocol call	Returned value	Cost
$\mathsf{Csort}(\langle\!\langle x \rangle\!\rangle, \langle\!\langle b \rangle\!\rangle)$	$\langle\!\langle x \rangle\!\rangle$ sorted by $\langle\!\langle b \rangle\!\rangle$	$\mathsf{ShareConv}_k^n \oplus \mathsf{Mult}_k^n \oplus \mathsf{Shuffle}_{2k}^n \oplus \mathsf{Declassify}_k^n$
$\mathsf{Rsort}(\langle\!\langle\langle x \rangle\rangle\!\rangle)$	sorted x	$(\mathsf{ShareConv}_k^n \oplus \mathsf{Mult}_k^n \oplus \mathsf{Shuffle}_{2k}^n \oplus \mathsf{Declassify}_k^n)^{\oplus k}$
$\mathsf{Qsort}(\langle\!\langle x \rangle\!\rangle)$	sorted x	$\mathsf{Shuffle}_k^n \oplus (\mathsf{LessThan}_k^n \oplus \mathsf{Declassify}_1^n)^{\oplus \log n}$
$\mathsf{PQsort}(\langle\!\langle x \rangle\!\rangle)$	sorted x	$(\mathsf{LessThan}_k^n \oplus \mathsf{Declassify}_1^n)^{\oplus \log n}$

Set Intersection and Difference. We describe the algorithms that we use for set intersection and set difference. The sets are represented as arrays of elements. We do not want to leak the precise set cardinality, and we assume that there is a known upper bound n on the number of elements. If the set has less than n elements, then the entries that represent missing elements are set to 0.

Let two sparse vectors be represented as sequences of index-value pairs, where the indices are encoded by ℓ bits, and values are encoded by k bits. We give an algorithm for a bit more general task, that allows to compute a pointwise product of values these vectors, matching their entries by indices. We then show how to instantiate it to set intersection and set difference.

Let u be a vector of length n. For each $u[i]$, let $u[i].\mathsf{idx}$ and $u[i].\mathsf{val}$ denote the index and value component respectively. The pointwise product algorithm is given in Algorithm 4. It concatenates the vectors u and v, obtaining a vector w. It then sorts the obtained vector w by indices, so that similar indices are now consequent. It then computes the products $w[i].\mathsf{val} \cdot w[i+1].\mathsf{val}$ and leaves only those $w[i]$ for which $w[i].\mathsf{idx} = w[i+1].\mathsf{idx}$ holds. In other words, it leaves exactly the products $v[i].\mathsf{val} \cdot u[j].\mathsf{val}$ such that $v[i].\mathsf{idx} = u[j].\mathsf{idx}$. In the end, the algorithm sorts the entries back to their initial positions (applying the sorting permutation inverse σ^{-1} and UnShuffle), so that the second half of the resulting vector (the entries that are 0 anyway) can be discarded. Counting the number of all used subprotocols of this algorithm, we get the cost

$$\mathsf{Shuffle}_{\ell+k}^{2n} \oplus \mathsf{PQsort}_\ell^{2n} \oplus (\mathsf{Equal}_\ell \oplus \mathsf{ShareConv}_k \oplus \mathsf{Mult}_k^2)^{\otimes 2n-1} \oplus \mathsf{UnShuffle}_k^{2n} \ .$$

This algorithm can be easily adjusted to set intersection and set difference. The summary of costs of these set operations is given in Table 3. Let the set elements be the indices of u and v. We show how to assign the values.

Set Intersection. For the set intersection task, we set $u[i].\mathsf{val} = u[i].\mathsf{idx}$ and $v[i].\mathsf{val} = 1$ for all $i \in \{1, \ldots, n\}$. As the result, Algorithm 4 returns exactly those indices of $\langle\!\langle u \rangle\!\rangle$ that are present in $\langle\!\langle v \rangle\!\rangle$.

Set Difference. To compute the difference between two sets, we need to keep exactly those elements of $\langle\!\langle u \rangle\!\rangle$ that are *not* present in $\langle\!\langle v \rangle\!\rangle$. If we flip the bit $\langle\!\langle b[i] \rangle\!\rangle = \mathsf{Equal}(\langle\!\langle w[i].\mathsf{idx} \rangle\!\rangle, \langle\!\langle w[i+1].\mathsf{idx} \rangle\!\rangle)$ and keep only those elements $\langle\!\langle w[i].\mathsf{idx} \rangle\!\rangle$ for which $\langle\!\langle b[i] \rangle\!\rangle = 0$, we will also get elements of $\langle\!\langle v \rangle\!\rangle$ that are not present in $\langle\!\langle u \rangle\!\rangle$, and we do not need them. To get rid of these elements, for all $i \in \{1, \ldots, n\}$,

Algorithm 4. Pointwise product of two sparse vectors

Data: Shared sparse vectors $\langle\!\langle u \rangle\!\rangle$, $\langle\!\langle v \rangle\!\rangle$
Data: n — the number of non-zero elements in a vector
Result: The vector $\langle\!\langle d \rangle\!\rangle$ such that $d[i] = u[i] \cdot v[i]$
$\langle\!\langle w \rangle\!\rangle \leftarrow \langle\!\langle u \rangle\!\rangle \| \langle\!\langle v \rangle\!\rangle$;
$\langle\!\langle w \rangle\!\rangle \leftarrow \mathsf{Shuffle}(\langle\!\langle w \rangle\!\rangle)$;
$\langle\!\langle w \rangle\!\rangle \leftarrow \mathsf{PQsort}(\langle\!\langle w \rangle\!\rangle, \langle\!\langle w \rangle\!\rangle.\mathsf{idx})$;
Let σ be the public sorting permutation of PQsort ;
foreach $i \in \{1, \ldots, 2n-1\}$ **do**
$\quad\quad \langle\!\langle b[i] \rangle\!\rangle \leftarrow \mathsf{ShareConv}(\mathsf{Equal}(\langle\!\langle w[i].\mathsf{idx} \rangle\!\rangle, \langle\!\langle w[i+1].\mathsf{idx} \rangle\!\rangle), k)$
$\quad\quad \langle\!\langle d[i] \rangle\!\rangle \leftarrow \mathsf{Mult}(\langle\!\langle b[i] \rangle\!\rangle, \mathsf{Mult}(\langle\!\langle w[i].\mathsf{val} \rangle\!\rangle, \langle\!\langle w[i+1].\mathsf{val} \rangle\!\rangle))$
$[\![d]\!] \leftarrow \sigma^{-1}([\![d]\!])$
$[\![d]\!] \leftarrow \mathsf{UnShuffle}([\![d]\!])$
return $\langle\!\langle d[1] \rangle\!\rangle, \ldots, \langle\!\langle d[n] \rangle\!\rangle$

Table 3. Set operations (k-bit elements, $|a| = |b| = n$)

Protocol call	Returned value	Cost
$\mathsf{Set}_\cap(\langle\!\langle a \rangle\!\rangle, \langle\!\langle b \rangle\!\rangle)$	$\langle\!\langle c \rangle\!\rangle = \langle\!\langle a \cap b \rangle\!\rangle$	$\mathsf{Shuffle}_{2k}^{2n} \oplus \mathsf{PQsort}_k^{2n}$
$\mathsf{Set}_\setminus(\langle\!\langle a \rangle\!\rangle, \langle\!\langle b \rangle\!\rangle)$	$\langle\!\langle c \rangle\!\rangle = \langle\!\langle a \setminus b \rangle\!\rangle$	$\oplus(\mathsf{Equal}_k \oplus \mathsf{ShareConv}_k \oplus \mathsf{Mult}_k^2)^{\otimes 2n-1} \oplus \mathsf{UnShuffle}_k^{2n}$

we may initially set $u[i].\mathsf{val} = u[i].\mathsf{idx}$ and $v[i].\mathsf{val} = 0$, and take $\langle\!\langle d[i] \rangle\!\rangle \leftarrow \mathsf{Mult}(1 - \langle\!\langle b[i] \rangle\!\rangle, \langle\!\langle w[i].\mathsf{val} \rangle\!\rangle))$ as the final result. Only those entries $\langle\!\langle d[i] \rangle\!\rangle$ that correspond to u can now be nonzero.

Algorithm Costs on Sharemind. We assume that the algorithms are run on secure multiparty computation system Sharemind [3]. In Table 4 we present the costs of basic operations that are used in our algorithms. The numbers are taken mainly from [3, 10]. We take the cost of set intersection (and set difference) from Table 3, and substitute the costs of basic operations with values from Table 4. The summary of protocol costs on Sharemind platform is presented in Table 5.

Table 4. Basic operation costs of Sharemind

Sharing	Operation	Rounds	Communication
additive	Sum_k^n	0	0
	Mult_k	1	$6k$
	$\mathsf{OuterProd}_k^{n,m}$	1	$3(n+m)k$
	$\mathsf{ShareConv}_k$	2	$5k+4$
bitwise	$\mathsf{LessThan}_k$	$\log k$	$30k$
	Equal_k	$\log k$	$12k-9$
both	$\mathsf{Shuffle}_k^n$, $\mathsf{UnShuffle}_k^n$	3	$6nk$
	$\mathsf{Declassify}_k$	1	$6k$

Table 5. Auxiliary algorithm costs of Sharemind

Sharing	Protocol	Rounds	Communication
bitwise	PQsort_k^n	$\log n(\log k + 1)$	$\log n(30nk + 6n)$
	$\mathsf{Set}_{\cap k}^n$, $\mathsf{Set}_{\backslash k}^n$	$9 + (\log 2n + 1)(\log k + 1)$	$60nk \log n + 154nk$
both	Csort_k^n	7	$30nk$

Counting Sort: The sorting assumes that the secret-shared input bits, according to which the sorting is done, are already given. The cost of counting sort is $\mathsf{ShareConv}_k^n \oplus \mathsf{Mult}_k^n \oplus \mathsf{Shuffle}_{2k}^n \oplus \mathsf{Declassify}_k^n$. The total communication is $n \cdot (5k + 4) + n \cdot 6k + 6n \cdot 2k + n \cdot 6k = n(29k + 4) \approx 30nk$ bits. The total number of rounds is $2 + 1 + 3 + 1 = 7$.

Quicksort: The cost of PQsort is $(\mathsf{LessThan}_k^n \oplus \mathsf{Declassify}_1^n)^{\oplus \log n}$. Since $\mathsf{LessThan}$ is more efficient using bitwise sharing, the entire PQsort is also more efficient using bitwise sharing. For PQsort, the total number of rounds is $(\log k + 1) \cdot \log n$, and the total communication is $\log n \cdot (30nk + 6n)$ bits.

Set Intersection and Difference: let us assume that the values are bitwise shared, since the comparisons and the quicksort are faster in this case. The number of bits for a single instance of intersection is $6 \cdot 2n \cdot 2k + (30k + 6)2n \log 2n + (12k - 9 + 5k + 4 + 12k) \cdot (2n - 1) + 6 \cdot 2n \cdot k = 24nk + (60nk + 12n)(\log n + 1) + (29k - 5)(2n - 1) + 12nk = 154nk + 60nk \log n + 12n \log n + 2n - 29k + 5$ bits, which is ca $60nk \log n + 154nk$. The number of rounds is $3 + \log 2n(\log k + 1) + (\log k + 2 + 2) + 3 = 9 + (\log 2n + 1)(\log k + 1)$.

3.2 Comparing Bit Matrix and Set Based Approaches

We will now compare the bit representation and the sparse representation for FIM task. As mentioned in Sect. 3, for the bit representation, the intersections are found by multiplying two bit vectors pointwise, and the set difference can be computing analogously, by taking the negation of the bit vector that is being subtracted. For the sparse representation, the set operations can be found by using the algorithms defined in Sect. 3.1. The size of the sparse sets is m', and the set elements are encoded with $\log m$ bits. The numbers m' and n are defined as in the beginning of Sect. 3.

Cost of Intersection for Bit Representation. First of all, we estimate the rounds and communication of the bit matrix intersections, based on the operation costs of Table 4.

According to the multiplication protocol [10] of Sharemind, this is $6m \log m$ for multiplying pointwise two $\log m$ bit vectors of length m. Another possibility to do the same thing is to keep all the bits in $\mathbb{Z}_2$, doing the share conversion *after* the multiplication. Now the multiplication of m bit pairs has cost $\mathsf{Mult}_1^m = 6m$, and the share conversion $\mathsf{ShareConv}_{\log m}^m = m \cdot (5 \log m + 4)$, which is in total $5m \log m + 10m$. This approach is more efficient for $m \geq 2^{10}$.

Note that, if we need to compute $\mathsf{Mult}(\langle\!\langle a_i\rangle\!\rangle, \langle\!\langle b_j\rangle\!\rangle)$ for all $i \in \{1, \ldots, n_a\}$, $j \in \{1, \ldots, n_b\}$, then we could apply $\mathsf{OuterProd}(\langle\!\langle a_1\rangle\!\rangle | \ldots \| \langle\!\langle a_{n_a}\rangle\!\rangle, \langle\!\langle b_1\rangle\!\rangle \| \ldots \| \langle\!\langle b_{n_b}\rangle\!\rangle)$ instead, which has the same operation cost as $\mathsf{Mult}_1^{n_a+n_b}$. However, the share conversion would still have to be applied to all products, having cost $\mathsf{ShareConv}_{\log m}^{n_a \cdot n_b}$, and it does not scale well with n_a and n_b. Hence, we use only the first approach in this paper. The number of rounds in the first approach is 1, compared to the 3 rounds of the second approach.

Cost of Intersection for Sparse Representation. Now we estimate the rounds and communication of the sparse intersections, based on the operation costs of Table 4. We compare them to the analogous cost metrics of bit matrix approach.

In the sparse representation, we have m' elements in the sets, each encoded using $\log m$ bits. Suppose that we are going to find the intersections of some n_a sets with some n_b sets.

Round Advantage: An intersection takes $9 + (\log 2m' + 1)(\log\log m + 1)$ rounds instead of 1 round of bit representation. This is an obvious disadvantage, but we hope to win in memory consumption and communication.

Communication: One set intersection requires $60m'\log m \log m' + 154m'\log m$ bits of communication, compared to $6m\log m$ of the bit based approach. Hence, for a single intersection, the advantage is non-negative iff $m \geq 10m'\log m' + 26m'$. However, while the total cost of all intersections is $(n_a + n_b)\cdot 6m\log m$ for the bit approach, it is $n_a \cdot n_b \cdot (60m'\log m \log m' + 154m'\log m)$, so the sparse approach scales badly. The advantage of set intersection is non-negative iff

$$m \geq \frac{n_a \cdot n_b}{n_a + n_b}(10m'\log m' + 26m') \ .$$

Comparisons of the bit and the set representations is given in Table 6.

Table 6. Multiple set algorithm costs of Sharemind

Type	Bit communication cost
Bit representation	$(n_a + n_b)\cdot 6m\log m$
Set representation	$(n_a \cdot n_b)\cdot 6m'\log m \cdot (10\log m' + 26)$

Caveats of Sparse Representation. The best choice of m' depends on the values of n_a and n_b, which in turn depends on particular input data and the parameters, and these values are in general not known beforehand. In general, we would like to fix m' already in the beginning, since making m' dependent on data may leak more about it. On the other hand, we can make some further intersections worse if we underestimate the values of n_a and n_b.

Another problem is that, even if the data is sparse, they may be some single columns that have too many elements to make sparse approach applicable.

Algorithm 5. Bit vector to a set Bits2Set

Data: A bit vector $[\![b]\!]$ of length m with at most m' nonzero entries
Result: A bitwise shared set representation $\langle\!\langle c \rangle\!\rangle$ of $\langle\!\langle b \rangle\!\rangle$

1   $\langle\!\langle b \rangle\!\rangle = [\![b]\!] \bmod 2$;
2   **foreach** $i \in \{1, \ldots, m\}$ **do**
3     $\lfloor \quad \langle\!\langle c[i] \rangle\!\rangle = \langle\!\langle b[i] \rangle\!\rangle \cdot i$;
4   $\langle\!\langle d \rangle\!\rangle = \mathsf{CSort}(\langle\!\langle c \rangle\!\rangle, \langle\!\langle b \rangle\!\rangle)$;
5   **return** $\langle\!\langle d[1] \rangle\!\rangle, \ldots, \langle\!\langle d[m'] \rangle\!\rangle$;

We cannot just remove excessive transactions since we would have to decide which transactions exactly should be removed, and that choice may affect the final result significantly. On the other hand, finding an intersection between a dense column and a set of sparse columns is even worse than if sparse columns were treated as bit columns, regardless of the advantage that intersections of sparse columns give themselves.

If we agree to leak whether the number of nonzero entries has become at most m' after finding the intersection of two dense columns, then we may turn the resulting column into sparse. We convert bit columns of some branch of Diffset and Eclat to set columns only after *all* columns of that branch become sparse, and only if conversion still makes sense for the number of intersections that is going to be done on the next step.

Converting a Bit Matrix Column to a Set Matrix Column. The protocol Bits2Set transforms a column of a bit matrix to m' bitwise shared row identifiers, where m' is a known upper bound on the number of nonzero entries of a sparse column. This protocol is given in Algorithm 5. Assuming that m is a power of 2, even though the input bit vector is additively shared in $\mathbb{Z}_m$, it is easy to convert it to a bit vector shared in $\mathbb{Z}_2$ by locally truncating each entry up to the least significant bit. In practice, at least using Sharemind system, the entries should be shared over $\mathbb{Z}_{2^{\lceil \log m \rceil}}$ anyway.

Computing the multiplications is a local operation since we are multiplying by a public value j. The bit $b[i] \in \{0, 1\}$ should be multiplied (in $\mathbb{Z}_2$) with each bit of j, and this is a local operation. The cost of Csort is 7 rounds and $30mk$ communication. For fixed m and m', the total cost of the protocol is denoted Bits2Set$_{m, m'}$, which is 7 rounds and $30m' \log m$ communication.

3.3 Combining Dense and Sparse Representations

We still assume that we are using the standard Eclat and Diffset algorithms without modifying them in general. The algorithms should now additionally decide, which columns should be represented as sets, and which columns as bit vectors. As an example, we describe the new privacy preserving Eclat algorithm, and Diffset would be analogous, just using set difference instead of set intersection. Let m' be the bound for which set based approach is applicable. Each iteration of privacy preserving Eclat (depicted in Algorithm 6) works as follows.

Algorithm 6. Privacy Preserving Eclat

Data: $\mathcal{X}$ is a set of n itemsets of size $k-1$ with the same prefix
Data: $\langle\!\langle\mathbf{M}\rangle\!\rangle$ is the $m \times n$ matrix of supports of the itemsets
Data: threshold t
Result: Frequent itemsets of size at least k with the same prefix

1 **if** $\langle\!\langle\mathbf{M}\rangle\!\rangle$ *has a bit representation* **then**
2     $\langle\!\langle\mathbf{C}\rangle\!\rangle \leftarrow \mathsf{OuterProd}(\langle\!\langle\mathbf{M}\rangle\!\rangle, \langle\!\langle\mathbf{M}\rangle\!\rangle)$;
3     $\langle\!\langle s\rangle\!\rangle \leftarrow \mathsf{Sum}(\langle\!\langle\mathbf{C}\rangle\!\rangle)$;

4 **else**
5     $\langle\!\langle\mathbf{C}\rangle\!\rangle \leftarrow \mathsf{Set}_\cap(\langle\!\langle\mathbf{M}\rangle\!\rangle, \langle\!\langle\mathbf{M}\rangle\!\rangle)$;
6     $\langle\!\langle s\rangle\!\rangle \leftarrow \mathsf{Sum}(1 - \mathsf{Equal}(\langle\!\langle\mathbf{C}\rangle\!\rangle, 0))$;

7 $b \leftarrow \mathsf{Declassify}(\langle\!\langle s\rangle\!\rangle \geq t)$;
8 **foreach** $i \in \{1, \ldots, n\}$ **do**
9     $F_i \leftarrow \emptyset$; $\langle\!\langle\mathbf{M_i}\rangle\!\rangle \leftarrow [\,]$;
    foreach $j \in \{i+1, \ldots, n\}$ **do**
10       $R = X_i \cup X_j$;
11   **if** $b[i \cdot n + j] \geq t$ **then**
12         $F_i = F_i \cup \{R\}$;
13         $\langle\!\langle\mathbf{M_i}\rangle\!\rangle = \langle\!\langle\mathbf{M_i}\rangle\!\rangle \| \langle\!\langle\mathbf{C}\rangle\!\rangle[i \cdot n + j]$;

14   **if** $F_i \neq \emptyset$ **then**
15       $n' = |F_i|$;   //number of all columns
16       $n'' = \mathsf{Sum}(\langle\!\langle s\rangle\!\rangle \leq m')$;   //number of sparse columns
17       $cost_{bit} = (\mathsf{OuterProduct}_{\log m}^{n',n'})^{\otimes m}$;
18       $cost_{set} = (\mathsf{Bits2Set}_{m,m'} \oplus \mathsf{Equal}_{\log m}^{m'})^{\otimes n'} \oplus \mathsf{Set}_{\cap \log m}^{(n'^2 - n')/2}$;
19   **if** $(n' \neq n'')$ *or* $(cost_{bit} < cost_{set})$ **then**
20         $F_i' = \mathsf{Eclat}(F_i, \langle\!\langle\mathbf{M_i}\rangle\!\rangle, t)$;

21   **else**
22         $F_i' = \mathsf{Eclat}(F_i, \mathsf{Bits2Set}(\langle\!\langle\mathbf{M_i}\rangle\!\rangle, m), t)$;

23 **return** $\bigcup_i F_i'$;

The itemsets that are found to be frequent are public: similarly to [1], they will be declassified in the end anyway and hence do not leak any additional information. Let the itemsets of the current iteration of Eclat be represented by $\mathcal{P}$, as in Algorithm 2. The invariant is that, the secret shared supports of the itemsets $\mathcal{P}$ are either all in bit representation, or are all in set representation. It is not possible that some supports of the same prefix are bit columns, while others are set columns, since computing intersections between columns of different representations would be too inefficient.

The representation determines the algorithm used to compute the supports of the next iteration, which are the intersections of current supports. The resulting bit columns that have at most m' elements are converted to set columns using

Bits2Set protocol. To keep the invariant, this is being done only if *all* columns of the current prefix have at most m' elements. It is important that, even if a column is sparse enough, it make sense to convert bit columns to set columns only if it indeed gives advantage on the next step. This is done by estimating the cost of both approaches and comparing them.

Let n' be the current number of columns for the given prefix. The cost of one conversion is $\mathsf{Bits2Set}_{m,m'}^{n'}$. In addition, set representation makes counting ones a bit harder, requiring more comparisons of cost $\mathsf{Equal}_{\log m}^{m'n'}$. This overhead should be added to the cost of sparse intersections $(\mathsf{Set}_{\cap \log m}^{m'})^{(n'^2-n')/2}$. The resulting cost $cost_{set}$ is compared to the cost that we would have without converting bit columns to sparse columns, which is $cost_{bit} = (\mathsf{OuterProd}_{\log m}^{n',n'})^m$. The bit columns are converted to set columns iff $cost_{set} \leq cost_{bit}$.

We note that comparison with m' can be done only if all other conditions are satisfied, and it is not needed for the bit representation. Hence, the cost $\mathsf{LessThan}_{\log m}^{n'} \oplus \mathsf{Declassify}_1^{n'}$ of the comparison itself can be added to $cost_{set}$ as well, but its contribution is very small.

4   Benchmarks

We have implemented our algorithms in Sharemind and tested them on some public datasets that are available e.g. in [7]. We tested Diffset on the denser dataset Chess (3196 rows, 75 columns, 49,3% density), the medium density dataset Mushroom (8124 rows, 119 columns, 19,3% density), and we tested Eclat on the sparse dataset Retail (88162 rows, 16470 columns, 0.06% density). Since Retail is a very large set, and we had to take a very small threshold t to get use of sparse columns, we have taken only the first 5500 of its rows for our tests. We ran the FIM task them with different thresholds t and different upper bounds m' on sparse column size. If $m' = 0$, then only the bit representation was considered. The results are given in Table 7. In addition to time, we measured the memory usage and the bit communication. For these two metrics, we have three columns "sparse", "dense" and "total", denoting how much overhead was coming from the set operations of sparse columns, the dense columns, and in total. The communication cost of converting a bit column to a set column is treated as the cost of sparse representation.

We see that the advantage of sparse representation is very small. The reason is that there are too few columns for which set representation was more efficient. Sometimes, the results are even slightly worse than for pure bit representation. One reason for that is the data types of Sharemind are of fixed size, and it is not possible to encode value in k bits for an arbitrary k. Moreover, the set representation increases the number of rounds, and even some local computations, which still affect the efficiency, even though they are less significant than the communication.

Table 7. Benchmarks on Sharemind

dataset, FIM alg.	t	m'	Memory			Communication			Time (s)
			Sparse	Dense	Total	Sparse	Dense	Total	
chess, Diffset	3000	36	288 B	968 KB	968 KB	22 KB	16.5 MB	16.6 MB	2.0 s
		18	360 B	929 KB	930 KB	22.9 KB	16.4 MB	16.4 MB	2.5 s
		9	72 B	968 KB	968 KB	15.2 KB	16.5 MB	16.6 MB	3.0 s
		4	32 B	965 KB	965 KB	14 KB	16.5 MB	16.5 MB	2.5 s
		0	0	991 KB	991 KB	0	16.5 MB	16.5 MB	2.6 s
	2800	36	1.4 KB	8.5 MB	8.5 MB	154 KB	140 MB	141 MB	18 s
		18	396 B	8.6 MB	8.6 MB	122 KB	140 MB	140 MB	19 s
		9	144 B	8.6 MB	8.6 MB	115 KB	140 MB	140 MB	19 s
		4	64 B	8.6 MB	8.6 MB	113 KB	140 MB	140 MB	19 s
		0	0	8.6 MB	8.6 MB	0	140 MB	140 MB	18 s
	2600	36	8 KB	38.6 MB	38.6 MB	710 KB	595 MB	596 MB	83 s
		18	2.1 KB	38.9 MB	38.9 MB	528 KB	595 MB	595 MB	83 s
		9	576 B	39.1 MB	39.1 MB	487 KB	595 MB	596 MB	80 s
		4	208 B	39 MB	39 MB	476 KB	595 MB	594 MB	79 s
		0	0	39 MB	39 MB	0	594 MB	594 MB	77 s
	2400	36	220 KB	130 MB	130 MB	2.2 MB	2.0 GB	2.0 GB	265 s
		18	857 B	130 MB	130 MB	1.8 MB	2.0 GB	2.0 GB	267 s
		9	738 B	132 MB	132 MB	1.57 MB	2.0 GB	2.0 GB	255 s
		4	328 B	131 MB	131 MB	1.55 MB	1.95 GB	1.95 GB	254 s
		0	0	132 MB	132 MB	1.5 MB	1.95 GB	1.95 GB	253 s
mushroom, Diffset	2600	92	180 B	30.4 MB	30.4 MB	701 KB	516 MB	517 MB	83 s
		46	136 B	29.4 MB	29.4 MB	546 KB	510 MB	511 MB	81 s
		23	7.64 KB	29 MB	29 MB	641 KB	507 MB	508 MB	80 s
		11	3.43 KB	29.2 MB	29.2 MB	430 KB	506 MB	507 MB	80 s
		0	0	31.7 MB	31.7 MB	0	512 MB	512 MB	64 s
	2400	92	35 KB	45.5 MB	45.5 MB	1.29 MB	769 MB	769 MB	117 s
		46	25.3 KB	43.9 MB	43.9 MB	958 KB	757 MB	758 MB	124 s
		23	13.9 KB	43.5 MB	43.5 MB	952 KB	754 MB	755 MB	123 s
		11	6.67 KB	43.4 MB	43.4 MB	556 KB	753 MB	753 MB	123 s
		0	0	48.2 MB	48.2 MB	0	761 MB	761 MB	97 s
	2200	92	53.5 KB	62.7 MB	62.7 MB	1.94 MB	1.10 GB	1.10 GB	183 s
		46	38 KB	60.3 MB	60.4 MB	1.43 MB	1.08 GB	1.08 GB	172 s
		23	36.5 KB	59.9 MB	59.9 MB	7.97 MB	1.07 GB	1.08 GB	180 s
		11	18 KB	59.3 MB	59.4 MB	3.70 MB	1.07 GB	1.07 GB	179 s
		0	0	66.9 MB	66.9 MB	0	1.08 GB	1.08 GB	151 s
	2000	92	82 KB	101 MB	101 MB	3.00 MB	1.73 GB	1.73 GB	279 s
		46	74 KB	94.8 MB	94.9 MB	2.59 MB	1.70 GB	1.70 GB	260 s
		23	76.5 KB	92.2 MB	92.2 MB	16.8 MB	1.67 GB	1.69 GB	289 s
		11	92.5 KB	88.4 MB	88.5 MB	28.4 MB	1.63 GB	1.66 GB	300 s
		0	0	108 MB	108 MB	0	1.72 GB	1.72 GB	214 s
retail (trimmed), Eclat	15	62	744 B	25.0 MB	25.0 MB	22.9 MB	49.8 GB	49.8 GB	2660 s
		31	1.05 KB	24.9 MB	24.9 MB	22.9 MB	49.8 GB	49.8 GB	2640 s
		15	0	25.0 MB	25.0 MB	0	49.7 GB	49.7 GB	2690 s
		7	0	25.0 MB	25.0 MB	0	49.6 GB	49.6 GB	2690 s
		0	0	25.0 MB	25.0 MB	0	49.6 GB	49.6 GB	2620 s

5   Conclusion

We have presented two basic FIM algorithm for sparse datasets, an Eclat/Apriori based one, and a Diffset based one, where Diffset may be useful also for non-sparse matrices. The algorithms turn out to be not as efficient as we wanted. The main challenge is that the algorithms for sparse representation are not as linearizable as the bit vector algorithms are. Nevertheless, since our protocols can be easily integrated into the bit based approach, we may choose to apply them only on those steps where they indeed give advantage. Also, while sparse representation has not improved efficiency for the benchmarked tables, it allowed to reduce the local memory usage, which may be important for large datasets.

References

1. Bogdanov, D., Jagomägis, R., Laur, S.: A universal toolkit for cryptographically secure privacy-preserving data mining. In: Chau, M., Wang, G.A., Yue, W.T., Chen, H. (eds.) PAISI 2012. LNCS, vol. 7299, pp. 112–126. Springer, Heidelberg (2012). doi:10.1007/978-3-642-30428-6_9
2. Bogdanov, D., Laur, S., Talviste, R.: A practical analysis of oblivious sorting algorithms for secure multi-party computation. In: Bernsmed, K., Fischer-Hübner, S. (eds.) NordSec 2014. LNCS, vol. 8788, pp. 59–74. Springer, Cham (2014). doi:10.1007/978-3-319-11599-3_4
3. Bogdanov, D., Niitsoo, M., Toft, T., Willemson, J.: High-performance secure multi-party computation for data mining applications. Int. J. Inf. Sec. 11(6), 403–418 (2012)
4. Chen, R., Reznichenko, A., Francis, P., Gehrke, J.: Towards statistical queries over distributed private user data. In: Proceedings of the 9th USENIX Symposium on Networked Systems Design and Implementation, NSDI 2012, San Jose, CA, USA, 25–27 April 2012, pp. 169–182. USENIX Association (2012)
5. Cheng, X., Su, S., Xu, S., Li, Z.: Dp-apriori: a differentially private frequent itemset mining algorithm based on transaction splitting. Comput. Secur. 50, 74–90 (2015)
6. Erlingsson, Ú., Pihur, V., Korolova, A.: RAPPOR: randomized aggregatable privacy-preserving ordinal response. In: Proceedings of the 2014 ACM SIGSAC Conference on Computer and Communications Security, Scottsdale, AZ, USA, 3–7 November 2014, pp. 1054–1067. ACM (2014)
7. Frequent itemset mining dataset repository. http://fimi.ua.ac.be/data/. Accessed 16 Sept 2017
8. Hamada, K., Kikuchi, R., Ikarashi, D., Chida, K., Takahashi, K.: Practically efficient multi-party sorting protocols from comparison sort algorithms. In: Kwon, T., Lee, M.-K., Kwon, D. (eds.) ICISC 2012. LNCS, vol. 7839, pp. 202–216. Springer, Heidelberg (2013). doi:10.1007/978-3-642-37682-5_15
9. Kantarcioglu, M., Clifton, C.: Privacy-preserving distributed mining of association rules on horizontally partitioned data. IEEE Trans. Knowl. Data Eng. 16(9), 1026–1037 (2004)
10. Kerik, L., Laud, P., Randmets, J.: Optimizing MPC for robust and scalable integer and floating-point arithmetic. In: Proceedings of WAHC 2016 - 4th Workshop on Encrypted Computing and Applied Homomorphic Cryptography (2016)

11. Lee, J., Clifton, C.W.: Top-k frequent itemsets via differentially private fp-trees. In: Proceedings of the 20th ACM SIGKDD International Conference on Knowledge Discovery and Data Mining, KDD 2014, pp. 931–940. ACM (2014)
12. McSherry, F.: Privacy integrated queries: an extensible platform for privacy-preserving data analysis. Commun. ACM **53**(9), 89–97 (2010)
13. Sun, C., Fu, Y., Zhou, J., Gao, H.: Personalized privacy-preserving frequent itemset mining using randomized response. Sci. World J. **2014** (2014). 10 pages. Article ID 686151. doi:10.1155/2014/686151
14. Zaki, M.J., Gouda, K.: Fast vertical mining using diffsets. In: Proceedings of the Ninth ACM SIGKDD International Conference on Knowledge Discovery and Data Mining, KDD 2003, pp. 326–335. ACM (2003)
15. Zeng, C., Naughton, J.F., Cai, J.-Y.: On differentially private frequent itemset mining. Proc. VLDB Endow. **6**(1), 25–36 (2012)

Applications

Free Rides in Denmark: Lessons from Improperly Generated Mobile Transport Tickets

Rosario Giustolisi[✉]

IT University of Copenhagen, Copenhagen, Denmark
rosg@itu.dk

Abstract. The term *security ceremony* describes a technical system extended with its human users. In this paper, we examine the inspection ceremony for the mobile transport ticket in Denmark. We find several security weaknesses that are ascribable to both human and computer components of the ceremony. The main vulnerabilities are due to the design choices of how the visual inspection ceremony is organised and the lack of information that is stored into the 2D barcode. These vulnerabilities allow a ticket holder to travel up to 8 zones with a 2-zone subscription and enable several people to travel with the same subscription. The attack is significant as it can be automated, and rather modest skills are necessary to break the inspection ceremony. We state four principles that aim at strengthening the security of inspection ceremonies and propose an alternative ceremony whose design is driven by the stated principles.

1 Introduction

Denmark has a modern transport infrastructure. The Danish government has recently allocated substantial investments in infrastructure for railways, buses and new metro lines [1]. Copenhagen, the capital with just over 700,000 inhabitants, has one of the most advanced metro systems in the world that runs autonomously 24 h a day. Approximately one million people use the metro every week [2]. Several transportation companies operate bus, metro, and train services sharing the same ticketing system. Travellers can purchase three different ticket formats: paper, contactless smart card, and digital app. Each ticket format has its purchase and inspection procedures.

Travellers can purchase season tickets in digital format through a smartphone app, which in Danish is called *Mobilpendlerkort*. The app enables commuters to buy a digital subscription for a 30 up to 183 days in adjacent travel zones in Denmark. Season tickets allow commuters to take advantage of unlimited trips on buses, trains and metro for a discounted price. Hence, they are personal and should be used only by the person who is registered as the user in the app.

Introducing new payment technologies goes hand in hand with designing the corresponding process to check the validity of a ticket. There are many different ways of how such a process could be organised. London transit, for example,

© Springer International Publishing AG 2017
H. Lipmaa et al. (Eds.): NordSec 2017, LNCS 10674, pp. 159–174, 2017.
https://doi.org/10.1007/978-3-319-70290-2_10

has introduced the Oyster card, which is physically checked upon entering and before leaving buses or stations. In Denmark, the solution is different. Train guards are going around and checking the validity of the tickets. In long distance trains, every customer is asked to provide a ticket, in local trains and Metro trains inspections occur often, and in buses, inspections are permitted but rarely happen at all, although one has to display the proof of purchase to the bus driver before entering. Also, new mobile payment technologies brings along new challenges of adjusting existing and designing new ticket inspection ceremonies. Nowadays, the inspection ceremony must also work for digital tickets including those printed on paper or displayed on a mobile phone's screen. Inspectors may use barcode scanning technology or use other means to assess the validity of a ticket.

In this paper, we investigate the security ceremony that involves the inspection of the mobile transport tickets. According to Ellison [3], the term *ceremony* refers to a technical system extended with its human users. Security ceremony analysis, also known as socio-technical security analysis, is an area of research that has been initiated only recently although it is widely accepted that security incidents usually bootstrap from social engineering practices. Those practices target the weakest link in the security chain, namely, the user. The idea behind the socio-technical security analysis is to combine computer security and social sciences into an interdisciplinary approach that brings the human in the context of information security analysis. Recently, a socio-technical attack has been carried out against UK rail tickets [4]. BBC reporters bought forged rail tickets on the dark web. Although the tickets appeared genuine, magnetic strips were not accepted by the barriers. However, train guards let BBC reporters through the barriers at all occasions without asking any questions.

In a similar vein, we conduct our security analysis considering both human and computer components involved in the inspection ceremony. Our findings include an attack, which would allow a malicious commuter to exploit weaknesses in the app and the inspection ceremony to ride trains, buses, and metro in Denmark for free. If carefully orchestrated, the attack can elude post-processing inspection analysis attempting to detect fraudulent activities and may affect railway companies: the prices of a personal season ticket range from a minimum of 375 kroner (50 euros), for one month covering 2 zones, to a maximum of 32099 kroner (4315 euros), for three months covering 28 zones in first class. The cost depends on multiple factors such as period, travel class, the number of travel zones, and type of ticket. Our work in this paper makes the following contributions:

- We detail the security analysis of the inspection ceremony of digital season tickets and reports some security weaknesses that enable a concrete attack to the ceremony. Since there are no publicly available specifications of the ceremony, we build the ceremony via a three-phase analysis (observation, interaction, validation) of the procedures that train guards follow during an inspection. For a similar reason, we decode the barcode printed into the Mobilpendlerkort to gain the encoded data.

– We advance four principles to transport operators aimed at improving the security design of the tickets. We formulate our principles on the basis of the outcome of our ceremony analysis and findings. It follows that the principles are specifically devised for the ticket inspection ceremony but can be further generalised and applied to other socio-technical contexts.
– We propose an alternative inspection ceremony that is aimed at strengthening the security of the inspection procedure. The design of the alternative ceremony is driven by our principles. While it provides stronger security guarantees, it is simpler than the original ceremony.

Outline. Section 2 describes the Mobilpendlerkort app and the inspection ceremony; Sect. 3 details the steps that enable the attack to the ceremony; Sect. 4 outlines four principles for the design of ticket inspection ceremony; Sect. 5 details an alternative ceremony that mitigates the attack in the original one; Sect. 6 discusses related work; finally, Sect. 7 concludes the paper.

2 Ceremony Description

The inspection ceremony is pivoted on the elements provided by the Mobilpendlerkort app and on how the train guard interprets these elements. The user installs the app and purchases a season ticket. The train guard is expected to execute the inspection ceremony according to the details of the season ticket available in the app.

2.1 Description of Mobilpendlerkort

The Mobilpendlerkort app is available for mobile devices running iOS or Android operating systems. It is published by *DSB*, the largest train operating company in Scandinavia. We consider the latest available version (3.01) that is available at the time of writing this paper. Both iOS and Android versions of the Mobilpendlerkort implement the same functionalities. Hence, the security issues and the principles later described in this paper apply to both versions of the app.

Once the app is installed, it requires the commuter to enter the mobile phone number. The app generates a subscription number (*stamkortnr*) that is bound to both device and phone number. If the commuter changes device or phone number, the app generates a new subscription number, hence former season ticket cannot be restored, and the commuter should purchase a new season ticket. To purchase a season ticket, the commuter provides her personal details (i.e., name, surname, and birth date), chooses the starting and ending dates, and the desired travel zones. After payment, the app downloads and installs the ticket, which is represented via two screens. The *primary* screen is intended for visual inspection and includes an animated visual watermark. The *secondary* screen is intended for computer inspection and includes a 2D barcode that encodes a digital signature on the ticket data. An instance of a valid ticket is depicted in Figs. 1 and 2, which show the primary and secondary screens, respectively.

Fig. 1. A screenshot of the primary screen (personal details are hidden)

Fig. 2. A screenshot of the secondary screen

Fig. 3. A screenshot of the primary screen with an extra zone ticket

The center of the primary screen displays the valid travel zones and the time interval for which the ticket is valid. In the lower-right corner, the screen displays the personal details of the commuter, the type of the ticket (i.e., young, adult, or senior), the number of zones and the number of days included in the subscription, the purchase price, and the subscription number. In the lower-left area of the primary screen there is the animated visual watermark that is intended to provide visual assurance about the authenticity of the app. The secondary screen displays the phone number of the commuter, the ticket number, and a 2D barcode, specifically an Aztec type of barcode, which is the *de facto* standard for mobile transport tickets. A new ticket number is generated whenever the subscription is renewed. The Aztec barcode contains most of the data displayed on the primary screen and the respective digital signature. Section 3 provides a detailed description of the content of the barcode.

The app also allows commuters to purchase additional extra zone tickets that extend the number of travel zones temporarily. Extra zone tickets are visually *stapled* on the primary screen. For instance, the extra zone ticket in Fig. 3 allows the commuter to travel to one additional zone, namely, any of the zones that are adjacent to the zones already included in the subscription. It is possible to purchase extra zone tickets for up to 8 additional zones.

2.2   Building the Inspection Ceremony

There is no available public specification for the inspection ceremony of transport tickets in Denmark. Thus, we derived the steps that form the ceremony empirically. The procedure to build the ceremony included three phases. In the *observation* phase, we observed how train guards interact with the app, either when a valid or invalid ticket (i.e., expired, with wrong zones or personal details) is checked. We noted the behaviours of train guards on metro and trains in the Copenhagen metropolitan area. On buses, we observed the inspection done by bus drivers. The output of this phase was a preliminary draft version of the ceremony. The draft served as input to the *interaction* phase in which we interacted with train guards to refine the structure of the ceremony when they checked our valid tickets. For example, we asked the train guard to execute a full inspection. We found that the inspection ceremony varies according to the mean of transport and the cost of the ticket. For example, guards on metro and buses are not equipped with scanners that can check barcodes, hence those guards can only proceed with the visual inspection. Also, we realised that guards check ID documents on metro rides either randomly or if they believe that the personal details displayed on the primary screen of the app do not match with the mobile phone holder (e.g., the app reports a typical female first name but the holder is a male). Then, we moved to the *validation* phase, in which we asked the IT security department at DSB to confirm about the correctness of the ceremony. DSB personnel confirmed that steps outlined in the ceremony are correct, while they preferred not to comment whether the ceremony comprises other steps that we have not observed empirically. More details about DSB response are outlined in Sect. 4.

Figure 4 summarises the various steps required for mobile ticket inspection, which unfolds as follows. Upon request of the train guard, the commuter shows the primary screen of the ticket along with an authentic and valid ID document. The train guard visually checks the authenticity of the primary screen provided by the animated visual watermark, image background, and font of the text. The train guard then checks the validity of dates and zones reported on the ticket. For short rides, such as metro rides, or in rush hours, the train guard may decide to conclude the inspection and consider the ticket valid. Earlier successful conclusions of the ceremony are depicted with dashed lines in Fig. 4. Otherwise, the train guard may check that the personal details of the ID document match name, surname, and birth date reported in the primary screen. During the last step of the inspection ceremony, the train guard checks the validity of the barcode using an hand-held scanner, which is equipped with the verification key to validate the signature. The device emits a green light if the verification is successful and a red light otherwise.

It is clear that ticket inspection is a socio-technical procedure as it involves actions and decisions from both human and computer components. It follows that the security of such procedure depends on elements ascribable to human users (i.e., the train guard), computer technology (i.e. the app and the scanner), and the interaction among them. As we shall see later, this observation is important

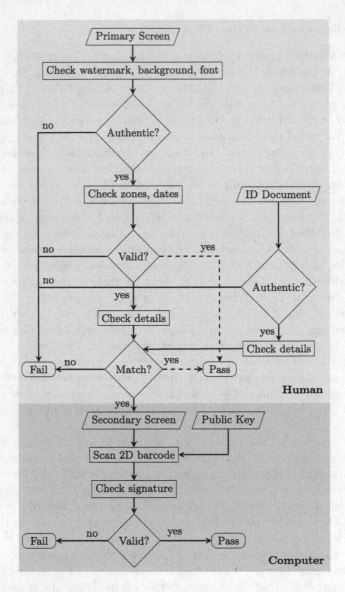

Fig. 4. The inspection ceremony executed by the train guard

as it leads to the finding of security weaknesses and provides grounds for the security principles.

3   Attack Demonstration

This section describes an attack that exploits the security weaknesses of the inspection ceremony. To describe the attack, we first report the procedure leading

to evidence of weaknesses on the generation of the barcode. Then, we discuss
how to forge the primary screen to break the inspection ceremony by combining
the forged primary screen with the barcode.

```
23 55 54 30 31 31 31 38   36 30 30 30 30 31 30 2c   |#UT011186000010,|
02 14 6d 48 6c 80 cb 75   95 79 58 bc 92 6a 8f 9b   |..mHl..u.yX..j..|
f7 ee 04 5d f8 7b 02 14   1c ad fe ef 43 4a 84 1c   |...].{.....CJ..|
4d ce 94 7a 18 6c 7b 55   ba 4a c5 c6 00 00 00 00   |M..z.l{U.J......|
30 32 37 35 78 da 65 91   cd 4e c3 40 0c 84 5f 65   |0275x.e..N.@.._e|
cf 55 a9 c6 fb e3 dd e4   80 54 9a a8 54 e9 a1 a2   |.U.......T..T...|
e9 a1 5c aa 48 c9 81 4b   40 81 f7 17 76 8a c2 22   |....H..K@...v.."|
7c fb 6c cf 78 b2 b9 dc   9e eb 6d 05 02 82 23 4a   ||.l.x.....m...#J|
4c d1 c7 00 07 93 15 a4   61 41 91 6c 48 be ef 86   |L.......aA.1H...|
f1 72 6b 8f db ab 88 7c   e1 5e 76 ad 05 6c 04 84   |.rk....|.^v..1..|
49 94 52 ae 3a 3f 09 26   10 91 f6 e9 fc d5 8d 7d   |I.R.:?.&.......}|
37 f5 e6 54 0b 49 1f ac   7b dc 1e 76 4d dd c2 ca   |7..T.I..{..vM...|
05 31 d3 96 d7 10 e2 22   5e 64 55 6b f7 d7 63 75   |.1.....".^dUk..cu|
d8 97 46 13 40 12 c8 28   e8 a6 48 cc c7 20 46 b4   |..F.@..(..H.. F.|
9c 01 6f e0 0c 50 aa 77   b0 4b 9f fc 06 de 80 b4   |..o..P.w.K......|
1f 83 9f 83 ea 60 85 78   3f b5 50 91 13 23 23 26   |.....'.x?.P..##&|
dd 2c 94 8a d7 f7 71 30   a4 d3 48 b9 9e f5 99 a0   |.,....q0..H.....|
b1 61 1f 1e 11 ff 22 3b   ff cf c1 c9 52 9e 87 39   |.a....";....R..9|
2d 64 25 41 ca 66 29 fb   52 d5 4f a5 a1 b5 71 bf   |-d%A.f).R.O...q.|
99 e7 5f 90 03 e5 30 67   09 3f e0 f4 75 c4 51 2a   |.._...0g.?..u.Q*|
9d a6 b7 4f 53 35 0d 39   d6 03 77 9b 95 94 4f 69   |...OS5.9..w...Oi|
0d 7c 03 b6 cf 72 04                                |.|...r.|
```

Fig. 5. Raw data encoded in the 2D barcode. The sequence 302c 0214 reveals the
header of the signature. The sequence 78da reveals the header of a compressed payload.

3.1   Barcode Analysis

The first step for the analysis of the barcode content is to get the raw data cod-
ified in the barcode. There are many barcode decoders available on the Internet
that can recover data from barcodes, e.g., ZXing, WebQR. Figure 5 shows the
raw data encoded in the barcode contained in the secondary screen of Fig. 2.
The first 14 bytes are reserved for the header of the ticket. The presence of the
string "#UT" in the header confirms that the structure of the barcode follows
the UIC 918-3 standard [5], which prescribes the use of public key cryptography. Information about the verification keys used by DSB and most popular
train operators in Europe are publicly available for download [6]. The header is
followed by a digital signature that is generated using DSA-SHA1 (1024-160).

Most of the data encoded in the barcode is the payload, which is compressed
into zlib format. Since only the content of the payload is signed, it is important
that it contains as much information as possible about the ticket to prevent its
forgery.

Since the structure of digital tickets is not open source, we derived it using
a differential analysis approach. We purchased different season tickets and studied how the respective payloads differ. In so doing, we were able to isolate the
elements that form the structure of the digital ticket, for example, the purchase
date, the price, etc. This approach led to an interesting finding. We considered two season tickets for the same travel zones but having different personal

```
U_HEAD0100531186 17475030 0503201712584daenU_TLAY010493RCT200270001010
300003DSB0018011100011Standard
PE0118010600006TICKET02050104000041186030101120012GYLDIG:
2017025801050000501pe06010111000110603.00:00065201110001114.0401:
000754010100001*0703010100001*0709010100001*0760010100001
*0613010900009Zone10010713010100001*0632010200002->
0732010200002->0634010900009Zone 10030734010100001
*0668010100001207680101000001*0801011100011Zoner: 1,
3090101000000010010100000001101010000000120102500000013520108000008
Pris DKK1361011000010****488,00


U_HEAD0100531186 17641124 1303201721194daenU_TLAY010493RCT200270001010
300003DSB0018011100011Standard
PE0118010600006TICKET02050104000041186030101120012GYLDIG:
2017025801050000501pe06010111000114.0300:00065201110001115.0401:
000754010100001*0703010100001*0709010100001*0760010100001
*0613010900009Zone 10010713010100001*0632010200002->
0732010200002->0634010900009Zone 10030734010100001
*0668010100001207680101000001*0801011100011Zoner: 1,
3090101000000010010100000001101010000000120102500000013520108000008
Pris DKK1361011000010****400,00
```

Fig. 6. The payload of the two tickets. The differences between the two payloads are in bold. There is no mention of the personal details of the commuters.

details, starting, and ending dates. We expected that the different key information between the tickets should be reflected on the payload. Figure 6 highlights the data of the two payloads. As expected, we found that the payload includes information about the ticket number, purchase date and time, ticket type, travel zones, and purchase price. Surprisingly, the personal details of the commuter cannot be found in the payload. This means that the authenticity of the personal details reported in the primary screen only relies on the unforgeability of that screen.

3.2 Primary Screen Forgery

The primary screen should be usable, namely the visual inspection should not take too long to evaluate the authenticity of the data reported on the screen. On the other hand, any alteration of the screen should be noticeable to the train guard at visual inspection. We note that the unforgeability of the primary screen mainly depends on fonts, the position of the text, background image, and animated visual watermark. As expected, the animation has complex elements that make the animation difficult to reproduce.

Unfortunately, the primary screen can be forged in few steps and, most importantly, once the background and the visual watermark are reproduced, the forgery can be automated. We observe that no text overlays the visual watermark, hence the latter can be obtained by recording a video of the primary screen on the phone (Quicktime can capture iPhone screens). Then, the resulting video can be converted into a GIF image to facilitate editing. We also observe that only name, surname, and birth date need to be edited because other information is sealed into the signature encoded in the barcode. It follows that only a small portion of the background needs to be edited and replaced with the desired

Fig. 7. A screenshot of a complete empty forged primary screen as template

Fig. 8. A screenshot of the forged primary screen as template

Fig. 9. A screenshot of the forged primary screen filled with a placeholder name

name, surname, and birth date, although a complete empty primary screen can be easily obtained (Fig. 7). This makes the forgery procedure even easier than expected: there is no need for a dedicated app, a GIF image is sufficient to make a successful forgery.

The last step needed to complete the forgery is to use the correct font. Any online font identifier can recognise the font, which in this case is *RobotoSlab Regular*. Figure 8 shows a screenshot of a forged primary screen that can be used as a template to create several primary screens with different personal details automatically. Figure 9 depicts the template filled with placeholder details.

The attack is carried out by combining the secondary screen with a forged primary screen. In fact, even if the train guard checks the ticket holder's ID document, she may end up accepting the forged digital ticket. The attack breaks the inspection ceremony in both its human and computer components. It takes advantage of poor security choices in the design of the primary screen breaking the visual inspection on the screen (human component). The attack exploits the naive generation of the signature encoded in the barcode to pass the electronic check (computer component) of the secondary screen.

One may note that post-processing systems will eventually identify the concurrent use of a forged digital ticket, and annul it. Still, a forged digital ticket can be successfully used at different times of the day by different people.

The attack is very effective if orchestrated carefully. A more sophisticated version would be to purchase a season ticket with many travel zones and take advantage of the reduced price due to the incremental discount rate. The ticket would be distributed among several commuters that travel to different zones. In this case, the forged digital ticket can be implemented into a dedicated app that allows users to report when a forged digital ticket is scanned. This sophisticated version would be particularly virulent: fraudsters can coordinate the issue of forged digital tickets, minimising the risk that post-processing systems identify the concurrent or suspicious use of the tickets.

3.3   Extra Zone Ticket Forgery

Forging the extra zone ticket is even easier than the primary screen as there is no need to forge the app at all in this case. Less than one person-day of effort was needed to make an Android *floating app* that overlaps the primary screen of the Mobilpendlerkort. A floating app is an application that opens in a window and floats over all other applications allowing multitasking on a device. A popular use of floating apps are the chat heads of messaging apps like Facebook Messenger. In our case, the floating app consists of a fake extra zone ticket with a countdown timer that simulates the remaining validity time of the ticket. Our app does not float freely but sticks to the primary screen.

The extra zone ticket forgery is effective because the barcode does not change when extra zone tickets are purchased and takes advantage of the poor security design of the ticket in the primary screen to attack the visual inspection (human component).

4   Principles

Upon the basis of our findings, we formulate four principles that form the foundation for our proposed solution. The first principle focuses on differences between paper and mobile tickets.

Principle 1. *The security design of paper tickets should not influence the security design of electronic tickets.*

Often, look and feel of traditional systems tend to be copied in their electronic counterpart. This is a natural design choice because it takes advantage of the preexisting familiarity that stakeholders have with the system: it minimises end-user confusion caused by the introduction of electronic components; it allows system developers to get immediate correctness feedback from a system they already know well. With a security take, this approach should be practised more cautiously, prioritising secure by design principles when possible. For instance, it is mistakenly believed that a digital animation image gives the same degree of authenticity to mobile tickets as watermarks give to paper tickets. In our setting, the forged digital ticket is a clear example of how a pre-existing (working) approach in the traditional system influences the electronic one negatively.

Principle 2. *Computer inspection should be prioritised over visual inspection.*

An immediate consequence of Principle 1 is that computer components should not be seen as add-ons that follow the traditional inspection ceremony. Train guards are used to that a successful visual inspection signifies that the ticket is valid. Habits are hard to eradicate, especially when traditional and electronic systems coexist, as in the case for transport tickets. The human-then-computer ceremony may lead train guards to execute only the traditional inspection ceremony and consider mobile tickets valid when they are not. Designing a computer-then-human ceremony would result in train guards to diverge from the traditional ceremony, fostering awareness of the different inspection ceremonies.

When possible, it is also advisable to intertwine human and computer components in the inspection ceremony. A closer look at the inspection ceremony in Fig. 4 reveals an additional issue due to the separation of human and computer components. The last check validates the signature of a payload that is unintelligible to the human because the payload is encoded as a 2D barcode. The payload may contain different data respect to the information reported in the primary screen, and the output of the scanner (i.e., green or red light) is not sufficient to the train guard to check whether the payload actually encodes the same data as in the primary screen.

Principle 3. *The inspection ceremony should enable the verification of ticket key information either electronically or manually.*

Maintaining information in electronic form has many advantages, such as quick and human-errorless ticket verification. However, it is necessary that all key information is considered for checking. Complete information may not be available due to intentional or unintentional computer malfunctions. For example, scanners may not work properly during the inspection. It is desirable that the inspection ceremony provides the strongest possible security guarantees to both electronic and manual verification procedures. For example, the ceremony would benefit from practices such as data redundancy and data duplication. Both practices prompt verification procedures to have access to the necessary data and would help to avoid weaknesses such as the exclusion of commuter personal details from the payload of the barcode.

Principle 4. *Security should be preferred over usability in the design of visual inspection of an electronic ticket.*

The rationales underlying this principle are twofold. First, it comes from the observation that a ticket is not a receipt. The goal of a transport ticket is to prove *to the train guard* that the holder has a certain right, while the goal of a receipt is to prove *to the holder* that ticketing system received the money. A common mistake is not to separate concerns, for example, separate tickets with receipts. The design of a ticket demands security and usability towards the train guard while the design of receipt focuses on usability towards the ticket holder. The effort in balancing those conflicting requirements may lead to compromises and may introduce security weaknesses, which can be easily avoided by designing ticket and receipt separately.

Secondly, electronic tickets are more suitable for computer inspection than visual inspection, hence it is easier to guarantee the security of the inspection via computer components rather than human components. The security of visual inspection cannot be taken for granted, and visual assurance elements should be carefully designed to maximise security, sacrificing usability if necessary. The following design advice list aims at maximising the security of visual inspection.

- Prefer complex, large, and dynamic visual watermarks over simple, small, and static ones to mitigate forgery.
- Superimpose critical information and visual watermark to prevent data alteration.
- Display the current time to ensure liveness.
- Display visual watermarks across the screens to evidence screen correlation.

Of course, the list contains elements that may negatively affect the usability of the ceremony. However, we observe that this is not a real issue since train guards are expected to be specifically trained to perform a visual inspection. This is an additional element in support of prioritising security over usability in the design of visual inspection.

5   Alternative Inspection Ceremony

We propose a new inspection ceremony inspired by the principles outlined above. Where possible, we reuse the components of the actual ceremony. We believe that the reuse of existing components will reduce the need of training for train guards, hence will favour a faster adoption of the new ceremony.

Figure 10 shows the steps of the alternative inspection ceremony. A main pillar of the alternative ceremony is to prioritise computer inspection over visual inspection, as advocated in Principle 2. The primary screen is replaced with a *barcode screen*, that contains the Aztec barcode, which should encode zones, dates, and personal details of the commuter. This enables the verification of all ticket key information electronically as suggested in Principle 3. The hand-held scanner checks the barcode and, if the signature is valid, it shows the content of the payload (i.e., zones, dates, and personal details) on the screen of the scanner. The train guard now checks the validity of zones and dates by looking at the screen of the scanner. The last step consists of checking whether the personal details displayed on the screen of the scanner match with the details reported on the ID document.

The proposed inspection ceremony is simpler than the original one. The app needs only one screen to represent the ticket. This is possible by taking advantage of the screen on the scanner. The visual inspection now takes place on a device controlled by transport companies rather than on the commuter's device, which provides only the elements for computer inspection. Thus, the alternative inspection ceremony minimises the attack surface due to forged digital tickets

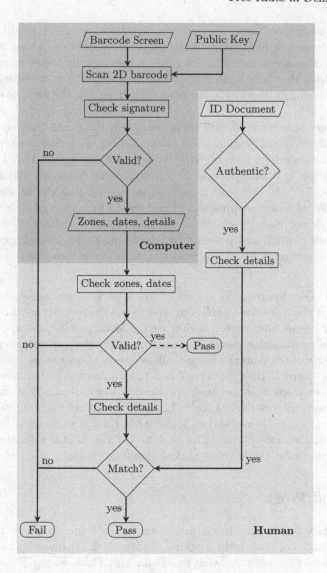

Fig. 10. The proposed inspection ceremony

on the human visual inspection. The proposed design provides more security guarantees in case of an earlier successful conclusion of the ceremony. The scanner does not emit light to signal the outcome of the verification anymore, hence the train guard needs to validate travel zones and dates, whose authenticity is provided by the digital signature encoded in the barcode. Although an earlier successful conclusion of the proposed ceremony does not ensure that the holder has the right to travel with that digital ticket, it still guarantees that the ticket is valid. Conversely, an early successful conclusion of the original ceremony does not give any guarantees about the validity digital ticket, which might be forged. An additional element in support of the proposed inspection is that it helps the post-processing analysis as each ticket inspection entails ticket scanning. The device can store the data collected on each scan, which is eventually sent to the post-processing systems as it happens with tickets in the form of contactless smart cards.

A secondary screen might still be necessary in the event that an hand-held scanner is unavailable. The animation image displayed on that screen should be replaced by a more robust form of digital watermark as suggested by the elements listed in Principle 1. It is important to stress that the ceremony should lead the train guard to look at this screen only under exceptional circumstances. This justifies the use of a more sophisticated and secure form of digital watermark at the cost of a slower visual inspection. The security of the alternative inspection ceremony is not proven, and it would be interesting to use formal approaches to appreciate the security of both original and proposed ceremonies.

Responsible Disclosure. We notified DSB as soon as we completed the security analysis of the inspection ceremony and found the security weaknesses. DSB examined our report and started reviewing the Mobilpendlerkort app. However, at the time of writing this paper, no new version of the app has been released. While DSB personnel confirmed that all the steps depicted in the inspection ceremony are correct, they preferred not to comment whether the ceremony comprises other steps that we have not observed empirically. For example, we would have been interested to know whether the scanner can remotely check a ticket online. If so, it is essential to know what data is exchanged between the scanner and the remote server. The presence of the digital signature into the barcode suggests that the scanner makes an offline check.

6   Related Work

Ceremony analysis has been increasingly studied as a way to better understand the security issues of real-world systems. Radke et al. [7] investigated strengths and weaknesses of ceremony analysis. Bella and Coles-Kemp [8] advanced a model in support of the socio-technical analysis of security termed *the ceremony concertina*. Johansen and Jøsang [9] proposed a probabilistic modelling of humans based on the ceremony concertina. Probst et al. [10] have recently discussed different approaches to modelling and analysing socio-technical systems formally. In our case, the approach to the ceremony analysis is informal. However, we believe that our case study would benefit from formal approaches and that new security issues might be found.

Garcia et al. [11] reverse engineered and made a cryptoanalysis of the MIFARE Classic card, a major player in contactless smart card market, with a strong presence in public transport systems. They found two attacks that allowed an intruder to get the secret key of a card. MIFARE Classic card could be cloned in under a second. Differently from our work, they exploit a security vulnerability of the cryptographic primitive of the authentication protocol. Murdoch et al. [12] attacked "Chip and Pin" cards issued by the EMV (Europay, MasterCard, Visa) consortium. They were able to make a successful transaction with a stolen card without knowing the PIN. They exploited the fact that the PIN verification step is never explicitly authenticated. In so doing, they were able to build a man-in-the-middle attack using a few hardware components and a fake card.

E-ticketing insecurity traces back to Schneier [13], which explained how easy was to fly on someone else's ticket by changing the name on the e-ticket boarding pass printed out at home. More recently, Jaroszewski [14] created a fake boarding pass app to enter airline lounges. Instead of generating the boarding pass, the app generates a barcode that is based on the flight number, and that can then be scanned at the entrance to lounges. The fake barcode worked because it was generated without any cryptographic authentication mechanism so unmanned automatic scanners could not check properly the eligibility of the passenger. This case is complementary to the case studied in our paper in which the involvement of human user (i.e., the train guard) might be harmful to the security of the system.

7   Conclusions

The socio-technical approach to the security analysis allowed us to find the security weaknesses that lead to the attacks. The contribution of this work is not in the attacks that we have identified but in the principles we have derived from our observation for the design of security ceremony. Building a security ceremony is not a straightforward task. There is no standard notation or known guideline that prescribes the right formalism to represent the flow of information in a security ceremony. The presence of human users further complicates the construction of the ceremony. In our case, we decided to use flowcharts because they provide a simple and clear graphical description, and naturally describe the heterogeneous protocol where users and computers are the players. The presence of the human component also complicates the design of a security ceremony. Human users tend to take shortcuts as in the case of earlier successful conclusions of the ceremony. With a security take, the shortcuts should be reduced as much as possible and, ideally, eradicated.

Acknowledgement. This work is supported in part by DemTech grant 10-092309 from the Danish Council for Strategic Research, Programme Commission on Strategic Growth Technologies.

References

1. Ministry of Transport of Denmark: Danish infrastructure investments (2012). https://goo.gl/irpQQR
2. Ministry of Foreign Affairs of Denmark: Transport infrastructure in Denmark (2012). http://denmark.dk/en/practical-info/work-in-denmark/transport-infrastructure-in-denmark

3. Ellison, C.: Ceremony design and analysis. IACR eprint (2007)
4. BBC News: Forged rail tickets sold on dark web, BBC investigation reveals (2016)
5. International Union of Railways: Uic 918–3: International rail ticket for home printing (2007)
6. International Union of Railways: the UIC public key management website (2017). https://railpublickey.uic.org/download.php
7. Radke, K., Boyd, C., Gonzalez Nieto, J., Brereton, M.: Ceremony analysis: strengths and weaknesses. In: Camenisch, J., Fischer-Hübner, S., Murayama, Y., Portmann, A., Rieder, C. (eds.) SEC 2011. IAICT, vol. 354, pp. 104–115. Springer, Heidelberg (2011). doi:10.1007/978-3-642-21424-0_9
8. Bella, G., Coles-Kemp, L.: Layered analysis of security ceremonies. In: Gritzalis, D., Furnell, S., Theoharidou, M. (eds.) SEC 2012. IAICT, vol. 376, pp. 273–286. Springer, Heidelberg (2012). doi:10.1007/978-3-642-30436-1_23
9. Johansen, C., Jøsang, A.: Probabilistic Modelling of Humans in Security Ceremonies. In: Garcia-Alfaro, J., Herrera-Joancomartí, J., Lupu, E., Posegga, J., Aldini, A., Martinelli, F., Suri, N. (eds.) DPM/QASA/SETOP -2014. LNCS, vol. 8872, pp. 277–292. Springer, Cham (2015). doi:10.1007/978-3-319-17016-9_18
10. Probst, C.W., Kammüller, F., Hansen, R.R.: Formal modelling and analysis of socio-technical systems. In: Probst, C.W., Hankin, C., Hansen, R.R. (eds.) Semantics, Logics, and Calculi. LNCS, vol. 9560, pp. 54–73. Springer, Cham (2016). doi:10.1007/978-3-319-27810-0_3
11. Garcia, F.D., Koning Gans, G., Muijrers, R., Rossum, P., Verdult, R., Schreur, R.W., Jacobs, B.: Dismantling MIFARE classic. In: Jajodia, S., Lopez, J. (eds.) ESORICS 2008. LNCS, vol. 5283, pp. 97–114. Springer, Heidelberg (2008). doi:10.1007/978-3-540-88313-5_7
12. Murdoch, S.J., Drimer, S., Anderson, R., Bond, M.: Chip and pin is broken. In: 2010 IEEE Symposium on Security and Privacy, pp. 433–446 (2010)
13. Schneier, B.: Flying on someone elses airplaine ticket (2003). https://www.schneier.com/crypto-gram/archives/2003/0815.html#6
14. Jaroszewski, P.: How to get good seats in the security theater? Hacking boarding passes for fun and profit. In: DEF CON 24 Hacking Conference (2016)

Using the Estonian Electronic Identity Card for Authentication to a Machine

Danielle Morgan[1] and Arnis Parsovs[2,3]([✉])

[1] Tallinn University of Technology, Tallinn, Estonia
[2] Software Technology and Applications Competence Center, Tartu, Estonia
[3] University of Tartu, Tartu, Estonia
arnis@ut.ee

Abstract. The electronic chip of the Estonian ID card is widely used in Estonia to identify the cardholder to a machine. For example, the electronic ID card can be used to collect rewards in customer loyalty programs, authenticate to public printers and self-checkout machines in libraries, and even unlock doors and gain access to restricted areas. This paper studies the security aspects of using the Estonian ID card for this purpose. The paper shows that the way the ID card is currently being used provides little to no assurance to the terminal about the identity of the cardholder. To demonstrate this, an ID card emulator is built, which emulates the electronic chip of the Estonian ID card as much as possible and is able to successfully impersonate the real ID card to the terminals deployed in practice. The exact mechanisms used by the terminals to authenticate the ID card are studied and possible security improvements for the Estonian ID card are discussed.

1 Introduction

The state of Estonia issues several types of credit card-sized identity documents that contain a contact-type smart card chip. These are the identity card, the residence permit card, the digital identity card and the e-resident's digital identity card [24]. The common term "ID card" is used in this paper to refer to all these chip cards.

The main purpose of the electronic chip embedded in the ID card is to perform cryptographic operations with two RSA 2048-bit keys stored on the chip. One key is used for authentication and the other for digital signatures. The authentication key can be used to sign TLS client certificate authentication challenges and to decrypt encrypted documents sent to the cardholder, while the digital signature key can be used by the cardholder to create eIDAS-compatible [29] qualified electronic signatures (QES). Cryptographic operations require that the cardholder authenticate using a PIN.

In Estonia, it is a widespread practice to use the ID card as a credential to electronically identify oneself to a machine. Several large merchants in Estonia allow the ID card to be used as a customer loyalty card [6], providing access to rewards once the ID card is inserted in the merchant's terminal. Similarly,

© Springer International Publishing AG 2017
H. Lipmaa et al. (Eds.): NordSec 2017, LNCS 10674, pp. 175–191, 2017.
https://doi.org/10.1007/978-3-319-70290-2_11

the ID card can be used to authenticate to self-service printing machines and self-checkout machines in libraries. Pharmacies use the ID card chip to look up the drugs prescribed using the digital prescription system. In some public and less public security installations the ID card can be used as an entrance card to unlock the door and gain access to restricted areas [5]. Using the ID card is convenient, as every Estonian resident is supposed to have one, and using such a universal identification token means that people do not need to carry around a large number of service provider-specific identification tokens.

However, one common characteristic observed in this type of identification is that the ID card is authenticated without the cardholder being required to enter a PIN. This means that the cryptographic capabilities provided by the card are not used to authenticate the chip. The card terminal simply reads a cardholder identifier stored on the card and uses it to decide if access to the service should be granted.

This paper studies how the smart card terminals deployed authenticate the ID card chip in practice. This is done by building an ID card emulator, which emulates the chip of the real ID card as much as possible and logs the commands received from the terminal. We discuss the security aspects of using this type of chip authentication and also analyze the risks faced by the cardholder when inserting their ID card in an untrusted terminal. We acknowledge that for certain ID card use cases discussed in this paper, the risk of fraud is so low that a secure authentication solution may not be needed. The analysis of fraud feasibility, however, is not in the scope of this study.

The paper is structured as follows. Section 2 describes the current ID card chip authentication mechanism and the related issues. Section 3 analyzes the security risks faced by the cardholder when inserting their card in a malicious terminal. Section 4 describes the design of the ID card emulator. Section 5 describes the results of using the ID card emulator in the terminals deployed in practice. Section 6 discusses possible improvements to the current card authentication mechanism. Finally, Sect. 7 concludes the paper.

2   Card Authentication

To identify the cardholder, the terminals deployed in practice read the publicly readable personal data file that resides on the chip of the Estonian ID card. The records contained in the personal data file are shown in Table 1.[1] To read the records, the terminal has to send several Application Protocol Data Unit (APDU) commands to the smart card and read the responses. An example of reading the 5th record (nationality code) from the personal data file is shown in Table 2. In practice, the process of reading the whole personal data file takes half a second.

For the purpose of identifying the cardholder, the personal ID code (record No. 7) is the best option. The personal ID code does not change during the

---

[1] The digital identity cards issued before December 2014 only have the document number (field No. 8) filled. These cards will expire by December 2017.

Table 1. Contents of a personal data file stored on an ID card [30, Sect. 10]

No	Content	Example	Length (bytes)
1	Surname	ŽAIKOVSKI	Max 28
2	First name line 1	IGOR	Max 15
3	First name line 2		Max 15
4	Sex	M	1
5	Nationality code	POL	3
6	Date of birth	01.01.1971	10
7	Personal ID code	37101010021	11
8	Document number	X0010536	8 or 9
9	Expiry date	13.08.2019	10
10	Place of birth	POOLA / POL	Max 35
11	Date of issuance	13.08.2014	10
12	Permit type		Max 50
13	Notes line 1	EL KODANIK / EU CITIZEN	Max 50
14	Notes line 2	ALALINE ELAMISÕIGUS	Max 50
15	Notes line 3	PERMANENT RIGHT OF RESIDENCE	Max 50
16	Notes line 4	LUBATUD TÖÖTADA	Max 50

Table 2. APDU commands for reading the 5th record from the personal data file

Command	Command APDU	Response APDU	Description
SELECT FILE	00 A4 01 0C 02 EE EE	90 00	Select EstEID DF
SELECT FILE	00 A4 02 0C 02 50 44	90 00	Select personal data file
READ RECORD	00 B2 05 04 00	61 03	Read 5th record
GET RESPONSE	00 C0 00 00 03	50 4F 4C 90 00	Retrieve 3-byte response

cardholder's lifetime and is the standard identifier used by Estonian information systems to uniquely identify a person. The personal ID code, however, is not just a unique identifier – it also reveals the person's date of birth, sex, and the district where the code was issued.

2.1   Document Expiration and Revocation Checks

To verify that the card has not expired and is not revoked, the terminal can check the expiry date (record No. 9) and use the document number (record No. 8) to run an online check against the public online document validity service provided by the Estonian Police and Border Guard Board [7]. Alternatively, the terminal can check the validity of X.509 certificates stored on the card using the OCSP service provided by the CA free of charge. However, since the certificates on the card can be revoked without revoking the identity document, the validity status of certificates may not reflect the validity status of the document. For example, the cardholder could have revoked the certificates only to disable the card's cryptographic functionality. On the other hand, in case a card is lost or stolen the validity of certificates is often suspended temporarily in hopes of finding the card later, at which time the suspension can be terminated. The validity life-cycle of the identity document, however, does not allow temporarily suspending the validity of the document.

2.2 Card Impersonation

The data records stored in the personal data file are not cryptographically protected, therefore the terminal has to trust that the data received has not been modified and is read from an authentic ID card. With cheap programmable smart cards available on the market, this assumption of trust does not hold in practice. In Sect. 4 we demonstrate the design of a fake ID card chip that is able to trick the terminals into accepting the fake chip as a genuine ID card, and respond to the terminal with arbitrary data contained in the personal data file. This makes the schemes relying on chip authentication vulnerable to cardholder impersonation attacks.

3   Attacks by Malicious Terminals

The lack of cryptographic assurance in the chip authentication process allows a malicious cardholder to deceive the terminal. However, the cardholder also faces security risks if the ID card is inserted in a malicious terminal. These risks will be discussed in this section.

3.1   Compromising the Cardholder's Privacy

While their personal ID code is the only record needed to identify the cardholder, a malicious terminal can read other records from the personal data file, such as place of birth or information about a residence permit. There may be a legitimate reason for reading the expiry date and document number, if the document expiration and revocation checks are to be performed (see Sect. 2.1). Without the data stored in the personal data file, there is also other publicly readable information available on the chip, such as X.509 public key certificates, private key usage counters and PIN retry counters.

The X.509 certificates stored on the ID card [30, Sect. 13] do not contain any personal information other than the personal ID code and name of the cardholder. The certificates can also be obtained from the public LDAP directory maintained by the CA. Instead of reading the personal data file, the ID code to authenticate the cardholder could be extracted from the certificate. This approach, however, is not used in practice, since it is faster and easier to read a single record from the personal data file. On the other hand, by verifying the signature on the certificate, the terminal could at least obtain a cryptographic assurance that a person with such a name and personal ID code exists.

Private key usage counters [30, Sect. 12.4] show how many private key operations have been performed with a particular private key. The terminal can use this information to find out how active the cardholder is in using the ID card for authentication and digital signatures.

The PIN and PUK code retry counters [30, Sect. 12.2] show the remaining PIN tries, which lets the terminal find out how many times the particular PIN or PUK code has been entered incorrectly. Note that the retry counter is reset to 3 after each successful PIN/PUK verification, and hence these counters are unlikely to have any other value except 3.

Privacy Risk for Residence Permit Card Holders. Residence permit cards are ID cards issued to Estonian residents who are not citizens of the European Union [8]. These cards contain an additional contactless smart card chip that runs the ICAO compliant ePassport application storing digitally signed cardholder data, including biometric data. However, to read that information wirelessly, the terminal has to authenticate to the ePassport chip using the Basic Access Control (BAC) mechanism. To create the BAC key, the reader has to optically read the machine-readable zone (MRZ) and extract the document number, expiration date and date of birth. However, since the fields comprising the BAC key are also stored on the contact chip in the personal data file, a malicious terminal – if equipped with an additional contactless reader – can read the data stored on the ePassport chip without needing to scan the MRZ. The additional personal information that can be obtained this way are the cardholder's facial image and two fingerprints[2].

We note that the reading of the facial image may be useful for the purpose of cardholder verification (see Sect. 6.2), although the wireless reading of 20 KB 480×640 pixel facial image from the residence permit card takes around 15 s.

3.2   Denial-of-Service Attacks

A malicious terminal can execute denial-of-service attacks against the card, leaving the card in an impaired state. The most straightforward attack is to decrease the PIN/PUK retry counters to 0. This will block the cardholder's access to cryptographic operations, forcing the cardholder to visit the ID card customer service point to obtain a new PIN envelope. Such a service does not exist for holders of the e-resident's digital identity card, which means that their only option is to apply for a new ID card. Similarly, a malicious terminal can block the GlobalPlatform [10] applet management key, which will prevent the cardholder from renewing the applet in ID card customer service points or over the Internet.

---

[2] Fingerprints on cards issued after 3 November 2014 are additionally protected using the Extended Access Control (EAC) mechanism, which requires terminal authentication.

One could argue that these logical denial-of-service attacks against the ID card should not be of concern, since a malicious terminal can always cause damage, for example, by supplying excessive voltage to the electronic chip. In practice, however, the attacker may have gained only a logical control over the terminal, leaving application-level attacks against the smart card the only option available.

3.3 Unauthorized Use of Private Keys

A malicious terminal could also try to perform private key operations by guessing the 4-digit PIN1 protecting the authentication key or the 5-digit PIN2 protecting the digital signature key. The probabilities of guessing a random 4-digit or 5-digit PIN in 3 tries are 0.03% and 0.003%, respectively. In practice, however, the probability of successfully guessing the PIN can be several times higher, since some of the cardholders may have updated the random PIN generated by the card issuer with a PIN of their choice. Bonneau and others in [1, Table 3] show that compared to randomly generated PINs, for human-chosen PINs the probability of successfully guessing a 4-digit PIN increases from 0.03% to 5.52% if 3 guesses are allowed.

While this type of an attack has quite a low success probability for a targeted attack, an opportunistic attacker in contact with thousands of cards can succeed in guessing the PIN for some of the cards. Instead of performing 3 guesses per PIN (which will cause the PIN to be blocked), the attacker can perform only one try per PIN. The cardholder is unlikely to notice that after inserting the card in a terminal, the PIN retry counter has decreased. In fact, the malicious terminal can continue the attack once the cardholder returns with the PIN retry counter reset.

Nevertheless, the probability of guessing both PINs for the same cardholder is negligible. Therefore, high-risk transactions should always involve both the authentication as well as the digital signing operation.

4   Design of ID Card Emulator

In this section, we show the design of the ID card emulator that we used in the experiments performed in Sect. 5. The purpose of the ID card emulator is to emulate the chip of a real ID card as closely as possible. Since the private keys from the real chip cannot be extracted, the operations performed with the private keys cannot be emulated. However, as discussed before, for the purpose of card authentication, the emulation of private key operations is not needed. As the experiments will show, in practice the terminals only use the personal data file feature.

To implement ID card functionality in a smart card, a programmable smart card supporting JavaCard technology was chosen. The JavaCard technology allows smart card applications to be written using a subset of the Java programming language. Nowadays, most of the smart cards on the market run applications written using the JavaCard technology. The Estonian ID cards issued starting from 2011 also use the JavaCard technology to implement the required functionality. The source code of the EstEID JavaCard applet, which is installed on the ID cards in the card personalization stage, is the intellectual property of the card personalizer Trüb Baltic AS and is not public [3]. However, a detailed specification for terminal and chip communication is provided in [30] and [31]. Furthermore, there is an open source implementation of the EstEID applet called FakeEstEID [19], on which we based our ID card emulator. The FakeEstEID applet was modified to implement the APDU command logging functionality, add support for the arbitrary applet application identifier (AID), implement the GET RESPONSE emulation for T=0 Case 2 APDUs, and to emulate the EstEID v3.5 personal data file format.

4.1 Card ATR Adjustment

Whenever the power or the reset signal is supplied to a card, the card responds with a sequence of bytes called Answer To Reset (ATR). These bytes identify communication parameters supported by the card and may contain historical bytes that typically hold some kind of a card identifier. Several generations of Estonian ID cards in circulation respond with different ATRs. Each generation of ID card responds with two different ATRs, cold and warm ATR, depending on whether the power or the reset signal has been supplied to the card.

Since the ATR returned by the card can be used by the terminal to verify if the inserted card is the Estonian ID card, the ID card emulator had to be adjusted to respond with the ATR of the real ID card. The historical bytes of ATR can be changed within the applet using the JavaCard API call GPSystem.setATRHistBytes(). However, the ATR prefix, which encodes the electrical communication parameters of the card, cannot be changed. The solution was to find a blank JavaCard whose ATR prefix would match the prefix of the Estonian ID card.

Fortunately, a blank JavaCard "G&D SmartCafe Expert 6.0 80K Dual" [9] sold for under 15 pounds in small quantities by a UK-based seller [28] was found to have the ATR[3] whose ATR prefix 3B F? 18 00 00 80 31 FE 45 matched the ATR prefix of Estonian identity cards issued since 2011. Since this blank JavaCard returns a single ATR value for both the cold and warm ATR, we chose to configure the ID card emulator to respond with the cold ATR[4] of the Estonian identity card version issued from October 2014. The inability to emulate both ATRs at the same time is a deficiency of our ID card emulator. The terminals deployed in practice, however, are unlikely to validate both ATRs.

4.2 APDU Logging Functionality

The purpose of the APDU logging functionality is to study how the terminals deployed in practice interact with the ID cards. The logging functionality of the

---

[3] 3B FE 18 00 00 80 31 FE 45 53 43 45 36 30 2D 43 44 30 38 31 2D 6E 46 A9 (ATR of SmartCafe Expert 6.0).

[4] 3B FA 18 00 00 80 31 FE 45 FE 65 49 44 20 2F 20 50 4B 49 03 (cold ATR of EstEID v3.5 (10.2014)).

emulator applet writes the received APDU commands on the card's EEPROM storage and later releases them when a specific command is received.

Since in the JavaCard each APDU received from the terminal is passed to the selected applet's `process()` method, this method is the central place where all the APDU commands received are logged. The smart card technology allows several applications to reside on one card; however, only one applet on the card can be set as the implicitly selected (default) applet. To communicate with another applet, an explicit `SELECT FILE` APDU has to be sent, specifying the application identifier (AID) for the applet that should be selected. For the ID card emulator, the emulator applet is set as the default applet. This means that all the commands, including applet selection commands containing non-existent AIDs, will be received and logged by the emulator applet. In addition to logging the received APDUs, the emulator applet logs invocations of the applet's `select()` and `deselect()` methods.

The applet's `select()` method is invoked automatically when the first APDU is received from the terminal and before it is passed to the applet's `process()` method. The logging of `select()` invocation allows detecting if the card has been reset in the middle of an APDU trace. Note that if the terminal powers up the card just to read the ATR, this fact will not be logged, because the `select()` method is invoked only when the first APDU is to be processed. Since the smart card does not have a built-in clock, the timing of the APDUs received cannot be logged either.

The Estonian ID card being emulated supports both electrical transport protocols T=0 and T=1 defined by ISO 7816. To find out which communication protocol the terminal prefers, the applet's `select()` method logs the protocol used in the communication. The protocol used is obtained using the `APDU.getProtocol()` JavaCard API call.

The invocation of the applet's `deselect()` method is also logged. This method is invoked whenever the terminal selects the emulator applet or some other applet residing on the card. The possibility that the applet's deselection is caused by the terminal's explicit selection of the emulator applet itself can be ruled out, because the AID of the emulator applet is set to a random value. The only other selectable applet on the card is the Issuer Security Domain (ISD) with the standard AID `A000000003000000`, which is used for GlobalPlatform's [10] applet management purposes. A legitimate terminal should not communicate with the ISD. However, if it does, this fact is logged and detected.

From the 80 KB of the card's total EEPROM size, a 4 KB memory buffer was allocated to store logged APDU commands. The amount allocated is more than enough to store the APDU trace of a usual interaction between terminal and card.

4.3   Visual Imitation of ID Card

Since we wanted to avoid drawing attention to our fake ID card when it was used in supervised terminals, the white blank of the fake ID card had to be disguised to imitate the design of the real ID card. To avoid the potential legal problems associated with imitating a physical identification document, we decided to imitate the visual design of the digital identity card. The digital identity card does not serve the purpose of physical identification and hence has neither a facial image nor any security features on it.

We used a scanner to scan the original digital identity card and printed the scan on a sticker paper, which was then glued onto both sides of the fake ID card. Visually, the results were not bad, but the added layer created issues when the card was inserted into some terminals and the paper got wet and dirty very fast. A much better result was obtained using the Zebra ZXP Series 3 card printer to print the scanned image on both sides of the white plastic.

Chip Transplantation. A perfect visual imitation of the real ID card could be obtained if the chip from the fake ID card was transplanted onto the plastic of the real ID card. This way even a thorough inspection of the card's security features along with the verification of the cardholder's facial image would give no signs that the card's electronic communication was inauthentic. The only way to detect the inauthentic behavior of the card would be to compare the data read from the chip with the data printed on the surface of the card.

Since we did not wish to experiment with a real identity card, we tested the feasibility of chip transplantation using the ID card test cards obtained from SK ID Solutions AS [27]. The test card fully replicates the visual appearance of the identity card, including all the security features on it. The only difference from the real identity card is that the test card has the word "SPECIMEN" placed diagonally on the front of the card and the identity information on the card is that of a fictitious cardholder. We removed the fake ID card chip from the white blank card by heating the back of the card with a lighter for a few seconds. To remove the chip from the test card, we used a utility knife to carefully cut out the chip without damaging the plastic around the chip. The fake ID card chip was then glued to the test card. We found the chip transplantation process (Fig. 1) to be straightforward and reliable, and the end result (Fig. 2) had no visible traces of chip replacement other than the different contact pad layout of the fake chip.

5   Card Authentication in Practice

To study the exact mechanisms used by the terminals to authenticate a card, we used the ID card emulator in the most popular public deployments where the ID card is used for authentication to a machine. The protocol trace logged by the ID card emulator was later retrieved from the card and analyzed. Each terminal was tested using four slightly different fake ID cards.

1. The first card was a perfect ID card emulator described in Sect. 4. The card was used to test if the terminal accepts the fake ID card and to obtain APDU commands received from the terminal.

Fig. 1. Cards with the chips removed   **Fig. 2.** Original vs. transplanted card

2. The second card was the same as the first card, but with the ATR historical bytes set to random values. This card was used to test if the terminal validates the ATR against the list of ATRs for the ID cards in circulation.
3. The third card was the same as the first card, but with a document expiry date that has already passed and an invalid document number. This card was used to test if the terminal performs document validity checks described in Sect. 2.1.
4. The fourth card was used to test if the terminal supports both ISO 7816 electrical transport protocols. If the terminal preferred the T=0 protocol to communicate with the first card, the fourth card was the card with the arbitrary ATR supporting only the T=1 protocol, and vice versa. It is not particular important which protocol the terminal supports, since all ID cards in circulation support both protocols.[5]

The results of the tests are summarized in Table 3. In total 15 terminals were tested from May to July 2017. A more detailed description of the terminals analyzed and the raw APDU traces obtained from the terminals are provided in the extended version of this paper [16].

As expected, our ID card emulator was accepted by all the terminals tested. This shows that the terminals are vulnerable to card impersonation attacks. The results show that all the terminals perform cardholder identification based on data from the personal data file and not from the certificates. Most of the terminals read more records from the personal data file than required for cardholder identification. While we do not know if the personal data read is retained by the systems, this practice of excessive personal data reading is troubling.

As we can see from Table 3, not all of the terminals check the ATR of the ID card. ATR validation makes ID card forgery more challenging (see Sect. 4.1), however, in the past it has resulted in newer-generation ID cards being rejected temporarily [22]. Our tests with the third card show that none of the terminals tested perform the ID card expiration and revocation checks described in

---

[5] The exception is the digital identity cards issued before 2014, which support T=0 only.

Table 3. The results of using the ID card emulator in terminals deployed in practice

Terminal	Records read	ATR check	Protocol
Apotheka (PC reader)	First nine records	No	T=0 pref
Apotheka (prescr. lookup)	Name, ID code, doc. No	No	T=1 pref
Ektaco ARGOS	Doc. No	Yes	T=0 pref
Ingenico iPP320	All records	Yes	T=0 pref
National Library	All records	No	T=1 pref
Pilveprint	Doc. No	No	T=0 only
TUT library entrance	All records	No	T=1 pref
TUT library checkout	ID code	Yes	T=1 pref
VeriFone Vx805/Vx820	Name, ID code, doc. No., expiry date	No	T=0 pref
Worldline YOMANI	ID code, doc. No., expiry date	Yes	T=0 pref

Sect. 2.1. Almost all the terminals support both smart card communication protocols, but T=0 is widely preferred.

Use of ID Cards in Payment Terminals. Using the ID card as a loyalty card is a popular authentication method in Estonia.[6] To avoid the need for a separate smart card reader, the merchants tested in this study communicate with the ID card chip using the point-of-sale (POS) terminal. Support for the ID card has been implemented in the firmware of payment terminals, and the merchant's systems only receive the predefined records of the personal data file. We found three payment terminal models that were used by Estonian merchants to communicate with an ID card. These are: Ingenico iPP320 (Apollo, Apotheka, Grossi Toidukaubad, Olerex), VeriFone Vx805/Vx820 (Lido, Rahva Raamat) and Worldline YOMANI (K-Rauta, Prisma).

Digital Prescription Lookup. In the Apotheka pharmacy chain, the card reader connected to the pharmacist's computer is used to both identify a loyal customer and to automate the lookup of drugs prescribed to the patient in the digital prescription system [4]. We noted that contrary to the legal requirement, in the process of prescription lookup the pharmacist used our ID card emulator without asking for any physical identification document. However, even if a physical identification document was verified, the identity read from the chip would likely not be verified with the information printed on the card, and thus the chip transplantation (Sect. 4.3) could allow impersonating another person.

Using the ID Card as an Entrance Card. The Estonian company Ektaco has developed an ARGOS-series access control system where the ID card can

---

[6] For various reasons, not all merchants in Estonia accept the ID card as a loyalty card [21]. These merchants provide their own loyalty cards, which are usually magnetic stripe cards or contactless chip cards [15].

be used as a key. The APDU traces were collected from unsupervised Ektaco terminals installed on the side gate of TTK University of Applied Sciences and the front door of Tudengimaja. The National Library of Estonia and the library of Tallinn University of Technology (TUT) allow entering the library using an ID card and a standard smart card reader connected to a computer.

6   Discussion: Improvements

As demonstrated in the previous sections, the chip authentication mechanism currently used can be abused by a malicious cardholder to execute card impersonation attacks. In this section, we discuss the ID card's possible technological improvements that could improve its security and usability, therefore enabling wider use of the ID card as a physical authentication token.

6.1   Cloning Prevention

To prevent a card impersonation attack, the card authentication process should verify that the unclonable private key objects are on the chip. To achieve this, the terminal should require the card to sign a random challenge and verify it using the certificate. To prevent the abuse of the authentication or digital signature keys, the ID card should contain a separate card authentication key and the corresponding certificate. The key should be used only for card authentication purposes and should be operable without the requirement to enter a PIN. The data in the personal data file or its hash should be embedded in the card authentication certificate to prove its integrity. The validity information of the card authentication certificate should correspond to the validity of the document, thereby enabling reliable document validity checking (Sect. 2.1). This chip cloning prevention feature is similar to the FIPS 201-2 (PIV card) "card authentication key (CAK)" [18], and Active Authentication in the ICAO ePassport [11]. The card authentication feature can be remotely deployed as an additional JavaCard applet on the ID cards that have already been issued. The use of a separate applet provides flexibility, since then the applet will not have to be Common Criteria certified, which is a requirement for applets used to create eIDAS-compatible QESs. To make use of the cryptographic feature, the terminal owners will have to invest in adapting the terminal software. The software, however, will have to be updated to some extent anyway, since the new-generation ID cards (to be issued starting from 2019) will have a new ATR, and due to the eIDAS certification requirements, the EstEID applet will have to be replaced with the internationally developed IAS-ECC applet [20, slide 14]. It is not yet known if it will support the Estonian personal data file feature in its current form.

Performance. To evaluate the performance of the suggested card authentication feature, we performed measurements using two ID card generations in circulation: the chip of ID cards issued since 2011 and the newest-generation chip in ID cards issued since October 2014. The results are shown in Table 4. To summarize, if the ECDSA is used, the cryptographic card authentication process takes around 1.5 s on the ID cards issued after 2011, but only 0.6 s on the cards issued after October 2014. By utilizing the certificate caching mechanisms, this time could be decreased to be under 200 ms on the latest generation of ID cards. This would considerably improve the user experience when using the ID card as an entrance card.

Table 4. The performance of two ID card chip generations

Measurement	2011 (ms)	2014 (ms)
Reading the personal ID code	150	150
Reading the entire personal data file	625	450
Reading the 1.5 KB certificate	440	380
RSA-2048 signing	1500	385
ECDSA with NIST P-256	1000	160

Relay Attacks. While the card authentication process would prevent the use of ID card forgeries, a more sophisticated relay attack where the fraudulent card relays card authentication commands to the legitimate card is still possible. Relay attacks are not easy to prevent. The EMV contactless payment card specification tries to prevent them using distance bounding protocols, which are far from easy to implement in the actual physical hardware [17].

6.2   Cardholder Verification

Verification Using a PIN. Basic protection against the unauthorized use of an ID card by a non-owner of the card can be implemented by requiring cardholder verification using an additional PIN3. In practice, however, the added security value may be too insignificant to compensate for the degraded user experience caused by memorizing and entering yet another PIN. Entering a PIN into a terminal and in an environment controlled by some third party would also greatly increase the risk of the PIN being compromised. In the EMV payment card rollout in the U.S., banks have chosen to abandon PIN verification, because most fraud cases involve counterfeit cards, while fraud related to lost and stolen cards is minimal [13]. The use of lost or stolen cards can instead be prevented by card revocation mechanisms.

Cardholder verification using a PIN does not prevent fraud where the owner of the card has authorized the use of their ID card by some other person. In some fraud schemes (e.g., in some customer loyalty programs) the owner may have a direct or indirect interest in their card being used by someone else.

Verification Using Biometrics. To completely eliminate the use of an ID card by someone who is not the cardholder, the identity of the card user has to be verified. Due to chip transplantation attacks (Sect. 4.3), not only the facial image printed on the identity document has to be verified, but also the personal ID code retrieved electronically should be compared with the personal ID code printed on the document. To automate the verification task as much as possible, the facial image of the cardholder should be stored on the card, indirectly signed by including its hash in the card authentication certificate. The person performing cardholder verification would then only need to compare the digital image retrieved from the chip with the facial features of the card user. This task could be further delegated to a face recognition system.

A similar feature is already provided by the ePassport chip on the residence permit cards (see Sect. 3.1). Therefore, as an alternative to cryptographic improvements for the ID card chip, all types of ID cards could be equipped with the ICAO ePassport chip. The advantage of this solution would be that an internationally standardized method would then be used for cardholder authentication, and in the case of the Estonian ID card, the BAC key would be read from the contact chip without the need for optical character recognition. The disadvantage would be the need for two readers, which complicates the deployment and slows down the speed of card authentication transactions.

6.3 Contactless Interface

With the exception of the residence permit card, which contains a separate contactless ICAO ePassport chip, the current ID cards in circulation do not have a contactless interface. The potential benefits of adding a contactless interface to the ID card have been discussed in [3,12]. A non-public pilot for using NFC-enabled digital identity cards with mobile phones has been described in [14,23].

While the traditional electronic use of ID cards does not benefit much from a contactless interface (except perhaps to interface with mobile devices), the convenience provided by a contactless interface would be especially useful for using the ID card as an entrance card. This would allow making the terminals more vandal-proof and enhancing the convenience of the process.

However, the introduced security risk is that the cardholder's identifiable information could be retrieved covertly from a distance, and covert access to a PIN-less card authentication key would make relay attacks easier to execute. The security risk could be largely solved by introducing an NFC antenna-enabling button to the card [25]. Cards with such buttons, however, are currently not available on the market. The U.S. Department of Defense, for example, has decided to enable contactless interface for CAC cards, but has issued radio frequency shielding sleeves to cardholders [26].

7 Conclusion

We have shown the design of an ID card emulator able to impersonate a real Estonian ID card to the terminals deployed in practice. Building such an ID card emulator today is both feasible and affordable, and therefore the current ID card chip authentication mechanism, which does not involve any cryptographic assurance, should not be used for high-risk transactions. By demonstrating the reliability of the ID card chip transplantation process, we have shown that the authenticity of the data read from the chip should not be trusted even if the chip is part of a visually authentic ID card.

The study of terminals deployed in practice shows that the terminals do not perform document expiration and revocation checks, and most of the terminals read more personal data from the ID card than required for cardholder identification.

We hope that this paper will raise awareness of the risks related to the current ID card chip authentication mechanism and will facilitate the development of a secure and universal authentication solution. Such a solution is highly needed in the current situation in Estonia where a variety of proprietary solutions vulnerable to cloning and replay attacks are ubiquitous [2,15].

Acknowledgements. We would like to thank Martin Paljak for his feedback and the technical support he provided for this study, and all the people who gave their feedback on this paper. This work was supported by the European Regional Development Fund through the Estonian Centre of Excellence in ICT Research (EXCITE) and the Estonian Doctoral School in Information and Communication Technologies.

References

1. Bonneau, J., Preibusch, S., Anderson, R.: A birthday present every eleven wallets? The security of customer-chosen banking pins. In: Keromytis, A.D. (ed.) FC 2012. LNCS, vol. 7397, pp. 25–40. Springer, Heidelberg (2012). https://doi.org/10.1007/978-3-642-32946-3_3

2. Cybernetica AS: Cryptographic algorithms lifecycle report 2016. In: Cryptographic protocols over radio connection. 22 June 2016. https://www.ria.ee/public/RIA/Cryptographic_Algorithms_Lifecycle_Report_2016.pdf

3. e-Governance Academy: Study on the functionality of documents in ID-1 format (in Estonian), December 2013. https://www.siseministeerium.ee/sites/default/files/dokumendid/Uuringud/Isikut_toendavad_dokumendid/2013_id-1_formaadis_dokumentide_funktsionaalsuse_uuring.pdf

4. Estonian Health Insurance Fund: Digital Prescription, July 2017. https://www.haigekassa.ee/en/digital-prescription

5. Estonian Information System Authority: Electronic Identity Application Guide: ID card as an entrance card, May 2014. https://eid.eesti.ee/index.php/ID_card_as_an_entrance_card

6. Estonian Information System Authority: Electronic Identity Application Guide: Using ID-card as a loyalty card, May 2014. https://eid.eesti.ee/index.php/Using_ID-card_as_a_loyalty_card

7. Estonian Police and Border Guard Board: Online identity document validity check, May 2017. https://www.politsei.ee/en/teenused/inquiries/

8. Estonian Police and Border Guard Board: Residence card, May 2017. https://www.politsei.ee/en/nouanded/residence-card.dot

9. Giesecke & Devrient: Sm@rtCafé Expert operating systems: Sm@rtCafé Expert 6.0, February 2013. https://www.gd.gd/gd_media/media/en/documents/brochures/mobile_security_2/nb/SmartCafe-Expert.pdf

10. GlobalPlatform Inc.: GlobalPlatform Card Specification, Version 2.1.1, March 2013. http://www.win.tue.nl/pinpasjc/docs/Card%20Spec%20v2.1.1%20v0303.pdf

11. International Civil Aviation Organization: DOC 9303. Machine Readable Travel Documents. Part 11: Security Mechanisms for MRTDs (2015). https://www.icao.int/publications/Documents/9303_p11_cons_en.pdf

12. Joandi, E., Kuusik, A., Tammet, T.: Analysis of potential RFID usage in the context of extending Estonian ID-card (in Estonian), January 2008. https://www.mkm.ee/sites/default/files/rfid_id_analyys_-_koopia.doc

13. Krebs, B.: Chip & PIN vs. Chip & Signature, October 2014. http://krebsonsecurity.com/2014/10/chip-pin-vs-chip-signature/

14. Lehmann, A.: New Generation of eID Smartcard, 06 November 2014. https://sk.ee/upload/files/AK2014_New%20Generation%20of%20eID%20Smartcard_Andreas%20Lehmann.pdf

15. Morgan, D.: Security of Loyalty Cards Used in Estonia. MSc thesis, Tallinn University of Technology (2017). http://kodu.ut.ee/~arnis/loyalty_thesis.pdf

16. Morgan, D., Parsovs, A.: Using the Estonian Electronic Identity Card for Authentication to a Machine (Extended Version). Cryptology ePrint Archive, Report 2017/880 (2017). http://eprint.iacr.org/2017/880

17. Murdoch, S.J.: Do you know what you're paying for? How contactless cards are still vulnerable to relay attack, August 2016. https://www.benthamsgaze.org/2016/08/02/do-you-know-what-youre-paying-for-how-contactless-cards-are-still-vulnerable-to-relay-attack/

18. NIST: FIPS PUB 201–2: Personal Identity Verification (PIV) of Federal Employees and Contractors, August 2013. http://nvlpubs.nist.gov/nistpubs/FIPS/NIST.FIPS.201-2.pdf

19. Paljak, M.: FakeEstEID JavaCard applet, 16 January 2015. https://github.com/martinpaljak/esteid-applets/blob/master/docs/FakeEstEID.md

20. Paljak, M.: Off-line ID card (in Estonian), 18 October 2016. http://kliendikaart.publicon.ee/userfiles/RIA/idkaart/Martin_Paljak.pdf

21. Postimees: No plans to connect Kaubamaja Partnercard with ID-card (in Estonian), 5 August 2011. http://www.postimees.ee/521494/partnerkaarti-id-kaardiga-uhendada-ei-kavatse

22. Postimees: The new ID-cards will be refused (in Estonian), 23 January 2015. http://tarbija24.postimees.ee/3067299/uued-id-kaardid-voivad-torkuda

23. Postimees: Contactless Estonian ID-card has been built (in Estonian), 5 March 2016. http://tehnika.postimees.ee/3607697/video-valminud-on-kontaktivaba-eesti-id-kaart

24. Riigi Teataja: Identity Documents Act (2000). https://www.riigiteataja.ee/en/eli/504112013003/consolide/current

25. Roland, M., Hlzl, M.: Evaluation of Contactless Smartcard Antennas, June 2015. https://arxiv.org/abs/1507.06427

26. SecureIDNews: Defense Department order RF shields from National Laminating, November 2010. https://www.secureidnews.com/news-item/defense-department-order-rf-shields-from-national-laminating/

27. SK ID Solutions AS: Cards for testing 01 July 2017. https://sk.ee/en/services/testcard/

28. Smartcard Focus: Giesecke & Devrient: SmartCafe Expert 6.0 80K Dual, 11 April 2017. https://www.smartcardfocus.com/shop/ilp/id~684/smartcafe-expert-6-0-80k-dual-/p/index.shtml
29. The European Parliament, the Council of the European Union: Regulation 910/2014 on electronic identification and trust services for electronic transactions in the internal market and repealing Directive 1999/93/EC (2014)
30. Trüb Baltic AS: EstEID v3.4 card specification, 11 June 2012. http://www.id.ee/public/TB-SPEC-EstEID-Chip-App-v3.4.pdf
31. Trüb Baltic AS: EstEID v3.5 card specification, 14 March 2017. http://www.id.ee/public/TB-SPEC-EstEID-Chip-App-v3.5-20170314.pdf

Data Aware Defense (DaD): Towards a Generic and Practical Ransomware Countermeasure

Aurélien Palisse[1]([✉]), Antoine Durand[2], Hélène Le Bouder[3],
Colas Le Guernic[1,4], and Jean-Louis Lanet[1]

[1] INRIA, Campus de Beaulieu, 263 Avenue Général Leclerc, Rennes, France
aurelien.palisse@inria.fr
[2] ENSEIRB - MATMECA, 1 Avenue du Dr Albert Schweitzer, Talence, France
[3] IMT Atlantique, 2 Rue de la Châtaigneraie, Cesson Sévigné, France
[4] DGA - Maîtrise de l'Information, Route de Laillé, Bruz, France

Abstract. We present the *Malware - O - Matic* analysis platform and
the *Data Aware Defense* ransomware countermeasure based on real time
data gathering with as little impact as possible on system performance.
Our solution monitors (and blocks if necessary) file system activity of
all userland threads with new indicators of compromise. We successfully
detect 99.37% of our 798 active ransomware samples with at most 70 MB
lost per sample's thread in 90% of cases, or less than 7 MB in 70% of
cases. By a careful analysis of the few false negatives we show that some
ransomware authors are specifically trying to hide ongoing encryption.
We used free (as in free beer) de facto industry standard benchmarks to
evaluate the impact of our solution and enable fair comparisons. In all
but the most demanding tests the impact is marginal.

1 Introduction

Ransomware is a type of malware that prevents legitimate users from accessing
their machine or files and demands a payment for restoring the functionalities of
the infected computer. There are two classes of ransomware: the "simple lockers",
which block the usage of the computer, and "cryptors", that encrypt files on the
computer. In the case of encryption-based ransomware, the user data can only
be restored with the secret key(s) used during the attack.[1]

This class of malware has existed for a few decades [32], but the number
of attacks has increased drastically over the past couple of years [25]. The latest notable examples are the WannaCry and the Nopetya attacks. However,
recent findings suggest that Nopetya is a wiper with ransomware-like appearance.
Microsoft is concerned by the ransomware threat and plans to add a controlled
folder access feature in the next operating system update [3].

Recent advances have contributed in the current proliferation of ransomware.
Command and control (C&C) servers can be protected through the use of domain

---

[1] Usually the encryption keys are themselves encrypted with an asymmetric cryptosystem, the ransom must be paid in order to get the corresponding private key.

© Springer International Publishing AG 2017
H. Lipmaa et al. (Eds.): NordSec 2017, LNCS 10674, pp. 192–208, 2017.
https://doi.org/10.1007/978-3-319-70290-2_12

generation algorithms or The Onion Router (TOR) network. Popular applications have been diverted from their legitimate usage: Imgur [27], Twitter API [1], and Telegram Bot API [12] have been used to implement a C&C. Bitcoin and other cryptocurrencies (e.g., Zcash, Ethereum) facilitate ransom processing and handling. Similarly to most modern malware, current ransomware, hinder detection and analysis, through packing, virtualization, Windows Management Instrumentation (WMI) queries or obfuscated API calls. On March 28, Trend Micro discovered a new technique used by the Cerber family to evade static machine learning solutions [28]. A generic efficient defense mechanism seems challenging, but encryption-based ransomware have a clear common semantic that can be taken advantage of: they encrypt user data.

Similarly to recent approaches [5,13,14,23], reporting satisfying detection rates, we propose to monitor file activity. Since it has already been proven a valid approach in terms of detection, our main goal in this paper is to show that a good detection rate can be achieved with little to no impact on system performances. To this end we limit our monitoring to a minimum. In order to reduce the impact on detection with a low rate of false positive, we use the chi-square goodness-of-fit test instead of Shannon entropy (i.e., sensitive to compressed chunks of data [17]). We also achieve system completeness and fine granularity by monitoring the whole file system for all userland threads. In order to evaluate our prototype implementation, *Data Aware Defense* (DaD), under realistic conditions, we developed a bare-metal analysis platform, *Malware - O - Matic* (MoM), and ran it on a large and heterogeneous (compared to the litterature) live ransomware collection. We used de facto industry standard benchmarks to get a pertinent and reproducible assessment of the performance penalties.

Related work are presented in Sect. 2. Section 3 compares different statistical tests and their effectiveness to detect encryption. DaD, our ransomware countermeasure is introduced in Sect. 4, a significant effort is made to tackle the performance bottleneck. We evaluate its impact on the protected system performance in Sect. 5 and its effectiveness in Sect. 6 together with a description of our bare-metal automated malware analysis platform, MoM, and a discussion on our findings. Section 7 concludes the paper.

2 Related Work

Detecting malware is of prime importance. The main deployed approaches are either static pattern-based signatures, or behavioral signatures dynamically checked in a sandboxed environment. Unfortunately they rarely cover new malware or even new variants of known malware. The challenge is to design fast detection schemes that cover many samples with no false positives.

The subfield of ransomware countermeasures is relatively young. On top of classical malware approaches, one can rely on the specific common behavior of ransomware: they encrypt the victim's files. Dynamic solutions found in the literature are divided in two parts as suggested by Kharraz et al. [14]: cryptographic primitives hooks (i.e., user space) and low level disk activity monitoring (i.e., kernel space). One targets the cryptographic primitives as an essential gateway for ransomware whereas the second focuses on the system consequences.

In 1996 Young et al. [33] implemented a proof of concept "cryptoviral extortion" based on the Microsoft's Crytographic API (CryptoAPI). Nowadays a significant number of ransomware do indeed use the CryptoAPI to perform file encryption. This API enables the use of specific cryptographic providers, Palisse et al. [20] implemented their own and forced its use to get a trace of all cryptographic operations. PAYBREAK [15] live solution makes use of dynamic and static cryptographic hooks. A key escrow system is implemented and allows complete file recovery. However only symmetric-key encryption is considered and some obfuscation techniques will defeat the hooks, according to the authors. Moreover the static hooks need a prior knowledge of the libraries. Both papers suffer from two critical limitations: any ransomware that will embed its own cryptographic primitives bypass the solution (e.g., AES-NI [22]) and nothing prevents the ransomware to detect the hooks or the redirection.

Other approaches, like ours, focus on disk activity thanks to a file system driver. UNVEIL [13] detects ransomware by computing the increase, in term of Shannon entropy, between data read and written to disk by the same process and is also able to detect desktop locker, a benign type of ransomware. CRYPTO-DROP [23] relies on entropy too, but also takes into account file type changes and a file similarity score. SHIELDFS [5] applies machine learning to the disk activity. Features are selected from millions of disk I/O requests gathered from normal usage and ransomware attacks. The solution is composed of three drivers. One in charge of file recovery: for each write and renaming operations the corresponding file is backed up. The second detects cryptographic materials embedded in processes. Finally the third performs detection thanks to random forests mixed with incremental models which takes into account the short and long life of a process.

The three report good detection rate, over 96%, and almost no false positive on their respective data sets. Concerning the impact on system performance UNVEIL and CRYPTODROP give little information. The former is presented as an analysis tool and not a live solution, the authors of the later "believe that with future optimizations, CRYPTODROP can be run on a live system with a small overhead." and report a 9ms overhead per write operation without specifying their evaluation procedure. There are more information on SHIELDFS performance overhead. The time taken to open then read, or open the write, files of increasing size (from 1 KB to 128 MB) is measured. Using hard disk drive they get a 180% to 380% overhead if files need to be backed up or 30% to 90% if not. A "typical" overhead is also reported, unfortunately the evaluation procedure is unclear and hardly reproducible: they extrapolated a typical overhead from IRP logs taken from five users, resulting in an average estimated overhead of 26%.

Dynamic monitoring of file system activity seems to be the most promising approach to defend against ransomware. Indeed, the Intel AES-NI instructions defeat the approaches centered on the cryptographic libraries proposed in [15, 20].

The remaining challenge lies in an efficient implementation on a live system. One approach would be to limit the level of monitoring and focus on a single efficient distinguisher, at least at first, only suspicious threads need to be closely monitored.

3 Statistical Tests for Ransomware Attacks Detection

Ransomware involve a large number of ciphertext going through the file system; to detect such behavior, statistical tests can be used. The main idea is that ciphertext content distribution should be uniform. A one sample goodness of fit (GOF) test measures how close an information source is to a theoretical probability distribution function, also known as the "model". In practice, one data set F is compared to a known distribution function G and disproves or not the null hypothesis $H_0 : \forall x, F(x) = G(x)$.

We do not prove that both data sets come from a single distribution function but rather that there is no significant difference between them. The objective is to figure out which indicators of compromise is the most relevant to detect ransomware attacks in real time.

Shannon Entropy. It is a measure of the uncertainty of a random variable. Lots of disorder raise high entropy and structured data low entropy. The entropy of X a discrete random variable from the alphabet $\Omega = \{x_1, x_2, \ldots, x_n\}$ with the probability distribution function $p(x_i)$ at x_i is: $H(X) = -\sum_{i=1}^{n} p(x_i) \log p(x_i)$.

Chi-Square. The chi-square goodness-of-fit test (χ^2) is a test of distributional accuracy, it measures how closely a set of numbers follows a particular distribution. It is a non-parametric statistical test, meaning that no assumption is done on the samples distribution. The observed sequence of data is considered as discrete and arranged in a frequency histogram $[0; 255]$ with the degree of freedom v equals to 255 (i.e., number of possible outcome minus one). Suppose that N_i is the number of hit observed for the bin i, and n_i is the expected number according to a known distribution function. The formula for calculating the one-sided χ^2 test is:

$$\chi^2 = \sum_i \frac{(N_i - n_i)^2}{n_i} \tag{1}$$

A large value indicates that the null hypothesis is not likely verified, the N_i can not be drawn from the n_i. The significance level of the test denoted α_{TW}, is the probability of rejecting the null hypothesis when it is true. Traditionally, experimenters have used the 0.05 level (e.g., biology), thus we choose the same. The observed test statistic is compared to a boundary value, called the critical value, uniquely determined from the degree of freedom (or equivalently the size of the alphabet) and the desired significance level. If the χ^2 result is more extreme than the critical value [10], the null hypothesis is rejected.

Discussion. The robustness of the tests need to be checked against real world conditions (i.e., small and large samples). Previous papers [5,13,23], use the plug-in method (i.e., discrete symbols in histogram bins) to estimate the Shannon entropy. Nevertheless a study on TorrentLocker [17] shows that the Shannon entropy is not a good distinguisher especially with respect to JPEG compression[2]. Achieving encryption detection on compressed files that already have high entropy is a non-trivial task. The χ^2 test on the contrary can distinguish randomness (or encryption) from some compression schemes and is thus a more relevant statistic as shown in appendix Tables 4 and 5. For that reason we use it to detect suspicious behaviors in the next sections. No extensive study of a ransomware solution embedding the χ^2 has been presented earlier, but this statistic is already used in numerous applications [6,31].

A practical issue remains, like all statistical tests, the χ^2 test is not accurate on small samples. A good practice is to have at least five elements in each bins of the histogram, but we can reasonably think that this will not happen plenty of time. In this case, the test statistic will only reflect the small magnitude of the expected frequencies. To fix this problem we made the choice that the solution favor false positives over false negatives, by also computing the χ^2 test for small data. Moreover, the number of bytes involved in the computation is limited to 10,000. Indeed, if we do not set a limit, a zip bomb [11] can be used to crash or slow down our solution.

4   Towards a Generic and Practical Ransomware Countermeasure

In this section the architecture of *Data Aware Defense* is detailed, a file system driver for Microsoft Windows. An important part of the contribution is the usability of the solution, furthermore it can be used against zero-day ransomware.

4.1   File System Activity Monitoring

Windows, as most modern operating systems, splits its memory in several regions with different privilege requirements. The kernel mode (ring 0) has a high privilege level and is responsible, among other things, for managing disk operations.

Standard applications run in userland (ring 3), they are much less privileged and cannot perform disk operations directly. Instead they call the corresponding service from the Windows kernel, let the kernel manage the privileged operations, before safely returning in the userland application code. This separation ensures that no userland code directly manage critical operations in the system. To complete this security feature the 64-bit versions of the OS requires all kernel mode code to be signed by Microsoft to be accepted.

Up to now, to the best of our knowledge ransomware live in userland. That is why a countermeasure in kernel space can not be tampered with by malicious

---

[2] They use the Kullback-Liebler divergence instead but do not introduce an implementation.

code and is fully transparent. Users interactions with files are ultimately mapped to operations in the kernel. On top of this stack stand the I/O manager and at the bottom the file system driver (ntfs.sys). Microsoft offers a file system drivers framework [18] that allows third party developers to add functionalities between the two previous layers. Such component is called a file system minifilter driver and it is managed by the filter manager. The position of each minifilter driver in the I/O stack is defined by its "altitude". A minifilter driver that performs full disk encryption is below an anti-virus filter and thus avoid false positive detection of ransomware-like behavior. A minifilter driver can inspect all the operations that target the disks, regardless of whether the requested operation is an I/O request packet (IRP) or a fast I/O. In this context we are able to monitor write, read operations and so on. However a clever usage of such functionalities has to be done, otherwise a significant performance penalty occurs [23] and the solution can not be deployed in real world. Our file system minifilter driver has been extensively tested on Windows 7 and 10, for the 32 and 64-bit versions and follows the Windows Driver Model (WDM).

4.2 Implementation Design

In order to catch malicious behaviors efficiently we restrict our monitoring to write and information operations. We want to demonstrate that detecting ransomware behaviors with only two callbacks on the I/O requests is possible, while previous solutions [5,13,23] have at least twice as many callbacks. The so-called information operations allow to change various information about a file object, create a hard link, change the file position or the file name. We block all the information operations once a thread is marked as malicious to prevent aggressive files renaming. But without malicious activity, the information operations are always accepted and are not used to construct our compromise indicators. For each intercepted write operation the time spent in the callback function is minimized by collecting only the essential information. We are only interested in the content of the buffer that is passed through the file system stack, its size, offset, the corresponding absolute file name, process name, process id and thread id. This information is then copied to nonpaged memory and passed over to a new dedicated thread that is not part of the file system stack. Only the indispensable data is copied, not the thread context coming from the operation. As a consequence, write operations are immediately authorized to go through. Such features allow us to monitor all file system trees without excluding some assumed trusted threads. So far, we are the first to inspect the I/O operations with a thread granularity. It is a significant improvement in case of malicious code injection into a benign process. The Cerber ransomware already used a particular code injection technique, named process hollowing [28]. All costly computations are deferred to threads running at the lowest kernel priority level. This model solves the time restriction for the statistical computations and the synchronization deadlocks on shared resources.

Once the compromise indicator (e.g., χ^2) is obtained on the corresponding data we update an internal structure related to this thread behavior in nonpaged memory. This structure is stored for each thread of each process. To reduce the memory footprint, our own garbage collector has been implemented. All tracked information in memory can be exported to disk as a JSON file.

4.3   A Single Indicator of Compromise

To build an efficient filter, a sliding median on the last fifty write operations is computed. The goal of this basic statistic is to capture, for each thread, the ongoing file system behavior: it gives us a measure of central tendency. No assumption is made on the I/O patterns observed. This elementary statistic is low cost and does not involve complex calculations. Furthermore by monitoring the whole file system we have more chances to only lose files that did not belong to the user's important paths (e.g., Windows Defender, Python libraries, $Recycle.Bin).

At runtime if the χ^2 median go beyond the threshold, we suspend the corresponding process and collect information for postmortem analysis. We take advantage of the process suspension (i.e., malicious code will not notice) to dump from RAM the Portable Executable (PE) file and the committed pages of memory that belong to this process threads. All the process threads are stopped to ensure consistency of the memory dump, especially in case of self-modifying code. Finally, the thread that triggered the dump is tainted as malicious and all subsequent write and renaming operations are blocked for this particular thread.

Section 3 presents the χ^2 test. The significance level α_{TW} is set to be small (i.e., 0.05). It corresponds to the probability of rejecting the null hypothesis when it is true (i.e., type I error). This parameter also called the "testwise" alpha is relevant for one given hypothesis test. The sliding median is based on fifty consecutive χ^2 tests. Thus, the "experimentwise" alpha is the probability of having one or more errors of type I within the hypothesis tests. The experimentwise error rate can be calculated using the Bonferroni threshold [2] as follows:

$$\alpha_{\mathrm{EW}} = 1 - (1 - \alpha_{\mathrm{TW}})^K \tag{2}$$

K is the number of uncorrelated hypotheses being tested at the α_{TW} level. However, the experimentwise error rate is the same than the testwise error rate when only one hypothesis is tested for a given dataset. Moreover, it is likely that for each thread the write operations are correlated to each others. Therefore, the experimentwise error rate is equal to 0.05.

5   Experiments: Performance Evaluation

An important contribution in this paper is to address the performance penalty. This section shows that our solution is the first live ransomware coutermeasure based on a file system driver able to tackle the performance bottleneck. The workload model is office work. In this configuration the solution is almost imperceptible. In the following parts, the global impact of the solution on the system is investigated with de facto industry standards software[3]. No formal studies of the in-memory footprint have been conducted, it can be considered negligible on consumer computers (i.e., less than 50 Mb). All test were run under Windows 7 with 4 Gb of RAM and a SSD[4]. In order to focus on performance and not interfere with the benchmarks, we deactivated the blocking mechanism of our solution.

5.1 Disk Performance

To precisely measure the driver impact on disk, we used the Windows Performance Toolkit (WPT) [19]. It produces in-depth performance profiles of Windows operating systems. We set the minifilter I/O activity scenario for two hours and a half and obtained 11348 writes that give us $\approx 133\,ms$ spent in total in the file system driver. Thus, an average of $11.7\,\mu s$ per operation. During the profiling, normal work activities were simulated, with office suite use, downloads and compression. CRYPTOLOCK [23] produces an overhead of 9 ms per write operation, even without knowing their testing procedure, it is probably safe to say that performance has been significantly enhanced by a factor of a few hundreds.

Additionally a second test based on a free software, CrystalDiskMark [7], has been performed. Two types of write operations are considered: "Sequential" correspond to large contiguous blocks of data and "Random" focus on small (4K) blocks to random locations. As shown in Table 1, with the Random 4K tests our solution lead to a significant loss of performance but the resulting bandwidth of 9.5 MB/s is still reasonable. On an hard disk drive (HDD) the seek time and the rotational latency create a bottleneck. Consequently, the Random 4K tests will be bounded by the HDD mechanical limitations (i.e., 2.5 MB/s for the fastest HDD[5]), not the file system driver. On the other hand no performance loss is observed when dealing with big chunks of data. Pushing the limit of the solution on disk access do not create undesirable effects (e.g., freeze).

Table 1. CrystalDiskMark tests (in MB/s) configured with 5 tests passes, file size of 2 GiB, random data generation and 3 min interval time.

Write test	Solution off	Solution on	Impact
Sequential	136.3	134.2	−1.5%
Sequential $Q32T1$	135.7	135.8	+0.07%
Random $4K$	55.28	9.581	−82.66%
Random $4KQ32T1$	122.0	65.48	−46.32%

---

[3] We restricted ourselves to free (as in free beer) softwares used to assess performance of personal computers to ensure pertinence and affordable reproducibility.
[4] Windows 7 SP1 6.1.7601, Intel Xeon W3550, NVIDIA Quadro FX 1800, 4 Gb DDR3, Intel SSD 120 Go SATA III.
[5] http://hdd.userbenchmark.com/WD-Black-6TB-2015/Rating/3519.

5.2   CPU Performance

Due to the high number of threads that complete asynchronous jobs, the system load needs to be considered. For this we used Geekbench 4 [9] that contains an all in one test with different workload models and PCMark 8 [21] that is recognized as an industry standard benchmark. For more details on the underlying tests performed by the software please refer to their technical reports. We measure initially (without our protection) a Geekbench overall score of 6625 (multi-core) and 5841 after enabling the protection, for a performance loss of 11.83%. Contrary to the previous test, with PCMark the system impact is less palpable. We executed two built-in scenarios, "work" that measures basic work tasks on office machine and "home" that focus more on media capabilities (e.g., gaming, photo, chat). As shown in Table 2, only a very small difference occurs (less than 1%).

Table 2. PCMark 8 benchmark score.

Test	Solution off	Solution on	Performance loss
Work conventional 2.0	2859	2845	−0.49%
Home conventional 3.0	2728	2705	−0.84%

5.3   Discussion

Precise comparison with other contributions [5,23] is difficult: we can not reproduce their evaluation procedure (when they have one) and we were not able to apply our own[6]. Still, with a significant impact only on the most demanding test that is only exhibited when using solid state drives, we can safely assume a significant improvement over previous work and report that DaD is practical.

6   Experiments: Ransomware Detection

In this section, the experiments demonstrate that *Data Aware Defense* (DaD) detects and blocks in real time, most of the ransomware in the collection (i.e., 99.37%) with a low number of bytes lost. Then we discuss about the ransomware-like behaviors that lead to false positives, and finally review two false negatives that bypass the countermeasure using mimicry attacks.

Before delving into the experimental evaluation of our approach and an analysis of ransomware behavior in the next section, let us present our malware analysis platform that was used to conduct those experiments.

---

[6] We solicited the authors and got a negative answer from [5], and no answer from [23] as of submission.

6.1   Malware - O - Matic

We designed and built *Malware - O - Matic* (MoM), an automated analysis platform that does not use a virtual machine, while keeping all the main features of a regular analysis framework. Such fully bare-metal platform is built on top of two open source software, Clonezilla [4] and Viper [29], which makes it reproducible. The platform is made of a single master server and several slaves, each one running the analysis loop in parallel. The whole system is on a dedicated network under the supervision of the network autonomous system (AS) and directly connected to the Internet, to emulate a typical home network. The loop itself consists in a few simple steps: setup of the monitoring environment, malware execution, results gathering and storage, cleanup. In the first step the slave download a script from the master, that will act as instructions about how to conduct the next analysis. Once the procedure is completed, the slave sets its next environment and reboot for cleanup. The cleanup process simply consists of flashing a clean disk image onto the slave's drive. So far, MoM is able to analyze up to 360 malware per-day with only 1 server and 5 slaves. The end goal of this platform is to run uninterruptedly and thus automate the analysis of samples.

6.2   Experimental Setup

MoM is used in two distinct modes for the experiments: "passive" or "active". The first one, downloads a sample, executes it and according to a cryptographic hash already determined on the user files, labels the sample as active if the hash changes or discards it. The second mode, evaluates DaD ransomware countermeasure (i.e., file system driver) with the samples marked as actives. With such scenario, once the analysis is complete a set of information (e.g., PE file) is sent to a remote server. To avoid evasion during the analysis, a corpus of files that looks like a plausible user environment is built, thanks to the Digital Corpora corpus [8] and manual additions. In the same way, a set of user interactions is emulated (e.g., mouse, keyboard). Each run is fifteen minutes long. A Windows 7 SP1 32-bit snapshot is chosen as the operating system to be infected, the user is logged in as administrator with the User Account Control (UAC) disabled.

The experiments are based on a long-term collection gathered from August 2016 to March 2017. A VirusShare archive dedicated to ransomware was used in combination with daily crawling on online repositories [16,30]. Such mixing allows us to have an heterogeneous ransomware collection of 798 active ransomware (i.e., they encrypt the user's files), decomposed in more than twenty families, with numerous singletons. The previous studies due to their virtualized analysis environment were unable to run the Cerber samples[7]. Our dataset is not limited by the anti-virtualization techniques. Samples labeling is achieved through the *Avclass* tool [24]. Detailed information about the collection can be found in appendix, Table 3. For each sample, a manual analysis has been performed in accordance with its JSON log file to highlight irrelevant samples, but also false positives and negatives.

---

[7] PAYBREAK did, might be samples mislabeling.

6.3   Detection Results

DaD is only interested in the write operations on disk, with a thread granularity and irrespectively of any signature. Such feature makes the solution agnostic which is necessary to tackle zero-day ransomware. The solution successfully detects 99.37% of our ransomware. Solely five circumvent the countermeasure. The following activities are simulated during the evaluation: mouse move, keyboard input and web browser usage. Only one false positive is encountered as seen Fig. 1. Up to 238K threads have been monitored during the samples evaluation. A very important point is that DaD's classification error rate is very low: $7.08e-05$.

Prediction outcome

		1	0
Actual value	**1**	True Positive 1870	False Negative 16
	0	False Positive 1	True Negative 238098

Fig. 1. The confusion matrix related to the suspicious (1) and non suspicious (0) threads monitored by *Data Aware Defense* (DaD) during the samples evaluation.

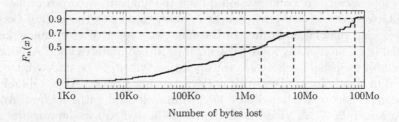

Fig. 2. The cumulative probability of the malicious threads for each thread number of bytes lost.

To assess the effectiveness of DaD, an estimation of the number of bytes lost across the ransomware collection is displayed Fig. 2. One can notice that for 70% of the samples' threads, at most 6.5 megabytes (MB) are encrypted, which is acceptable for most of the users. Unfortunately, considering 90% of the collection, 70 MB is lost per-thread in the most extreme scenario. Depending on the user needs, such loss can be tolerated, but it might be unacceptable for businesses. Most of the samples are single threaded, respectively 76.88%.

The detection is affected by the nature of the explored paths, and may be more or less prompt to block a malicious thread. Indeed, the folders with a few number of encrypted files set the sliding median far beyond the detection threshold. The epidemic of ransom notes is to blame. Even when this scenario occurs the malicious thread is successfully stopped. It demonstrates that monitoring the whole file system makes our solution resilient. Moreover, the compromise indication, a χ^2 sliding median on the last fifty write operations can be circumvented if less than half of the thread activity is dedicated to files encryption.

An important observation that we made, is that different χ^2 "layers" can be distinguished on the disk. Each one corresponding to a specific behavior, such as ransom notes and metadata appended to files. These patterns suggest a criterion to distinguish reversible from non-reversible ransomware. Indeed, metadata appended to files during encryption are visible on the file system and suggests a chance to get the data back (e.g., authors implement the decryption routine). Furthermore, no dissociation into separate threads has been observed for all the three following tasks: files encryption, ransom notes and the metadata.

6.4   Ransomware-Like Applications

Experimental results point that the solution is effective to stop infection but we did not discuss about limitations, in particular false positives and negatives. The primary purpose was to design DaD as an practical and efficient first line of defense against the ransomware.

False Positives. Looking at the χ^2 sliding median with a significance level, α_{EW} of 0.05, allows us to eliminate a significant number of false positives among most of the traditional applications. DaD monitors about dozens of processes and hundreds of thread while an user interacts with its machine. Few applications are blocked, in such cases, it disables a particular task (e.g., update), not the entire process. Moreover only very specific applications are able to obtain malicious file system behavior: files compression or encryption, secure files deletion, browsers startup and so on as shown Fig. 3.

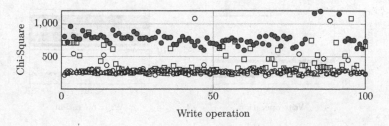

Fig. 3. χ^2 of the 100 first write operation of Mogrify ($\bullet$), 7-zip ($\square$), GPG4Win ($\triangle$), and μTorrent ($\circ$).

Indeed, the solution is not yet able to distinguish compression from encryption and a false positive is raised with 7-Zip, GPG4Win, or μTorrent. Still, as mentioned in Sect. 3 we can distinguish JPEG compression. The χ^2 statistic corresponding to the images rewritten by the Mogrify software is far away from the critical value (i.e., 293.24). In any cases, a major observation can be made: only a single "layer" is present. The χ^2 statistic alone is not sufficient to avoid false positives. The ransomware business model is based on extortion, in order to be paid, they need to make the ransom notes as visible as possible. Future works should focus on this idea.

False Negatives. The *Data Aware Defense* ransomware countermeasure focus on very specific activities on disk that belong to ransomware behavior but not exclusively. The underlying mechanics that comes with the ransomware to date is well known and documented: they encrypt files. However, the ransomware industry is very prolific and no one is immune to more stealthy behaviors, we are faced with an arms race. For example, the specific problem of boot sectors encryption (e.g., master boot record) is not addressed in this paper, a solution is proposed by the Talos Group [26]. In addition, as outlined in Mbol et al. [17], if an encryption algorithm preserving the distribution of the original files is used, it will evade the solution because randomness is the root of the detection. The ransomware which interleave malicious write operations with loops of unnecessary or redundant operations that look non random will go through DaD, as shown Fig. 4. Prior to block a malicious thread, DaD need to have a windows on the last fifty write operations. A multi-threaded ransomware where each file is encrypted by a unique thread can exploit this limitation. Finally, a kernel exploit is a potential breach that ransomware might use to unload DaD and more, in this scenario, the system is completely compromise.

Five ransomware samples among the collection (i.e., 0.62%) bypass the solution. Three Xorist samples used weak encryption algorithms (e.g., Tiny Encryption Algorithm). The last two samples behaviors are different than all we have previously observed. Figure 4, illustrates such statements.

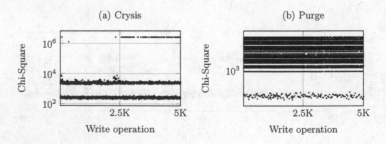

Fig. 4. Two false negatives samples: crysis and purge. Both interleave each malicious write operation with numerous garbage operations.

The Crysis sample does not write ransom notes, and after approximately 3,000 write operations begin to perform large write operations of 2^{18} bytes. Such operations with a zeroed buffer are repeated multiples times. The Purge sample hide his malicious behavior through a set of heterogeneous write operations. For each encrypted files, the ransom note is rewritten with chunks of 128 bytes. In both cases, the sliding median is inefficient to detect the underlying encryption process, the willingness of the ransomware authors is to hide the primary purpose of the application behind useless operations (i.e., mimicry attack). The ransomware can not be seen any more like a simplistic version of malware, in the future they will pretend to do something else than just encrypting files, which was not the case so far to the best of our knowledge.

7 Conclusions

The *Data Aware Defense* is based on file system monitoring and no assumptions is done concerning the malicious I/O patterns. In addition we achieved a thread granularity control on the system and do not restrict files protection on a particular folder. The χ^2 test by its own can replace the Shannon entropy and catch up some of its weaknesses. Moreover our countermeasure is efficient and can be deployed on Windows 7/10 machines with a reasonable performance hit, with an average delay of $12\,\mu s$ per write operation on disk, a few hundred times smaller than previous approaches. Our extensive experiments show that the more sophisticated ransomware already use mimicry attacks. However we successfully detect 99.37% of the samples with at most 70 MB lost per sample's threads in 90% of cases and less than 7 MB in 70% of cases.

These promising results in terms of performance and detection rate were obtained thanks to single simple metric computed for all threads of all processes running on the system, allowing us to track code injection attacks in particular. False positives seem inherent to the approach: we are detecting large write operations of random data. But its speed and low negative rate makes it a good candidate as a first line of defense. Once a thread is deemed malicious, instead of blocking disk accesses, other more costly metrics can be used to improve the false positive rate without impacting performance, since it would not be computed for all other threads. As an example, future work should focus on the distribution of random (encrypted files) and constant (ransom notes) data. Once false positive rate is small enough, an interaction with the user to eliminate the last ones seem reasonable. Indeed ransomware have a very specific behavior and the average user should know if she is encrypting all its files on purpose or not. Future work should investigate which information to report to a user and if the approach is practical.

Appendix 1: Ransomware Collection

Table 3. An overview of the active ransomware families used in the experiments (i.e., 87.98%). More details at: http://people.rennes.inria.fr/Aurelien.Palisse/DaD.html.

Family	Samples	Family	Samples	Family	Samples
Teslacrypt	195 (24.43%)	Yakes	25 (3.13%)	Shifu	9 (1.12%)
Cerber	135 (16.91%)	Deshacop	19 (2.38%)	Fsysna	8 (1%)
Xorist	125 (15.66%)	Locky	17 (2.13%)	Shade	7 (0.87%)
Bitman	101 (12.65%)	Gpcode	13 (1.62%)	Dalexis	5 (0.79%)
Zerber	27 (3.38%)	Gamarue	9 (1.12%)	Usteal	5 (0.79%)

Appendix 2: Empirical Tests

Table 4. Shannon entropy values with 10K files for each file type.

File types	Minimum	Average	Maximum	Variance
PNG	0.14	7.87	7.99	0.33
PDF	1.45	7.74	7.99	0.16
ZIP	3.21	7.93	7.99	0.07

Table 5. Chi-Square (χ^2) values with 10K files for each file type.

File types	Minimum	Average	Maximum	Variance
PNG	275.72	1.69e+6	3.76e+9	2.74e+15
PDF	306.86	1.50e+6	5.07e+8	1.30e+14
ZIP	220.44	4.74e+5	9.11e+8	1.23e+14

References

1. Bisson, D.: C&C servers? too risky! Android botnet goes with Twitter instead. https://www.bleepingcomputer.com/news/security/candc-servers-too-risky-android-botnet-goes-with-twitter-instead/
2. Bonferroni, C.E.: Teoria statistica delle classi e calcolo delle probabilita. Libreria internazionale Seeber (1936)
3. Cimpanu, C.: Microsoft announces controlled folder access to fend off crypto-ransomware. https://www.bleepingcomputer.com/news/microsoft/microsoft-announces-controlled-folder-access-to-fend-off-crypto-ransomware/

4. Clonezilla: The free and open source software for disk imaging and cloning. http://clonezilla.org/

5. Continella, A., Guagnelli, A., Zingaro, G., De Pasquale, G., Barenghi, A., Zanero, S., Maggi, F.: ShieldFS: a self-healing, ransomware-aware filesystem. In: Proceedings of the 32nd Annual Conference on Computer Security Applications, pp. 336–347. ACM (2016)

6. Craig: Differentiate   encryption   from   compression   using math,   June   2013.   http://www.devttys0.com/2013/06/differentiate-encryption-from-compression-using-math/

7. Crystal Dew World: CrystalDiskMark is a disk benchmark software. http://crystalmark.info/software/CrystalDiskMark/index-e.html

8. Corpora, D.: Producing the digital body. http://digitalcorpora.org/

9. Geekbench: New benchmarks, redesigned interface. http://geekbench.com/

10. Octave, G.N.U.: Scientific programming language. https://octave.sourceforge.io/octave/function/chi2inv.html

11. Haschek, C.: How to defend your website with ZIP bombs. https://blog.haschek.at/2017/how-to-defend-your-website-with-zip-bombs.html

12. Ivanov, A., Sinitsyn, F.: The first cryptor to exploit telegram. https://securelist.com/blog/research/76558/the-first-cryptor-to-exploit-telegram/

13. Kharraz, A., Arshad, S., Mulliner, C., Robertson, W., Kirda, E.: UNVEIL: a large-scale, automated approach to detecting ransomware. In: Proceedings of the 25th USENIX Security Symposium, Austin Texas, pp. 757–772. Usenix (2016)

14. Kharraz, A., Robertson, W., Balzarotti, D., Bilge, L., Kirda, E.: Cutting the Gordian knot: a look under the hood of ransomware attacks. In: Almgren, M., Gulisano, V., Maggi, F. (eds.) DIMVA 2015. LNCS, vol. 9148, pp. 3–24. Springer, Cham (2015). doi:10.1007/978-3-319-20550-2_1

15. Kolodenker, E., Koch, W., Stringhini, G., Egele, M.: PayBreak: defense against cryptographic ransomware. In: Proceedings of the 2017 ACM on Asia Conference on Computer and Communications Security, pp. 599–611. ACM (2017)

16. Malekal: Malware repository. http://malwaredb.malekal.com/

17. Mbol, F., Robert, J.-M., Sadighian, A.: An efficient approach to detect TorrentLocker ransomware in computer systems. In: Foresti, S., Persiano, G. (eds.) CANS 2016. LNCS, vol. 10052, pp. 532–541. Springer, Cham (2016). doi:10.1007/978-3-319-48965-0_32

18. Microsoft: File system minifilter drivers. https://msdn.microsoft.com/en-us/windows/hardware/drivers/ifs/file-system-minifilter-drivers

19. Microsoft: Windows performance toolkit. https://msdn.microsoft.com/en-us/windows/hardware/commercialize/test/wpt/index

20. Palisse, A., Le Bouder, H., Lanet, J.-L., Le Guernic, C., Legay, A.: Ransomware and the legacy crypto API. In: Cuppens, F., Cuppens, N., Lanet, J.-L., Legay, A. (eds.) CRiSIS 2016. LNCS, vol. 10158, pp. 11–28. Springer, Cham (2017). doi:10.1007/978-3-319-54876-0_2

21. PCMark 8: The complete benchmark for Windows 8.1, Windows 8 and Windows 7. https://www.futuremark.com/benchmarks/pcmark

22. PolarToffee: Found a sample of the AES-NI ransomware, April 2017. https://twitter.com/PolarToffee

23. Scaife, N., Carter, H., Traynor, P., Butler, K.R.: Cryptolock (and drop it): stopping ransomware attacks on user data. In: 2016 IEEE 36th International Conference on Distributed Computing Systems (ICDCS), pp. 303–312. IEEE (2016)

24. Sebastián, M., Rivera, R., Kotzias, P., Caballero, J.: AVCLASS: a tool for massive malware labeling. In: Monrose, F., Dacier, M., Blanc, G., Garcia-Alfaro, J. (eds.) RAID 2016. LNCS, vol. 9854, pp. 230–253. Springer, Cham (2016). doi:10.1007/978-3-319-45719-2_11

25. SonicWall: Annual threat report. Technical report, SonicWall (2017). https://www.sonicwall.com/docs/2017-sonicwall-annual-threat-report-white-paper-24934.pdf

26. The Talos Group: MBR filter driver. https://github.com/vrtadmin/MBRFilter

27. Micro, T.: CryLocker uses Imgur as C&C. http://www.trendmicro.com/vinfo/us/security/news/cybercrime-and-digital-threats/ransomware-recap-sept-2-2016-crylocker-uses-imgur-as-c-c

28. Micro, T.: Cerber starts evading machine learning. http://blog.trendmicro.com/trendlabs-security-intelligence/cerber-starts-evading-machine-learning/

29. Viper: Binary management and analysis framework. http://viper.li/

30. VirusShare: Malware repository. https://virusshare.com/

31. Wardle, P.: Towards generic ransomware detection. https://objective-see.com/blog/blog_0x0F.html

32. Young, A., Yung, M.: Cryptovirology: extortion-based security threats and countermeasures. In: 1996 IEEE Symposium on Security and Privacy, Proceedings, pp. 129–140. IEEE (1996)

33. Young, A.L., Yung, M.M.: An implementation of cryptoviral extortion using Microsoft's crypto API. CiteSeerX (2005)

A Large-Scale Analysis of Download Portals and Freeware Installers

Alberto Geniola[1], Markku Antikainen[2(✉)], and Tuomas Aura[1]

[1] Aalto University, Espoo, Finland
[2] Helsinki Institute for Information Technology,
University of Helsinki, Helsinki, Finland
markku.antikainen@helsinki.fi

Abstract. We present a large-scale study of Windows freeware installers. In particular, we look for potentially unwanted programs (PUP) and other potentially unwanted modifications to the target system made by freeware installers. The analysis is based on almost 800 installers gathered from eight popular software download portals. We measure how many of them drop PUP, such as browser plugins, or make other modifications to the system. In addition to these results, we find that most installers that download executable files over the network are vulnerable to man-in-the-middle attacks, which in the worst cases may be used to execute arbitrary code with elevated privileges on the target system. Moreover, serious man-in-the-middle vulnerabilities are found in application managers provided by download portals.

1 Introduction

Most computer users download and install some freeware applications from the Internet. The source is often one of the many download portals, which aggregate software packages and also offer locations for hosting them. It is common concern that the downloaded software might be infected with malware or have other unwanted side effects. Freeware installers are also known for dropping potentially unwanted programs (PUP) to the user's computer. PUP and other unwanted system modifications to desktop computers can be considered a security threat [5, 21]. This phenomenon is partly caused by the *pay-per install* (PPI) business model where freeware software developers monetize their software by bundling it with other third-party applications or by promoting some software and services by changing the user's default settings. This business model is not always illegal as the application installer may inform the users about the third-party software and even allow them to opt-out. However, this is often done in a way that the user is not fully aware of the choices made.

In this paper, we set out to analyze nearly 800 popular software installers from download portals. We do this with an automated analysis system that downloads and installs the applications in a sandbox while monitoring the target system. The sandbox emulates the behavior of a lazy user who tries to complete the installation process with the default settings of the installer. That is, we assume

H. Lipmaa et al. (Eds.): NordSec 2017, LNCS 10674, pp. 209–225, 2017.
https://doi.org/10.1007/978-3-319-70290-2_13

that the user wants to finish the installation as fast as possible and is habituated to accept the default settings and to bypass warnings.

Our study differs from earlier research [4,20] in several respects. First, we try to better understand the prevalence of any problems by gathering large quantities of software from the most popular download portals. Second, we do not differentiate between legitimate and malicious actions, which would easily lead to complicated legal and moral arguments, but instead try to cover all potentially unwanted changes to the system. Thirdly, our research methodology provides insights to software installers and download portals in general.

The most important findings from our study are following. We find that, while the most popular download portals do not distribute malware, some (1.3%) of the studied installers drop a well-known PUP to the target system. Furthermore, nearly 10% of the installers came with a with a third-party browser (e.g. Chrome) or a browser extension. On the positive side, we find no evidence that download portals would themselves bundle significant amounts of potentially unwanted content to the downloads – the PUPs seem to come from the original freeware authors. When analyzing the installers, we also find prevalent vulnerabilities. The installers often download the application binaries over HTTP, and over half of the installers that do so, do not verify the integrity of the binary and are thus vulnerable to man-in-the-middle (MitM) attacks. We also spot serious MitM vulnerabilities in update managers of two major download portals, which allow an attacker to underhandedly advertise malicious binaries as software updates.

The rest of this paper is organized as follows. Section 2 reviews related work. In Sect. 3, we describe the methodology and then briefly explain the analysis system. Analysis results are presented in Sect. 4 and further discussed in Sect. 5. Section 6 concludes the paper.

2 Background

This section describes the related work and ideas on which our research is based.

Downloading applications from the Internet can be dangerous, and this also applies to download portals [9,10]. The applications might come with unwanted features that range from clearly malicious, such as bundled malware and spyware, to minor nuisances like changing the browser's default search engine. Such software is often referred to as *potentially unwanted programs* (PUP)[1]. We use the broad definition of Goretsky [8], which states that a PUP is an application or a part of an application that installs additional unwanted software, changes the behavior of the device, or perform other kinds of activities that the user has not approved or does not expect. PUP often functions in a legal and moral gray area. The threat of legal action from PUP authors has been suggested as the reason why anti-malware labels it as "potentially unwanted" rather than "malicious" [2,13], and this was also confirmed by anti-malware developers who gave feedback on our research.

---

[1] Potentially Unwanted Application (PUA) is another often used term.

Recent studies have shown that freeware installers only rarely come bundled with critical malware [11]. More often, the system modifications are just unnecessary and unexpected. The user may even be informed about them, e.g., in the EULA, or the installer may allow a careful user to opt out of unwanted features. Users, however, do not always read EULAs and may be habituated to accept default settings and ok any warnings [1,17]. This *rushing-user* behavior leads the user to giving *uninformed consent* to the system modifications. While solutions have been proposed, they have not been widely adopted [2]. Moreover, PUP installers often come with a complex EULAs [7], which users are likely to accept blindly [3].

One root cause for the problem of unwanted software is the pay-per-install (PPI) business model. PPI is a monetization scheme where a software developer or distributor gets payed for dropping unrelated third-party applications to the target computer. This may be done with or without the user's consent. Recent research publications have studied the PPI business model [4,11,20]. The PPI application installer typically downloads the third-party software from a PPI distributor. Caballero et al. [4] reverse engineered protocols used by PPI distributors and found that the choice of applications depends on the target computer's geolocation. Another result is that, while PPI distributors do spread some known malware, this is not a very prevalent phenomenon—probably because blacklisting by anti-virus vendors would hurt the PPI business [11]. Another related paper analyzed black-market PPI that installs third-party applications silently in the background [20]. In the current research, we consider commercial PPI that does not necessarily try to hide its actions but rather takes advantage of the rushing user behavior to maximize the number of installs. We also analyze other unwanted side effects of the installers even if not part of the PPI business.

In summary, while there is plenty of anecdotal evidence showing that download portals distribute PUP [10,18], probably due to the PPI business model, the true extent of this problem has not been studied methodically. We aim to fill this gap by providing a comprehensive analysis of nearly 800 application installers retrieved from the most popular download portals. While the PUP phenomenon is not limited to a single operating system or platform, we focus purely on Microsoft Windows, which still is the most popular OS on desktop and laptop computers (84% market share at the time of writing [19]).

3   Methodology

This section describes the methodology and the analysis system used in our study. While the analysis system is rather complex, we describe them only briefly because the focus of this paper is on the analysis results.

3.1   Analysis System Overview

Our goal is to implement automated analysis of large numbers of Windows freeware installers. For this, we need an infrastructure that automatically downloads, executes and analyzes the application installers. On a high level, the analysis system (1) crawls selected download portals for Windows freeware installers, (2) automatically runs them in guest machines with emulated user interaction, (3) monitors the modifications made to the guest machine as well as network communication, and (4) saves the results for later use.

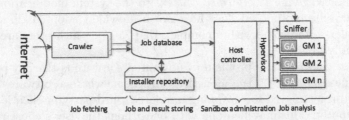

Fig. 1. Analysis system architecture

The architecture of the analysis system is shown in Fig. 1. First of all, we implemented crawlers for the download portals. The actual analysis is orchestrated by the host controller. It handles the life cycle of the guest machines, in which the installers are executed. This essentially means that the host controller is responsible for (1) fetching a job from the database, (2) initializing a guest machine and serving it the installer binary, (3) pre-processing and storing data about the installation process, and finally (4) cleaning up the guest machine. In each guest machine, there is a guest agent that pulls the installer from the host controller and drives its execution by launching it and interacting with its UI. The agent also monitors any filesystem and registry modifications and reports these to the host controller. The network traffic to the guest machines is routed through a network sniffer, which captures it. While the analysis system is modular and can support any guest OS, we have so far implemented the guest agent only for 32-bit Windows 7 guest machines.

The installers require user attention. Therefore, we implemented a heuristic interaction system which emulates the behavior of a lazy user during the installation process. When the installer runs, the guest agent tries to detect when it is waiting for user input and then sends the input event that is most likely to cause progress. The guest agent does this by observing screenshots that are taken periodically from the installer UI: the installer is likely to wait for user input if it is not performing any I/O operations and if the screenshots are stable for some time. The next input is chosen using heuristics that, for example, prefer rectangular shapes containing text such as "OK", "Install", or "Next". The graphical screenshot approach was taken because most installers do not make use of the standard Windows UI components. The UI interaction heuristics in the guest agent were optimized for Windows; however, they could easily be adapted to other operating systems.

3.2   Installer Crawling

We chose eight download portals based on their Alexa rankings (Table 1). While some of these sites also provide other content than application downloads, the ranking gives a rough picture of their popularity and perceived trustworthiness.

Each studied download portal promotes a list of the most popular applications on its front page, except Softpedia which promotes recent downloads. We decided to focus on the promoted applications and set a crawler to download up top 200 installers from each portal. When possible, it applied a filter for 32-bit Windows or Windows 7 freeware. With some portals, there were fewer than 200 actual downloads, mainly because of the limitations of the web interface. Table 1 summarizes portals chosen for our study and the number of downloaded files.

In addition to crawling, we also manually downloaded installers for the most popular freeware applications directly from the developers' websites. We used Alexa rankings of top freeware applications as well as Google Trends for the most popular searches that include the words "software download". The manual download was done to compare the behavior of the installers published directly by the shareware authors with those distributed through the download portals. However, it should be emphasized that we only downloaded 20 installers manually. More extensive comparison between the portals and "original" software would not scale because it cannot be automated, and it would also be complicated by the fact that many authors use one of the portals as their main distribution point.

Table 1. Download portals studied in this paper

Download portal	Alexa rank Oct.2016	Filters	Downloaded files	Successfully analyzed
download.cnet.com	159	Win,free	200	146
softonic.com	285	Win7,free	170	126
filehippo.com	662	Win	90	64
informer.com	881	Win,free	200	117
softpedia.com	1732	Win,free	200	148
majorgeeks.com	6077	Win,free	55	37
soft32.com	7279	Win,free	200	113
brothersoft.com	8600	Win,free	41	26
manual download	–	–	20	15
			1177	792

We were not able to automatically analyze every installer. First, almost 10% of the crawled files failed either because the application was not an installer in the first place (e.g. a stand-alone application) or because of missing hardware, software dependency, product key, or a similar reason. Additionally, 23% of the installers failed because the automated UI interaction was not smart enough. The reason was mostly complex interaction, such as selecting the directory to which the program should be installed. Another reason was that the installer used some other language than English. Nevertheless, a relatively high percentage of the installers (67%) completed. This was the result of iterative improvements to the UI automation heuristics and other parts of the analysis system.

The results discussed in the rest of this paper were obtained from the 792 installers completed successfully. Of these files, 751 were unique. We nevertheless consider even the installers with the same hash as distinct because some download portals have in the past served identical installers for several applications[2]. In these cases, the installer executable determines the further files to download and install based on its own filename.

4   Results

We present the results of our analysis in two parts. Section 4.1 describes what we can learn simply by looking at the files served by the download portals. Then, Sect. 4.2 presents the results of dynamic analysis. All the results are based on the 792 installers that were successfully executed. Some of the results are not directly related to security but are of general interest and serve as background information.

4.1   Static Properties of the Installers

This section describes some of the basic properties of the analyzed installers.

Analyzed Applications: We first compare the applications promoted on different portals. This helps to understand the data and is interesting in itself. We manually grouped the different versions of the same applications. Table 2 shows the overlap in applications at different portals. The number of distinct applications served by each portal is on the diagonal.

Our first observation is that the portals serve quite different sets of applications. Those promoted by CNET, FileHippo, Informer and Soft32 overlap the most. On the other hand, Softonic and Softpedia tend to promote applications that are not on the other portals. In the case of Softpedia, the reason may be that it does not promote the most popular software but the latest downloads. Finally, some portals use only the last week's downloads for the popularity ranking. This metric is susceptible to manipulation and short-term fluctuation, e.g. when an update is published. For these reasons, one has to be very careful when comparing different download portals based on our data.

---

[2] CNET's downloader VT report available at https://virustotal.com/it/file/ 9961ebc9782037f68b73096bcff3047489039d6dc5c089f789b3dbff4109e21b/analysis/.

Table 2 also shows the median ages of the installers served by each portal. Software age may be an indication of how seriously the publisher or download portal take security. We obtained the application ages from VirusTotal. Our assumption is that popular software tends to be submitted to VirusTotal soon after release. Although the first-seen date obtained from VirusTotal does not precisely tell how old a binary file is, it gives an independent indication of when the software began spreading more widely.

The overall observation is that much of the popular freeware is not frequently updated, and many installers are several years old. This can be a cause for concern. The collected data also shows that CNET, MajorGeeks and Softpedia serve relatively recent software installers while the rest of the portals serve considerably older binaries. In addition to the actual age of the software, the results could be explained by differences in which software the sites promote and the type of software that each portal distributes. For instance, there may be value to archiving popular legacy software that is no longer updated. But even considering such alternative explanations, we can still assert that the most popular download site CNET distributes relatively recent software: its installer ages align closely to those of manually downloaded files, which can serve as a reference metric.

Table 2. Number of common applications served by each download portal pair (different versions of same application have been combined). *Age* shows the median age in days of the installers served by each portal.

	Brothersoft	CNET	FileHippo	Informer	MajorGeeks	Soft32	Softonic	Softpedia	manual	age
Brothersoft	26	1	3	2	0	0	1	0	0	953
CNET	1	144	19	22	6	21	7	0	4	111
FileHippo	3	19	64	18	6	15	4	1	4	160
Informer	2	22	18	117	7	14	3	0	5	604
MajorGeeks	0	6	6	7	35	3	1	0	1	8
Soft32	0	21	15	14	3	112	6	2	2	573
Softonic	1	7	4	3	1	6	125	2	0	723
Softpedia	0	0	1	0	0	2	2	148	0	18
manual	0	4	4	5	1	2	0	0	15	117
# distinct files	26	146	64	117	37	112	126	148	15	

Application Signing: Our first security-related question was whether the installer binaries are signed. Recent research showed that while malware is generally not signed, potentially unwanted programs are [12]. We wanted to know where software distributed by the download portals stands.

The application signature verification results can be seen in Table 3. While most of the analyzed binaries (64%) had a valid signature, 30 (3.8%) cases did not verify correctly. Publisher certificate expiration was the most common cause

Table 3. Signature verification of analyzed installers

Verification outcome	# .EXE	# .MSI	# Total
Signed and verified	486	23	509
Verification error	26	4	30
Unsigned	239	14	253

of failure (24 cases). Other causes were explicit revocation (1 case) and untrusted root CA (5 case). The remaining 32% of the analyzed installers were unsigned.

Interestingly, there were differences between the download portals. CNET, FileHippo and Informer had about 80% correctly signed code while Soft32, Softonic and Softpedia had lower rates (62%, 61%, 44%, respectively). The other portals appeared to belong to the latter group, but there were too few installers for a fair comparison. The high percentage of signed files in three of the four most popular download sites seems to indicate that there is value for the publishers in code signing even though the portals do not require it.

4.2   Dynamic Analysis of Installers

This section presents results from the dynamic execution and monitoring of the installers.

Network Traffic Analysis: The sniffed traffic was analyzed with the Tshark and Bro protocol analyzers. We also implemented custom Python scripts for extracting further information.

We begin the discussion by looking at the network protocols (Table 4). Most of the traffic is HTTP and HTTPS over TCP (99%). The most frequent UDP packets were for UPnP, SSDP and DNS. Our script was unable to classify some of the UDP packets. Manual investigation revealed that such traffic mainly belongs to the BitTorrent protocol, legitimately used by torrent-based installers. In three cases, we identified JSON encoded text over UDP, which is used by content-sharing applications for advertising themselves on the local network. In one case, the installer used a variant of the GVSP video streaming protocol, presumably to show a video to the user.

Next, we focus on HTTP, which constitutes the bulk of the network traffic. Figure 2 shows the domains that are contacted by most installers and from which the installers download most of the data. It can be seen that more than 80% of the HTTP downstream traffic is from well-known CNDs. Akamai and Google are the two most-contacted ones. The figure also reveals that quite a few installers contact Google but only download small volumes of data. A close investigation revealed that 23 installers made least one HTTP request for the Google Analytics web beacon and 29 installers downloaded the Google Analytics JavaScript library. This may indicate that many freeware authors benefit from its value-added services such as user tracking and demographic data.

Table 4. Breakdown of network traffic (inbound and outbound), total for all analyzed installers

Transport layer	Application protocol	MB	(%)
TCP	HTTP	6567.84	95.22
	TLS/SSL	328.63	4.76
	Others	1.25	0.02
		6897.72	99.25
UDP	UPnP	18.03	34.51
	SSDP	17.25	33.02
	DNS	4.39	8.40
	Others	2.90	5.55
	Unknown	9.67	18.52
		52.24	0.75
ICMP		0.01	0.00
Total		6949.96	100

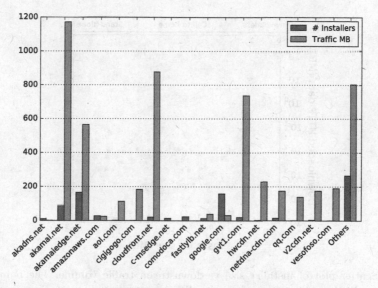

Fig. 2. HTTP downstream traffic breakdown by top domains. The blue bars (left) represent downstream traffic volume, while the cyan bars (right) indicate the number of installers that contacted the domain (Color figure online)

We reassembled and inspected the HTTP streams for binary content. Table 5 shows the results. Executable files and binary payloads constitute most of the traffic. This indicates that many of the installers (348) behave as install-time downloaders. To see if there is a clear distinction, we plotted the installer binary size and the downloaded traffic volume in Fig. 3. We have visually classified the

Table 5. HTTP downloads by MIME type

MIME type	Downloaded data (MB)	# Installers
application/x-dosexec	1879.17	96
application/octet-stream	1808.99	227
application/vnd.ms-cab-compressed	475.02	25
application/gzip	462.82	11
application/zip	267.32	32
application/vnd.ms-office	257.15	8
application/x-7z-compressed	228.09	4
application/x-bzip2	29.51	4
text/plain	16.90	787
text/html	12.33	138
others	18.13	155
Total	5455.42	788 (Distinct)

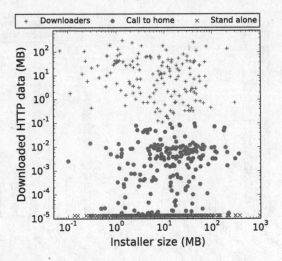

Fig. 3. Scatter plot of installer size vs downstream traffic volume. The points were divided to the three classes based on visually observed grouping

installers into three classes: downloaders, installers that call home but do now download significant amounts of data, and stand-alone installers. It can be seen from the figure that there is no apparent correlation between the installer size and the amount of data it downloads.

Man-in-the-middle Vulnerability: As seen in Table 4, installers tend to download binary files over insecure HTTP connections. These files are typically executed within the installation process, possibly with high system privileges. In such a context, it is essential that the installers authenticate the downloaded files, e.g. with a digital signature. To check if they do that, we implemented an automated MitM attack against the installers. This was done with a transparent HTTP proxy in the sniffer component of the analysis system, which replaced executable files in HTTP responses (but not HTTPS) with its own. The malicious binary (EXE or MSI file) that was fed to the installer simply took note of its running privileges and terminated. The malicious binary was injected in the following cases:

1. Request URL ended with .EXE or .MSI
2. Response MIME type matched executable or MSI
3. First bytes of the HTTP response body matched magic numbers for EXE or MSI.

Among the 792 analyzed installers, the MitM attack was triggered 100 times. Amazingly, more than half of the attempted attacks (55%) led to immediate execution of the attacker's binary file, meaning that no authentication or integrity check was done for the downloaded binaries. Only 8 installers refused to execute the tampered file and removed it right away. In the remaining 37 cases, the attacker's file was not executed, yet it was found on the disk after the installation. 17 of these were saved in temporary system folders (subject to later removal upon system cleanup) while 20 were stored in persistent file system locations, such as under the *Program Files* directory. The latter cases leave the system open to a delayed attack when the application is used.

The MitM attack is particularly dangerous because the attacker's file is executed with the same privileges as the installer application. In 75 out of the 100 successful attacks, the malicious binary was executed with elevated privileges.

There are straightforward ways of mitigating the MitM vulnerability. One approach would be to use HTTPS for the download. Another possibility is to use asymmetric signature verified by the installer application with a static publisher public key. Clearly, there is no good technical excuse to be vulnerable.

It is worth noticing that most of the download portals distribute installers via HTTP in the first place, and the installer itself could be fake. The user, however, has the opportunity to check that the installer binary is signed by the correct publisher. In comparison, the MitM attack succeeds even if the user takes care to only execute legitimate, signed installer binaries from the correct publisher.

File System Analysis for Malware Drops: We collect the hashes of all files that are temporarily or permanently stored on the guest machine as well as hashes of files reconstructed from network traces including HTTPS connections. We looked up all the collected file hashes in VirusTotal, which aggregates results from various virus scanners. This was done two months after running the installers to leave time for new malware variants to be detected.

The number of positive results was high (235 files) but most of these were reported by only one scanner and, most likely, were false positives. Only 1.3% of the installers contained files detected as malware by six or more scanners. More importantly, majority of such the positives were labeled as PUP. There was only one detected critical threat, and it was an Android rootkit that does not infect Win32 systems. The files with highest detection rates are listed in Table 6.

The analysis shows that download portals are not used for blatant malware distribution. The portals probably perform scans of the binaries before publishing them. On the other hand, the presence of PUP related to the well-known InstallCore PPI network indicates that download portals could still implement stricter countermeasures against grayware and bundled unwanted applications.

Table 6. Threats ranked by VirusTotal detection rate.

File name (truncated)	# Positive scanner(s)	# Inst	Description	Source portal
rootf.apk	30	1	Android rootkit	Soft32
fusion.dll	17	1	PUP InstallCore	CNET
videora.exe	15	1	PUP OpenCandy	Softonic
...Setup.exe	14	1	Adware Mobogenie	CNET
fusion.dll	14	1	PUP InstallCore	CNET
fusion.dll	12	1	PUP InstallCore	Soft32
...0061e.exe	12	1	PUP Montiera toolbar	Soft32
...B2C6B.dll	8	1	PUP Conduit	Brothersoft
...126C1.exe	6	1	PUP Zugo Toolbar	Softonic

Registry Modifications: We tracked the registry modifications made by the installers and analyzed changes to the following:

– Automatic program startup
– Default browser
– Browser plugins

There are many ways for a program to start automatically in Windows including registry keys [16] and specific file system folders. We found that 88 installers (12%) configured the installed software to automatically run at system startup.

Similar analysis was done on default browser changes: 26 installers replaced the default browser. Interestingly, 11 of these are not browser installers. Table 7 reports the details. Google Chrome turned to be the only third-party browser installed by non-browser installers. Manual investigation revealed that, in all cases, Google Chrome installation is optional but pre-selected by default.

Our analysis continued with the inspection of installed third-party browser modules. We focused on Internet Explorer, which was the only browser installed by default on the fresh Windows 7 guest machine. There were 69 registry modifications regarding browser extensions by 38 installers. As shown in Table 8, the

Table 7. Installers bundling unrelated third party browsers

Product name	Publisher	Bundled browser
Adobe Shockwave Player	Adobe Systems Inc	Google Chrome
CCleaner	Piriform Ltd	Google Chrome
Defraggler	Piriform Ltd	Google Chrome
PhotoScape	Mooii	Google Chrome
Recuva	Piriform Ltd	Google Chrome
Speccy	Piriform Ltd	Google Chrome
SUPERAntiSpyware Free	SUPERAntiSpyware	Google Chrome

Table 8. Third-party plugins dropped on IE

Portal	#Installers	Toolbar	Menu extensions	BHO	Total	%
Brothersoft	26	2	0	4	6	23.1
CNET	146	4	7	7	18	12.3
FileHippo	64	7	0	9	16	25.0
Informer	117	4	1	7	12	10.3
MajorGeeks	37	0	0	0	0	0.0
Soft32	113	2	0	4	6	5.3
Softonic	126	1	1	2	4	3.2
Softpedia	148	1	0	3	4	2.7
manual	15	1	0	2	3	20.0
Total	792	22	9	38	69	

predominant type of installed extension is the *Browser Helper Object* (BHO), which is the most powerful and potentially most dangerous IE component type because it runs in the same memory context as the browser and has access to the user's browsing data [6].

In the case of browser extensions, there were differences between the portals. MajorGeeks installers did not bundle any browser plugins, and Softpedia registered a total of just 4 dropped items over 148 installers (2.7%). Softonic and Soft32 had also relatively low rates. CNET and Informer, on the other hand, dropped considerably more browser plugins, and FileHippo topped the league with 25% drop rate. Interestingly, even manual downloaded installers bundled browser plugins. This could indicate that the plugins are bundled by the original software vendors and not by the portals.

Installer Best-Practices Compliance: Microsoft advises vendors to follow certain best practices for installers [15]. Firstly, each installed application should provide a consistent uninstall feature. For this, the installer should populate two registry keys on the system, one with the program's human-readable name and the other with a path to the uninstaller binary. If one of these two values is missing, removal of the application becomes cumbersome. Of the analyzed installers, 82 failed to specify both the program name and uninstaller path. Another 5 only stored the product name without specifying an uninstaller binary.

Secondly, Microsoft requires installers to specify valid *ProductName* property in their metadata, which is usually placed within the resource section of the executable file. It is exposed to the user in a properties dialog [14]. 174 of the analyzed installers failed to provide this information.

4.3   App Managers and Software Updates

Some download portals provide app-manager clients for simplifying software downloads and updates. By default, app managers run at system startup and regularly check for application updates. We analyzed three app managers (File-Hippo, Informer and MajorGeeks).

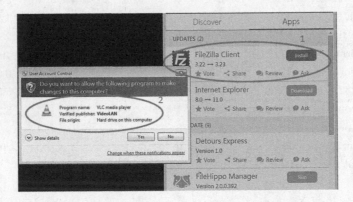

Fig. 4. Example of a successful MitM attack. User decides to update FileZilla (1), but is served a VLC installer instead (2)

Two of the analyzed app managers (FileHippo and Informer) are vulnerable to MitM attack. More specifically, they communicate with the portal using HTTP and do not check the integrity of the served binaries in any way. We were able to make modifications to the list of available updates, such as changing the download URL or adding new entries to the list of available software. Furthermore, both vulnerable app managers by default periodically check for software updates. This makes the attack even more serious in two respects: first, the attacker can push arbitrary binaries to the victim's computer without any initial user action and, second, the MitM attack can be mounted when a mobile user is visiting an untrusted access network such as wireless hotspot.

Figure 4 illustrates the MitM attack. Windows UAC still asks user's permission to run the attacker-selected application, which is either unsigned or has a different name from what the use should expect, but the user may not pay attention to such details—especially after explicitly deciding to install the update.

5   Discussion and Future Work

This research was motivated, in part, by the suspicion that download portals might distribute malware or bundle excessive amounts of unwanted programs to freeware downloads. The results of our analysis do *not* support these suspicions. No serious infections by known malware were detected, and the bundled PUPs seem to have been mostly included by the original freeware authors. Thus, freeware distribution does not appear to be such a Wild West as has been suggested in the past.

From the security viewpoint, the most important negative discoveries were:

– The median age of installers varies notably among portals, and the distributed freeware versions are often not the latest.
– Some portals host installers that bundle known PUP.
– The most common types of PUP bundled with freeware are third-party browser plugins and the Google Chrome browser.
– Many installers that download executable files are vulnerable to MitM attacks that enable code injection to the client machine.
– Some app managers provided by the portals are similarly vulnerable to MitM attacks and code injection.

While we make some comparisons between the portals throughout the paper, it would not be possible to make fair ranking of the portals regarding security or PUP. The portals differ in the types and quantity of software available. While Softpedia does well on all the metrics, it promotes a different set of software than the other portals (based on recent downloads rather than popularity), and thus the results may not be comparable. Further, our study is based on a snapshop and is limited to a single point in time. A longitudinal study would be needed for comparing different portals.

Our methodology does have some limitations. Firstly, our UI automation could still be more reliable. Manual inspection of installation screenshots confirmed that our UI engine was able to correctly automate 67% of the installers. There clearly is still scope for improvement. Secondly, our current analysis system only supports 32-bit Windows 7 on the guest machines. This is because of the free availability of an API hooking library for this platform. Thirdly, it is possible that some installers drop a plugin to an already installed third-party application (e.g. a Google Chrome plugin). Because the guest machines were initialized with a fresh OS, we have not spotted these. Fourthly, we have focused on grayware PUP and explicit changes to the system. We did not emphasize stealthiness in the design of our analysis system, due to which evasive malware may have escaped our analysis. We also cannot exclude the possibility of custom backdoors, spyware features, or other malicious behavior in the software.

6   Conclusion

We present analysis results from almost 800 freeware application installers obtained from download portals. The results indicate that portals do not actively distribute known malware. Many freeware installers, however, come bundled with unwanted programs and have links to PPI networks. The most dangerous detected security flaw is the lack of authentication for downloads made by installers and app managers. They are vulnerable to man-in-the-middle attacks that enable code injection to the client machines. We expect to extend the scope and coverage of the analysis and will share our data with the research community.

References

1. Böhme, R., Köpsell, S.: Trained to accept? A field experiment on consent dialogs. In: Proceedings of the SIGCHI Conference on Human Factors in Computing Systems, CHI 2010, pp. 2403–2406. ACM, New York (2010)
2. Boldt, M., Carlsson, B.: Privacy-invasive software and preventive mechanisms. In: International Conference on Systems and Networks Communications, ICSNC 2006, p. 21, October 2006
3. Bruce, J.: Defining rules for acceptable adware. In: Proceedings of the Fifteenth Virus Bulletin Conference (2005)
4. Caballero, J., Grier, C., Kreibich, C., Paxson, V.: Measuring pay-per-install: the commoditization of malware distribution. In: Proceedings of the 20th USENIX Conference on Security, SEC 2011, USENIX Association, Berkeley (2011)
5. Emm, D., Unuchek, R., Garnaeva, M., Ivanov, A., Makrushin, D., Sinitsyn, F.: It threat evolution in Q2 2016. Technical report (2016). https://securelist.com/files/2016/08/Kaspersky_Q2_malware_report_ENG.pdf
6. Esposito, D.: Browser helper objects: the browser the way you want it. https://msdn.microsoft.com/en-us/library/bb250436(v=vs.85).aspx. Accessed 29 Dec 2016
7. Good, N., Dhamija, R., Grossklags, J., Thaw, D., Aronowitz, S., Mulligan, D., Konstan, J.: Stopping spyware at the gate: a user study of privacy, notice and spyware. In: Proceedings of the 2005 Symposium on Usable Privacy and Security, SOUPS 2005, pp. 43–52. ACM, New York (2005)
8. Goretsky, A.: Problematic, unloved and argumentative: what is a potentially unwanted application (PUA)? Technical report, November 2011. Accessed 03 June 2016
9. Heddings, L.: Stop testing software on your PC: use virtual machine snapshots instead. http://www.howtogeek.com/206286/stop-testing-software-on-your-pc-use-virtual-machine-snapshots-instead/
10. Heddings, L.: Yes, every freeware download site is serving crapware (here's the proof). Technical report. http://www.howtogeek.com/207692/yes-every-freeware-download-site-is-serving-crapware-heres-the-proof/
11. Kotzias, P., Bilge, L., Caballero, J.: Measuring PUP prevalence and PUP distribution through Pay-Per-Install services. In: Proceedings of the USENIX Security Symposium (2016)
12. Kotzias, P., Matic, S., Rivera, R., Caballero, J.: Certified PUP: abuse in authenticode code signing. In: Proceedings of the 22nd ACM SIGSAC Conference on Computer and Communications Security, pp. 465–478. ACM (2015)
13. McFedries, P.: Technically speaking: the spyware nightmare. IEEE Spectr. **42**(8), 72–72 (2005)
14. Microsoft: VERSIONINFO resource. https://msdn.microsoft.com/en-us/library/aa381058.aspx. Accessed 05 Jan 2017

15. Microsoft: Windows installer and logo requirements. https://msdn.microsoft.com/en-us/library/windows/desktop/aa372825(v=vs.85).aspx. Accessed 30 Dec 2016
16. Microsoft: Run, RunOnce, RunServices, RunServicesOnce and Startup, November 2006. Accessed 08 Dec 2016
17. Motiee, S., Hawkey, K., Beznosov, K.: Do windows users follow the principle of least privilege? Investigating user account control practices. In: Proceedings of the Sixth Symposium on Usable Privacy and Security, p. 1. ACM (2010)
18. Slade: Mind the PUP: top download portals to avoid. Technical report, March 2015. http://blog.emsisoft.com/2015/03/11/mind-the-pup-top-download-portals-to-avoid/
19. Statcounter: Desktop operating system market share worldwide, June 2017. http://gs.statcounter.com/os-market-share/desktop/worldwide/#monthly-2017 06-201706-bar
20. Thomas, K., Crespo, J.A.E., et al.: Investigating commercial pay-per-install and the distribution of unwanted software. In: 25th USENIX Security Symposium (USENIX Security 2016), pp. 721–739. USENIX Association, Austin, August 2016
21. Wood, P., Nahorney, B., Chandrasekar, K., Wallace, S., Haley, K., et al.: Symantec internet security threat report trends for 2016. Technical report, April 2016

Access Control

GPASS: A Password Manager
with Group-Based Access Control

Thanh Bui$^{(\boxtimes)}$ and Tuomas Aura

Aalto University, Espoo, Finland
{thanh.bui,tuomas.aura}@aalto.fi

Abstract. Password managers make it easy for users to choose stronger and more random passwords without the burden of memorizing them. While the majority of our passwords should be kept secret, sharing passwords and access codes is necessary in some cases. In this paper, we present GPASS—a password manager architecture that allows groups to share passwords via an untrusted server. GPASS provides its own cryptographic access control mechanism in which all the information is transparent to the clients so that they can detect any misbehavior of the server. We implemented a proof-of-concept prototype to demonstrate the feasibility and effectiveness of the architecture.

1 Introduction

Despite the growing awareness of the security limitations of passwords, they remain the most widespread mean of user authentication. The main security issues in password authentication are that users tend to choose weak passwords [7,12] and re-use them in multiple services. Password managers offer a solution to these issues because they store passwords in an encrypted password vault, in which the encryption key is typically derived from a master password. Moreover, password managers enable users to select stronger and more random passwords, since they only need to memorize the master password that opens the vault.

While the majority of our passwords should remain secret, *sharing passwords* is useful in some scenarios. For example, third-party services may not allow creation of multiple user accounts, or the cost may lower with one shared user account. Also, short access codes are common in doors or photocopiers. If used as one-factor authentication, the short codes cannot be personalized because of the birthday paradox would make them easy to guess. In the workplace and at home, there may be need to manage dynamic groups of users and to share passwords and other access codes within those groups.

The application scenario that directly motivated this work is the following: *"In a company, employees can belong to multiple groups, such as projet teams or departments. Each group shares a set of access credentials. Each group has one or more leaders who can add or remove members."* This is the kind of system that we address in this paper: users form groups and share passwords among the group members, and a group might have several group leaders who have the authority to add new leaders and members and to remove existing ones.

© Springer International Publishing AG 2017
H. Lipmaa et al. (Eds.): NordSec 2017, LNCS 10674, pp. 229–244, 2017.
https://doi.org/10.1007/978-3-319-70290-2_14

Unlike the current popular password managers [1–3] that require considerable trust on the cloud server to enforce roles and permissions on the shared passwords, we aim to minimize the trust as much as possible. The reason is that with various incidents caused either by hackers [30] or governments [9,14,24], it is better not to place too much trust on any single party to protect such critical data. None of password managers presented in the literature features secure password-sharing. However, data sharing on untrusted cloud servers has been widely studied in the context of cryptographic distributed file systems [13,15,19,27]. We were inspired by these results, but our solution goes further by allowing group leaders to delegate their responsibility to others. The file owners in the cryptographic file systems generally cannot delegate their role to other users.

In this paper, we propose GPASS, a novel password manager architecture that enables passwords to be shared within groups and requires minimum trust on the storage server. The access control mechanism in the system is *key-oriented* in the sense that entities are represented by their public keys and authorizations are encoded into signed events. These events are transparent to the users so that they can detect server misbehavior. Like all other systems that do not depend on trusted entities, GPASS is unable to guarantee freshness of its data for users that are offline and do not observe the updates [22]. We propose an extension to GPASS that relies on trusted observers to tackle the issue. This solution is inspired by the recent proposals for monitoring the security of the web PKI [4,20,21,31]. To demonstrate the feasibility and efficiency of the solution, we implemented a prototype of the password manager and measured its performance from the user's perspective. We emphasize that our design is generic and can be used to implement similar group dynamics in other data-sharing applications, such as file systems.

The rest of the paper is structured as follows: Sect. 2 covers the related work. Section 3 describes the system and threat models and assumptions. Section 4 presents our password manager construction and Sect. 5 analyzes its security. Section 6 presents an extension of GPASS. In Sect. 7, we demonstrate its feasibility with a proof-of-concept implementation. Section 8 contains further discussion and Sect. 9 concludes the paper.

2   Related Work

Much of the literature on password managers focuses on solutions against offline brute-force attacks [6,8,16,26] and security analysis of existing password manager designs [23,28,29]. None of the effort has been put into secure password sharing. Existing work close to ours can be found in the area of cryptographic file systems and data access control on untrusted cloud servers. Access control in these systems usually rely exclusively on knowledge of cryptographic keys that are used to encrypt or sign the data.

The Cryptographic File System (CFS) [5] encrypts each file with a unique symmetric key, and access control is determined by possession of the key. Thus, only coarse-grained access control is supported. Cryptfs [33] is similar to CFS except that the keys are associated with groups of files. Cepheus [13] is also similar to CFS but supports group sharing. However, it relies on a trusted group database server to determine access rights.

SiRiUs [15] is the most closed to our proposal. In SiRiUs, each file is encrypted with a file encryption key. For access control, SiRiUs attaches to each file a metadata that contains the file access control list. Each entry in the list is the encryption of the file encryption key under an authorized user's public key. Only the file owner has the authority to change the metadata, while in GPASS, *our goal is to allow group leaders to be able to delegate their role to other users*. To guarantee freshness, each SiRiUs's user generates a hash tree of its files and directories and periodically timestamps the root of the tree using a freshness daemon. We consider having such a always-running daemon to be too strong assumption for a password manager application.

Plutus [19] is a similar cryptographic file system to SiRiUs. Each file in Plutus is divided into several blocks, each of which is encrypted with a unique symmetric file-block key. Files can be grouped into a filegroup and share the same set of file-block keys. These file-block keys are further encrypted with the lockbox-key of the filegroup. If the owner of a filegroup wants to share access, it distributes the lockbox-key of the filegroup to the intended users. These users can then redistribute the key to other users. We can see that an independent mechanism is required to identify members of a group with such access control mechanism. The paper also does not specify how it can be done.

Yu et al. [32] proposed a file system whose access control is based on key-policy attributed-based encryption (KP-ABE) [17] and proxy re-encryption. With these techniques, the system achieves fine-grain access control and data owners can delegate most computation tasks to the cloud server. However, like SiRiUs, the system does not allows delegation of administration rights.

3   Models and Assumptions

This section presents the system and threat models of GPASS. We also describe the system goals and the security properties that we aim to achieve.

3.1   System Model

At a high level, GPASS is composed of the following principals: *server*, *user*, and *device*. The users manage *accounts* on their set of trusted devices (since passwords are always stored with the respective usernames and domains, we will use "accounts" instead). Each device runs the password manager client software that performs all cryptographic operations to protect the user data before sending it to the server for synchronization purpose. The server is operated by the *service provider*. It stores the encrypted user accounts as well as access control information. It also allows update and synchronization requests from authorized clients.

3.2 Threat Model

We assume the following in our threat model. First, *the service provider might be malicious.* That is, it will try to discover as much secret information as possible based on its inputs. However, we assume that the service provider does not wish to attack its users in a public manner because it has a reputation to protect.

Second, *the password manager client is trusted.* While an attacker who has obtained user devices can crack the master password with brute-force attacks, we do not consider this threat. The password manager design presented in this paper can be combined with other proposals in the literature [6,8,16,26] to provide both secure group management and password-cracking resistance.

Third, *each entity is preloaded with a public-key pair* and the public key can be securely obtained by others when necessary.

3.3 System Goals

GPASS allows groups to share accounts. Users can arbitrarily form groups, and each user can belong to multiple groups. A user can have multiple roles in a group. We consider two roles in this paper: *leader* and *member.* These roles have the following properties:

Authorizations: Any member or leader can add, edit, or remove the shared accounts of the group, while only the leaders have the authority to add new leaders and members and to remove existing ones. For simplicity, we do not differentiate between read and write access in GPASS. Extending our proposal to support this privilege separation is trivial by using an additional key for signing data as in [15].

Separate Revocation: If a user has multiple roles in a group, each role can be revoked separately.

Role Persistence: A member's role is valid until it is explicitly revoked. It means that even when the leader who grants the role to the member is revoked, the role of the member remains valid.

3.4 Security Goals

GPASS aims to achieve the following security properties.

Confidentiality and Integrity of Shared Accounts: The shared accounts should be protected from the users that are not granted access. Even the administrators of the server or the attackers who have compromised the server should not be able to read them. Moreover, unauthorized modifications to the accounts must be detected by GPASS.

Secure Group Sharing: The server should enforce access control on all requests to the shared password vaults and allow only authorized operations. Any unauthorized operations must be detected by GPASS.

Freshness of Access Control Information: GPASS should guarantee the freshness of the access control information. This guarantee allows timely revocation and prevents rollback attacks [25] on access control information. However, it is practically impossible to guarantee freshness without online trusted parties. Thus, GPASS aims to achieve *fork consistency*, which is the strongest notion of freshness possible without trusted parties [22]. Fork consistency implies that if users can communicate or see each other's operations, they can detect the rollback attacks.

4   GPASS

We now get to present our password manager architecture. Before getting into the details, we introduce our notation: We denote an encryption using key K as $E_K(.)$, in which K can be either a symmetric key or the public part of an asymmetric key pair. We denote a hash function as $H(.)$. We denote a message authentication code (MAC) that is created with symmetric key K as $MAC_K(.)$ and a signature generated by the private key of an entity I as $Sig_I(.)$. For simplicity, we will use MAC_K and Sig_I to denote a MAC and a signature of all the data that the they are tagged with, respectively.

4.1   Overview

We take a key-oriented approach in GPASS. Specifically, each user is represented by an asymmetric key pair $[PK_U, SK_U]$. The public key PK_U identifies the user, while the private key SK_U is stored in the user's devices and never divulged. If the user has multiple devices, its key pair needs to be synchronized between them via a secure out-of-band channel. The user authenticates itself to the server with its key pair when requesting access or performing any operation. With this key-oriented approach, the group management in GPASS is as follows.

Any user can create a group by generating a key pair $[PK_G, SK_G]$ and giving the group a name. The group is identified by the combination of the public key PK_G and the name. We denote a group as G:

$$G = \langle PK_G, name \rangle. \tag{1}$$

We refer to PK_G as the *group's master key*. Initially, PK_G is the only leader of the group. The group owner will use it to add new members to the group.

The members of a group share access to a *password vault* on the server. The password vault contains two parts: the *encrypted data* and the *metadata*. The encrypted data contains the shared accounts in encrypted form, while the metadata contains the access control information.

Encrypted Data. The shared accounts are encrypted with a symmetric cipher. The encryption key must have sufficient entropy to make offline cracking attacks infeasible. We refer to the key as the *data encryption key* (*DEK*). The key is initially chosen by the group owner. When a member of the group is revoked, the *DEK* is changed and the shared accounts are re-encrypted with the new key. We assume that a user account includes: a unique ID, a username, a password, and the respective domain (e.g. a website's URL):

$$EncryptedData = E_{DEK}(account_1, \ldots, account_n), \ MAC_{DEK}$$
$$\text{where } account_i = \langle ID_i, name_i, password_i, domain_i \rangle, \text{ for } i = 1 \ldots n. \tag{2}$$

To provide integrity, the encrypted data is tagged with a message authentication code (MAC) calculated over a hash of the accounts using the *DEK*.

Metadata. The metadata contains access control information. It consists of a list of *lock boxes* and a *timeline of membership events*. Only the group leaders have the authority to edit the metadata.

. Each lock box corresponds to a member of the group. The lock box of user U is of the following form:

$$Lockbox_U = \langle KeyID_U, E_{PK_U}(DEK) \rangle \tag{3}$$

The lock box includes the *DEK* encrypted under the public key of the member and is tagged with the key ID corresponding to the public key (e.g. a hash of the public key). With these lock boxes, the group members can easily access the shared vault without storing the *DEK* locally. They also help to distribute the new *DEK* when it is changed.

The timeline is an *ordered* list of all membership events related to the group, which can be of either of two types: *add-member* and *revoke-member*. We construct the timeline as follows. Initially, it is empty. When a group leader issues a membership event, it appends the event to the end of the timeline. More specifically, after $n-1$ events, if a group leader PK_L wants to add role R to key PK_U, it appends the following event to the timeline:

$$E_n = \langle n, PK_L, PK_U, R, G, \text{``add-member''}, H(E_{n-1}), H(DEK), Sig_L \rangle. \tag{4}$$

A revocation is represented by a similar event:

$$E_n = \langle n, PK_L, PK_U, R, G, \text{``revoke-member''}, H(E_{n-1}), H(DEK), Sig_L \rangle. \tag{5}$$

These events function like signed certificates and revocations in public-key infrastructures (PKI) such as X509 [18] and SPKI [10, 11]. The difference is that they are stored on a central server for timestamping and accounting purposes, instead of being distributed to the intended users. By including in each event a hash of the previous event, no single entity, including the server, can tamper with their linear order if the signature algorithm is secure. The addition of events to the timeline is synchronized by the server so that no conflict is caused by multiple users updating the timeline at the same time. As we will see below, the timeline enables determination of each key's roles in the group. It also provides *transparency* to the access control information, which allows users to detect misbehavior of the server.

4.2   Fundamental Operations

Now we get to describe the fundamental operations of the system.

Role Determination. Role determination is a crucial process in GPASS. It enables the server to enforce access control on the password vaults. Also, it allows the users to verify the access control information stored in the metadata.

Algorithm 1. Constructing lists of leaders and members of a group

---

Input: PK_G, $Timeline_G$
1: $leaders \leftarrow \{PK_G\}$
2: $members \leftarrow \emptyset$
3: **for all** $E_i \in Timeline_G$ **do**
4:     $PK_I \leftarrow$ getIssuer(E_i)
5:     $PK_S \leftarrow$ getSubject(E_i)
6:     **if** $PK_I \in leaders$ & validSignature(E_i, PK_I) **then**
7:        **if** getType$(E_i) =$ "add-member" **then**
8:           **if** getRole$(E_i) =$ "leader" **then**
9:              $leaders \leftarrow leaders \cup \{PK_S\}$
10:           **else if** getRole$(E_i) =$ "member" **then**
11:              $members \leftarrow members \cup \{PK_S\}$
12:           **end if**
13:        **else if** getType$(E_i) =$ "revoke-member" **then**
14:           **if** getRole$(E_i) =$ "leader" **then**
15:              $leaders \leftarrow leaders \backslash \{PK_S\}$
16:           **else if** getRole$(E_i) =$ "member" **then**
17:              $members \leftarrow members \backslash \{PK_S\}$
18:           **end if**
19:        **end if**
20:     **end if**
21: **end for**
22: **return** $leaders$, $members$

Before presenting the process, we *define who is a group member or leader.* Let $G = \langle PK_G, name \rangle$ be a group, which is created at time t_0. At time t_0, PK_G is assigned role "leader" in G by default. A key PK_U has role R in a group at time $t > t_0$ if the following are satisfied: (1) there exists a key PK_1 that adds role R to PK_U at time t_1, such that $t_0 < t_1 < t$ and PK_1 has role "leader" in G at time t_1, and (2) there does not exist a key PK_2 that revokes role R of PK_U at time t_2, such that $t_1 < t_2 < t$ and PK_2 has role "leader" in G at time t_2.

According to the above definition, a role can only be granted to a key by a group leader and the role is persistent until it is explicitly revoked. Furthermore, if a key has multiple roles, revoking one does not invalidate the others. Thus, the definition satisfies all the properties that we specified in Sect. 3.3.

The role determination process is as follows. First, the verifier runs Algorithm 1 to construct the lists of leaders and members of the group. Initially, only the group's master key is in the list of leaders. The verifier then traverses through the timeline of the group. For each event in the timeline, it checks whether the event is *authorized*. An event is authorized if (1) the signature is cryptographically valid and (2) the issuer is in the list of leaders after the previous event. If the check succeeds, the subject of the event is added or removed from the lists depending on whether it is a add-member or revoke-member event, respectively. After the last event has been processed, the two lists of leaders and members indicate the current role assignments of the group. The verifier can then determine the roles of any key by finding it in the lists.

In practice, the server can maintain the lists of leaders and members together with the password vault of each group and update them when a new event is available. In the rest of the paper, *we assume that the server maintains these two lists for each group*.

Before describing other operations, we define two sub-processes, which will be used in all the operations.

Subprocess: Verifying Metadata Updates (run by the server). The server must verify all the updates to the metadata before accepting them. This verification is to prevent denial-of-service attackers from filling the timeline with invalid data. It is not needed for the correctness of membership management. The server verifies a new event as follows.

1. If the issuer of the event is not in the list of leaders, the verification fails.
2. Verify whether the signature of the event is cryptographically valid. If not, the verification fails. Otherwise, the verification succeeds.

If the verification succeeds, the server updates its local lists of leaders and members according to the event's information as in Algorithm 1.

Subprocess: Verifying the Metadata (run by any group member). Whenever accessing the shared password vault, the group members need to verify it first. This is to guarantee that the server does not accept any unauthorized membership events and the *DEK* is up-to-date. If the verification fails, it means that the service provider might have equivocated. Thus, the group members should raise an alarm about the honesty of the service provider in these cases (We do not specify how this is done in this paper).

To verify the metadata, the user remembers the last event that it has processed and maintains the lists of leaders and members locally as the server does. This way, the user only needs to download and verify the new events that it has not seen before. The verification process is as follows.

1. Traverse through the new events in their linear order. For each event, check whether it is authorized. If not, the verification fails. Otherwise, update the local lists of leaders and members as in Algorithm 1.
2. Calculate a hash of the *DEK*. If it is equal to that of the last event, the verification succeeds. Otherwise, the verification fails.

Accessing Shared Accounts (run by any group member). A group member can access the shared accounts with the following steps:

1. Authenticate itself to the server to download the password vault. Note that in all operations, a user does not need to download the lock boxes of all the members but only its respective one.
2. Decrypt the *DEK* and verify the metadata. Raise an alarm if the verification fails.
3. Decrypt the encrypted data and verify the MAC with the *DEK*.

Creating a New Account (run by any group member). A group member can create a new shared account with the following steps:

1. Run the process of accessing shared accounts above.
2. Append the new account to the existing list of shared accounts.
3. Encrypt the list and re-calculate the *MAC* using the *DEK*.
4. Submit the updates to the server.

Modifying an account is similar to this process. Thus, we will not present it in this paper.

Adding a Member (run by a group leader). Group leader PK_L adds user PK_U to role R in group G with the following steps.

1. Run the process of accessing shared accounts above.
2. Create a lock box for the new user:

$$Lockbox_U = \langle KeyID_U, E_{PK_U}(DEK) \rangle \qquad (6)$$

3. Create a new add-member event:

$$E_n = \langle n, PK_L, PK_U, R, G, \text{"}add\text{-}member\text{"}, H(E_{n-1}), H(DEK), Sig_L \rangle. \quad (7)$$

4. If R is "member", add PK_U to its local list of members. Otherwise, add PK_U to its local list of leaders.
5. Send the updates to the server. The server must verify the updates before accepting them.

Revoking a Member (run by a group leader). Group leader PK_L revokes role R of user PK_U in group G with the following steps.

1. Run the process of accessing shared accounts above.
2. Generate a new data encryption key DEK_{new} and re-encrypt the shared accounts with the new key.
3. If R is "member", remove PK_U from its local list of members. Otherwise, remove PK_U from its local list of leaders.

4. Create a new lock box for each remaining key in the lists of leaders and members.
5. Create a new revoke-member event:

$$E_n = \langle n, PK_L, PK_U, R, G, \text{``revoke-member''}, H(E_{n-1}), H(DEK_{new}), Sig_L \rangle.$$
(8)

6. Send the updates to the server. The server must verify the updates before accepting them.

The above process only prevents the revoked member from accessing accounts that are added after the revocation. It is naturally not able to revoke access of the member to the existed passwords. Thus, in practice, the group leader must change these passwords immediately besides revoking the member. This task could be done automatically with the password manager as some of the current password managers have already done[1][2].

5   Security Analysis

This section analyzes the security of GPASS. We can see that the role determination process is the basis of the security of the system. We will first present and prove a lemma about its correctness. We will then use the lemma to argue about the security properties of the system.

Lemma 1. *Algorithm 1 correctly identifies all the members of a group and their roles in the group.*

Proof. Let $G = \langle PK_G, name \rangle$ be a group. Suppose that the timeline of G contains n events $E_1, \ldots, E_n$, where $n \geq 0$.

Denote $leaders_k$ as the set of leaders and $members_k$ as the set of members after the kth iteration of the for-loop of Algorithm 1. We will show by induction that $leaders_n$ and $members_n$ are correct, i.e. they correctly represent the role assignments of the group after n membership events, for any $n \geq 0$.

1. Basis step: With $n = 0$, we have: $leaders_0 = \{PK_G\}$, $members_0 = \emptyset$. By definition, PK_G is the only leader of the group initially. Thus, the sets are correct.
2. Induction step: Assume that $leaders_n$ and $members_n$ are correct for some $n \geq 0$, we will prove that $leaders_{n+1}$ and $members_{n+1}$ are also correct. Denote the issuer, the subject and the role specified in E_{n+1} as PK_I, PK_S, and R, respectively.
   According to the algorithm, if $PK_I \notin leaders_n$ or E_{n+1} is not cryptographically valid, $leaders_{n+1} = leaders_n$ and $members_{n+1} = members_n$. Thus, they are correct by our assumption.

---

[1] https://www.dashlane.com/features/password-changer.
[2] https://blog.lastpass.com/2014/12/introducing-auto-password-changing-with.html/.

Otherwise, E_{n+1} is issued by a leader, meaning that it is authorized. In this case, if E_{n+1} is an add-member event, PK_S is assigned role R by a leader and the role has not been revoked at the moment. By definition, PK_S has role R after the event. Thus, adding PK_S to $leaders_n$ if $R =$ "leader" or adding PK_S to $members_n$ if $R =$ "member" is a valid change to the lists. It follows that $leaders_{n+1}$ and $members_{n+1}$ are correct. Similarly, if E_{n+1} is a revoke-member event, it also leads to a valid removal to one of the lists. Thus, $leaders_{n+1}$ and $members_{n+1}$ are also correct in this case.

By induction, $leaders_n$ and $members_n$ are correct for all $n \geq 0$. Therefore, the lemma always holds.

Confidentiality and Integrity of the Shared Accounts. Assuming that the symmetric encryption algorithm used to protect the shared accounts is secure (e.g., using standard algorithm such as AES) and the encryption key is sufficiently strong, only those with the correct DEK can decrypt the encrypted data. Since the DEK is initially created by the group owner and then securely distributed to the new members via the lock boxes, none other than the group members have access to the key. Furthermore, the DEK is changed for every revocation, the revoked members do not have access to the new DEK and thus, all the future updates to the shared accounts. A malicious user or server might attempt to add unauthorized events to the timeline so that it gets recognized as a valid group member. This attack will not work if the role determination process is correct, which can be deduced from Lemma 1.

GPASS achieves integrity of the encrypted accounts with the MAC created using the DEK.

Secure Group Sharing. According to Lemma 1, the role determination process correctly identifies all the members of a group and their roles. Thus, unauthorized operations, such as adding new timeline events without being a leader, will be detected by the users during the metadata verification process. In other words, the users act as auditors that monitor the server's behavior.

Freshness. As we mentioned in Sect. 3.4, freshness can never be guaranteed unless a trusted party is involved. In fact, the current construction of GPASS cannot prevent rollback attacks performed by a malicious server: Suppose that the latest event in the timeline is the revocation of Carol, which is issued by Alice, and Bob is not aware of the revocation. The server can conceal the event from Bob and roll the lock boxes back to their previous version accordingly. This way, Bob will not notice the revocation and keep using the old DEK. Consequently, Carol can read any updates made by Bob to the shared accounts.

GPASS does not prevent the attack, but it allows users to detect it. By including in each membership event a hash of the preceding, the events are linked to each other and the last event commits to the entire membership history of the group. If the server equivocates as in the above scenario, it will have to *fork* the timeline between Alice and Bob. Assuming that n events have occurred before the revocation of Carol, the forked timelines will share only these n events and never be the same again. As a result, the forked users will not see changes made by each other. Thus, if users can communicate or see each other's operations, they can detect the attack.

6 Extended GPASS

We propose an extended version of GPASS, which tackles the rollback attack discussed in the previous section. The idea is to have multiple observers monitoring the service provider's behavior. The observers may be its major clients or independent organizations. Periodically, the service provider publishes summaries of its data to the observers. Users can then compare their view with the observers' and be confident that their data is fresh. This architecture is similar to that of the proposals for monitoring the security of the web PKI [4,20,21,31].

The service provider can consolidate all of its data to one hash value by constructing a *Merkle prefix hash tree* as follows. (A Merkle prefix tree is basically a binary tree where each branch of the tree corresponds to a unique bit string x. Each bit in x represents either a left or right turn on the way down. We say that the leaf node that corresponds to x is *indexed* by x.) Each leaf node in the tree is indexed by a group's id (e.g. a hash of the group's master key and name) and contains a summary of the group's access control information. Since the last event of the timeline covers the whole membership history of the group, we can use a hash of this event as the summary. The value of a non-leaf node is the hash of its two children. This way, the value of the root of the tree summarizes the whole tree. The service provider periodically (e.g. once per hour) signs the root and sends it to the observers. The observers then publish this value for the benefit of the users.

With the new architecture, the metadata verification process run by the user needs to be changed. The user is required to perform an additional check of whether the timeline is fresh with the following steps.

1. Query a *proof-of-freshness* (POF) of the group from the server. The server constructs the POF by following the path determined by the group's id in the Merkle prefix tree. On the way down, it accumulates the proof as a list of the values of the siblings of the path. Together with a hash of the last timeline event, the POF allows the user to calculate the root of the tree.
2. Calculate the root of the tree using the received POF and compare it with the observers. The more observers the user consults, the more confident it is about the freshness of the metadata. If there is inconsistency between the root value that the user calculated and the observers', the verification fails. Otherwise, the verification succeeds.

We leave the analysis and evaluation of this extension as an interesting avenue for future work.

7 Implementation

We implemented a proof-of-concept of the basic GPASS. The system includes a server and a command-line client. They were written with Python (2.7.11) and the PyCryptodome cryptography library (3.4.5)[3]. We used AES-GCM for authenticated encryption, SHA-256 as the hash function and RSA-2048 keys as the user identities. The DEK was 256-bit long and chosen uniformly at random.

To evaluate the system, we setup groups that shared a fixed number of 50 accounts. The accounts were randomly generated so that the passwords were 15 characters long and the domains were in Alexa's top 500 websites[4]. We then simulated the membership management in each group and measured the overhead of the system from a user's point of view when the number of users of the group changed. More specifically, we measured the processing time and the bandwidth of the password manager client (for both downloading and uploading). In the simulations, we assumed that 20% of the membership events were revocations and that the client always saw $n * 0.1$ new events, where n is the number of users of the group, in the metadata whenever it accessed the shared password vault. (The client needed to download these new events only instead of the whole timeline.) We ran the simulation on a 3.4 GHz Intel(R) Zeon(R) E3-1231 machine with 32 GB of RAM.

Table 1. Average client's processing time and bandwidth of three operations, accessing accounts, adding members, and revoking members, when the numbers of users of a group changes (sampled over 1000 executions)

Number of users	100	1000	10000
Access accounts (processing time)	11.24 ms	92.5 ms	951.94 ms
Access accounts (bandwidth)	7.75 KB	37.97 KB	340.40 KB
Add member (processing time)	13.92 ms	94.57 ms	1.04 s
Add member (bandwidth)	7.83 KB	38.92 KB	341.42 KB
Revoke member (processing time)	139.86 ms	1.39 s	14.14 s
Revoke member (bandwidth)	42.61 KB	354.16 KB	3.48 MB

Table 1 summarizes the results that we obtained. We evaluated three main operations of the client: accessing shared accounts, adding a member, and revoking a member. To access the shared accounts, the client basically has to verify the metadata and the encrypted data, and decrypt the encrypted data using the DEK. The process of adding member is mostly the same, except that the client additionally needs to add a new lock box and a new add-member event to the metadata. We can see that the simulation results also illustrate this similarity. The processing time of revocations is significantly more than that of the others.

---

[3] https://github.com/Legrandin/pycryptodome.
[4] http://www.alexa.com/topsites.

It is due to the computationally expensive distribution of the new DEK. Assuming that there are n users in the group, revoking a member involves $n - 1$ public key encryptions. In fact, when there are 10000 users in the group, approximately 13.17 out of 14.14 s were devoted for the encryption of the new DEK.

Regarding the bandwidth, the encrypted data containing 50 accounts in our setting was approximately 4KB in size. The rest of the bandwidth is mainly for downloading the new membership events and if the operation is revoking a member, uploading the new lock boxes. Thus, the bandwidth will increase linearly with the number of users of the group.

It can be seen that the averaged processing time and bandwidth increase linearly with the number of users in the group. This makes GPASS not scale well when large groups of users share passwords. However, groups usually do not get too big in practice. Most of the largest organizations in the world have less than a million employees[5]. Furthermore, passwords are usually shared within a team in an organization, which make the number of members of a group even smaller.

8   Discussion

This section describes the lessons that we have learned from this research.

First, in a system that involves distribution of confidential data (the DEK in our case), accountability for access control information is required. In a distributed access control mechanism, such as SPKI, entities are usually represented by their public keys and access rights are encoded into certificates. These certificates are kept by the subjects instead of being stored with a central entity. When requesting access to a resource, a subject sends related certificates to the verifier, which then determines whether the subject is permitted to perform the requested operation by verifying the certificates. Such distributed access control models do not allow identifying all the subjects that have access rights to a resource. While it might be tempting to maintain a list of these subjects, such tasks cannot be left to the care of an untrusted party because the list could be tampered with. Therefore, if we apply a distributed access control model to our password manager application, it is difficult for a group leader to determine all the group members. Of course, it can locally maintain a list of the members that it has added, but it is not aware of those added by other leaders. As a result, it cannot distribute the DEK when the key is changed.

Second, we can entrust access control to an untrusted server as long as all access control information is transparent. Indeed, the transparency of the access control information in GPASS enables the users to audit the system. They are able to detect whether there are any unauthorized operations allowed by the server. The users, however, cannot detect the rollback attack without communicating with each other. In systems where communication between users is a

---

[5] https://www.statista.com/statistics/264671/top-20-companies-based-on-number-of-employees/.

natural part, such as messaging applications, the users themselves could detect all misbehaviors of the server without depending on trusted observers.

9 Conclusion

This paper presents GPASS—a password manager architecture that allows sharing passwords within groups. GPASS assumes that the storage server can be malicious and provides a cryptographic access control mechanism which allows users to monitor the behavior of the server. The system protects the confidentiality and integrity of shared accounts. The tradeoff for efficient sharing through the online server is that freshness of the stored data cannot be guaranteed. To tackle this, we propose an extended version of GPASS, which will be analyzed and evaluated in future work.

References

1. 1Password, December 2016. https://agilebits.com/onepassword
2. F-secure key, December 2016. https://www.f-secure.com/en/web/home_global/key
3. LastPass: Password manager, autoform filter, random password generator & secure digital wallet app, December 2016. https://lastpass.com/
4. Basin, D., Cremers, C., Kim, T.H.J., Perrig, A., Sasse, R., Szalachowski, P.: ARPKI: attack resilient public-key infrastructure. In: Proceedings of the 2014 ACM SIGSAC Conference on Computer and Communications Security, pp. 382–393. ACM (2014)
5. Blaze, M.: A cryptographic file system for UNIX. In: Proceedings of the 1st ACM conference on Computer and communications security, pp. 9–16. ACM (1993)
6. Bojinov, H., Bursztein, E., Boyen, X., Boneh, D.: Kamouflage: loss-resistant password management. In: Gritzalis, D., Preneel, B., Theoharidou, M. (eds.) ESORICS 2010. LNCS, vol. 6345, pp. 286–302. Springer, Heidelberg (2010). doi:10.1007/978-3-642-15497-3_18
7. Bonneau, J.: Guessing human-chosen secrets. Ph.D. thesis, University of Cambridge (2012)
8. Chatterjee, R., Bonneau, J., Juels, A., Ristenpart, T.: Cracking-resistant password vaults using natural language encoders. In: IEEE Symposium on Security and Privacy, pp. 481–498. IEEE (2015)
9. Electronic Frontier Foundation: National security letters, July 2016. https://www.eff.org/issues/national-security-letters
10. Ellison, C., Frantz, B., Lampson, B., Rivest, R., Thomas, B., Ylonen, T.: SPKI certificate theory. RFC 2693, IETF (1999)
11. Ellison, C.M.: The nature of a usable PKI. Elsevier Comput. Netw. **31**(9), 823–830 (1999)
12. Florencio, D., Herley, C.: A large-scale study of web password habits. In: Proceedings of the 16th International Conference on World Wide Web, pp. 657–666. ACM (2007)
13. Fu, K.E.: Group sharing and random access in cryptographic storage file systems. Ph.D. thesis, Massachusetts Institute of Technology (1999)
14. Gellman, B.: The FBI's secret scrutiny, July 2015. http://www.washingtonpost.com/wp-dyn/content/article/2005/11/05/AR2005110501366.html

15. Goh, E.J., Shacham, H., Modadugu, N., Boneh, D.: SiRiUS: securing remote untrusted storage. NDSS **3**, 131–145 (2003)
16. Golla, M., Beuscher, B., Dürmuth, M.: On the security of cracking-resistant password vaults. In: Proceedings of the 23rd ACM Conference on Computer and Communications Security. ACM (2016)
17. Goyal, V., Pandey, O., Sahai, A., Waters, B.: Attribute-based encryption for fine-grained access control of encrypted data. In: Proceedings of the 13th ACM Conference on Computer and communications security, pp. 89–98. ACM (2006)
18. Housley, R., Ford, W., Polk, W., Solo, D.: Internet X. 509 public key infrastructure certificate and CRL profile. RFC 2459, IETF (1998)
19. Kallahalla, M., Riedel, E., Swaminathan, R., Wang, Q., Fu, K.: Plutus: scalable secure file sharing on untrusted storage. In: Fast, vol. 3, pp. 29–42 (2003)
20. Kim, T.H.J., Huang, L.S., Perring, A., Jackson, C., Gligor, V.: Accountable key infrastructure (AKI): a proposal for a public-key validation infrastructure. In: Proceedings of the 22nd International Conference on World Wide Web, pp. 679–690. International World Wide Web Conferences Steering Committee (2013)
21. Laurie, B., Langley, A., Kasper, E.: Certificate transparency. RFC 6962 (2013)
22. Li, J., Krohn, M.N., Mazières, D., Shasha, D.: Secure untrusted data repository (SUNDR). In: OSDI, vol. 4, p. 9 (2004)
23. Li, Z., He, W., Akhawe, D., Song, D.: The emperor's new password manager: security analysis of web-based password managers. In: USENIX Security, pp. 465–479 (2014)
24. Lichtblau, E.: Judge tells Apple to help unlock iPhone used by San Bernardino Gunman, July 2016. http://www.nytimes.com/2016/02/17/us/judge-tells-apple-to-help-unlock-san-bernardino-gunmans-iphone.html
25. Mazires, D., Shasha, D.: Don't trust your file server. In: Proceedings of the Eighth Workshop on Hot Topics in Operating Systems, pp. 113–118. IEEE (2001)
26. McCarney, D., Barrera, D., Clark, J., Chiasson, S., van Oorschot, P.C.: Tapas: design, implementation, and usability evaluation of a password manager. In: Proceedings of the 28th Annual Computer Security Applications Conference, pp. 89–98. ACM (2012)
27. Miller, E., Long, D., Freeman, W., Reed, B.: Strong security for distributed file systems. In: IEEE International Conference on Performance, Computing, and Communications, pp. 34–40. IEEE (2001)
28. Silver, D., Jana, S., Boneh, D., Chen, E.Y., Jackson, C.: Password managers: attacks and defenses. In: Usenix Security, pp. 449–464 (2014)
29. Vigo, M.: Even the LastPass will be stolen, deal with it! February 2017. http://www.martinvigo.com/even-the-lastpass-will-be-stolen-deal-with-it/
30. Whitney, L.: LastPass CEO reveals details on security breach, December 2016. http://www.cnet.com/news/lastpass-ceo-reveals-details-on-security-breach/
31. Yu, J., Cheval, V., Ryan, M.: DTKI: a new formalized PKI with no trusted parties. IACR Cryptol. ePrint Arch. **2014**, 600 (2014)
32. Yu, S., Wang, C., Ren, K., Lou, W.: Achieving secure, scalable, and fine-grained data access control in cloud computing. In: INFOCOM, 2010 Proceedings IEEE, pp. 1–9. IEEE (2010)
33. Zadok, E., Badulescu, I., Shender, A.: Cryptfs: A stackable vnode level encryption file system. Technical report, Technical report CUCS-021-98, Computer Science Department, Columbia University (1998)

Towards Accelerated Usage Control Based on Access Correlations

Richard Gay[✉], Jinwei Hu, Heiko Mantel[✉], and Johannes Schickel[✉]

Department of Computer Science, TU Darmstadt, Darmstadt, Germany
{gay,hu,mantel,schickel}@mais.informatik.tu-darmstadt.de

Abstract. Low run-time overhead is crucial for the practicability of usage-control mechanisms. In this article, we propose an approach to accelerate usage control by exploiting access correlations. Our approach combines two main ingredients: firstly, a technique to compute decisions ahead of time and, secondly, a method to guide selection of usage events to pre-compute decisions for. For the first, we speculatively pre-compute decisions for usage events. For the second, we exploit access correlations to identify high acceleration potential. We implemented our approach and evaluated it in a case study of security policy enforcement in a distributed storage system. Our empirical results show that the speedup is substantial. More concretely, the speedup on average is up to 61.5%.

1 Introduction

Usage control [29] augments access control by protecting the access to resources as well as the subsequent use of the resources. For instance, usage control can ensure that a confidential document can only be accessed by authorized users and can also constrain the number of times the document is printed or propagated to other authorized users. Dynamic mechanisms are a popular approach to enforce usage control (e.g., [7,9,13,15,20]). Analysis of the system at run-time allows such approaches to precisely enforce usage control policies. However, by operating at the run-time of the system, dynamic mechanisms inevitably impose a performance overhead on the system.

Large performance overheads can easily deter the users of a system. How much overhead is acceptable in practice depends on the application domain. The question how much overhead is tolerable has been investigated, for instance, in the area of web services with the finding that delays of already a few hundred milliseconds can result in sales loss, reduced service use, and generally a competitive disadvantage [5,12,25,30,31].

A standard approach to reduce overhead of dynamic mechanisms is to reduce the number of program instructions that are instrumented to invoke the mechanism by static analysis (e.g., [1,3,10,11,22,27]). In this article we follow an orthogonal approach: We exploit domain knowledge to accelerate enforcement during run-time. Thus, our approach can be employed in addition to existing approaches based on static analysis.

© Springer International Publishing AG 2017
H. Lipmaa et al. (Eds.): NordSec 2017, LNCS 10674, pp. 245–261, 2017.
https://doi.org/10.1007/978-3-319-70290-2_15

In this article, we propose SPEED^{AC}, an approach to **sp**eculatively pre-compute decisions based on access correlations. Concretely, a usage-control mechanism following our approach computes and stores decisions for possible future usage events on the side and uses these pre-computed decisions when the events actually occur. Our approach exploits that a lookup of a decision can be more efficient than computing the decision. For example, in a distributed setting, computing a decision is expensive when it requires network communication [13, 20]. For selecting which decisions to pre-compute, SPEED^{AC} exploits access correlations on pieces of data such as files, database entries, and in-memory objects. By access correlations we refer to correlations between accesses to data in a program resulting from access patterns encoded in the program logic and from how the program is used. For instance, that an employee uses her company's storage service to access the agenda of a business meeting correlates with her accessing the meeting presentation.

We demonstrate our approach in a case study of a distributed storage system in which we employ usage control against conflicts of interest. For this scenario, we implemented our approach in a concrete usage-control mechanism and empirically evaluated the performance of our implementation based on the 6-hour MSN BEFS access trace by Microsoft [18]. Our evaluation showed speedup on average of up to 61.5% compared to not utilizing access correlations, which indicates that our approach is feasible and can significantly accelerate usage control.

In summary, the technical contributions of this article are:

- the SPEED^{AC} approach to accelerate usage control by speculatively pre-computing decisions, where selection of usage events is guided by access correlations
- an implementation of SPEED^{AC} against conflict of interest in distributed storage systems, and
- an empirical evaluation of the performance of our proposed mechanism based on a 6-hour access trace by Microsoft.

To our knowledge, our work is the first based on the idea of exploitation of *probabilistic* correlations *between* usage events to accelerate dynamic mechanisms. Our evaluation shows that the speedup can be substantial and hence that this is an interesting direction to counter the overhead caused by dynamic usage-control mechanisms. Our article constitutes a first step: Further research will enable a better understanding of the full design space for concrete SPEED^{AC}-based mechanisms and its potential to accelerate usage control.

2   The Problem

Performing usage control means ensuring that the usage events performed by a target program comply with given usage constraints. Dynamic approaches can utilize accurate usage histories for precisely enforcing usage constraints. This comes at the cost of run-time overhead that is perceivable by users of the system.

According to studies by Amazon, Bing, and Google Search [5, 12, 30], page load time increases of 100–200 ms already have a negative impact on sales and user experience. This suggests that the run-time overhead caused by a usage-control mechanism for a single page request should remain below 100 ms for the mechanism to be acceptable in practice. A single page request, however, can trigger a cascade of requests. For example, we observed that loading a single page of Dropbox's web interface for browsing photos[1] triggered Firefox to send 75 requests to Dropbox. That is, in this example a usage-control mechanism may take at most 1.33 ms per request to remain below 100 ms in total.

We looked at several usage-control mechanisms whose perceivable overhead has been measured experimentally: a non-distributed [15] and a distributed [19] mechanism for usage control on data and copies of data; an access control mechanism for grid computing [7]; and a generic mechanism to enforce security policies in distributed systems, which has been used to enforce Chinese Wall policies in distributed systems [13]. Table 1 summarizes our observations. The mechanism by Harvan et al. [15] exhibits overheads between about 1.5 ms and 1 s, depending on the test case. The mechanism by Kelbert et al. [19] introduces overheads of at least 106 ms for file transfers of size 100kB. The mechanism by Colombo et al. [7] exhibits overheads between 6 ms and 400 ms. CliSeAu, by Gay et al. [13], yields overheads between 1.9 ms and 16.1 ms, depending on the concrete experiment.

Table 1. Perceivable overhead caused by usage-control mechanisms

Reference	Experiments	Overhead of	Overhead
[15]	File copying; compilation	Dummy policy	1.5–33 ms
[15]	File copying; compilation	Policy monitoring	4–1000 ms
[19]	FTP and HTTP file transfers	Data-flow tracking (best-case), dummy policy	106–2931 ms
[19]	FTP and HTTP file transfers	Data-flow tracking (worst-case), dummy policy	131–53353 ms
[7]	Data storage via custom test program	Access control, trust and reputation management	6–400 ms
[13]	FTP file transfers	Local decision-making	1.9–3 ms
[13]	FTP file transfers	Cooperative decision-making	2.7–16.1 ms

The overhead of the different approaches is low already. However, in the Dropbox example, even the lowest overheads – 1.5 ms per access for a dummy policy and 1.9 ms for a Chinese Wall policy – accumulate to a total overhead of 112.5 ms for 75 requests. This raises the question how to further reduce overhead of dynamic mechanisms. The problem we therefore address in this article is how to further accelerate dynamic usage-control mechanisms.

---

[1] https://www.dropbox.com/photos.

3 Our Approach: SPEED^{AC}

The SPEED^{AC} approach for accelerating usage control is to speculatively pre-compute decisions ("SPEED") on the side and to use access correlations ("AC") to determine the speculative aspect of the pre-computation. Through the pre-computation on the side, SPEED^{AC} enables a reduction of the overhead that one can perceive when using a program that is subject to a usage-control mechanism.

3.1 Speculative Pre-computation of Decisions

With SPEED^{AC}, the decisions made by a usage-control mechanism are not all computed when they are needed but are, to some extent, pre-computed. That is, the mechanism computes decisions for some usage events already before the events actually occur. Since it is typically not known in advance which usage events occur, decisions are pre-computed speculatively for usage events that the mechanism can suspect to occur. Such a decision is then computed as if the usage event was actually about to occur. Rather than being used directly, the decision is stored in memory or in a database and might not be used at all when the respective usage event never occurs. For brevity, in the following we refer to speculatively pre-computed decisions simply as pre-computed decisions.

The pre-computation of decisions with SPEED^{AC} is performed "on the side" by a usage-control mechanism. That is, the mechanism need not suspend the target program for pre-computing decisions but uses, e.g., a concurrent thread for the pre-computation. The mechanism triggers the pre-computation of decisions after it has handled a concrete usage event. In particular, it allows the pre-computation to take the decision for this newest usage event into account.

Concretely, the decision-making with SPEED^{AC} integrates as follows into how the usage-control mechanism processes usage events. When the mechanism intercepts a usage event that the running target program is about to perform, the mechanism first performs a lookup for a pre-computed decision. In case of success, i.e., if a pre-computed decision for the intercepted event is available, this decision is enforced. For instance, if the decision is to permit the usage event, then the mechanism allows the program to perform the event. If the decision is to prevent the usage event, then the mechanism can, e.g., return an error code to the target such that it can afterwards resume its execution. When the lookup fails, i.e., when no pre-computed decision is available, then the mechanism computes a decision on the spot and enforces this decision. Unless a decision demanded to terminate the target program, the mechanism resumes the target after enforcing the decision and simultaneously triggers the pre-computation of decisions.

Figures 1 and 2 illustrate the cases of lookup success and, respectively, lookup failure. In the figures, time flows from left to right and shaded boxes represent functionality that is performed during the time. Notably, the target is blocked while the mechanism has intercepted an event and has not yet enforced a decision for the event. The target is not blocked while decisions are pre-computed.

We consider an attacker model of a malicious user of the target program. Concretely, the attacker can interact with the interface exposed by the program,

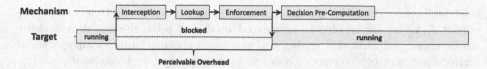

Fig. 1. Enforcement in case of lookup success

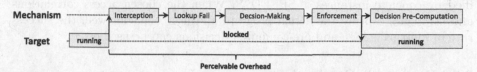

Fig. 2. Enforcement in case of lookup failure

such as a graphical user interface or a web interface, for trying to circumvent usage control. The attacker cannot directly observe or modify the mechanism. For the given attacker, neither soundness nor precision need to be sacrificed for increased performance when exploiting access correlations as well as parallelism for speculatively pre-computing decisions on the side. Preserving soundness and precision requires careful design and implementation of concrete mechanisms that use SPEED^{AC}. For instance, pre-computed decisions should not be utilized by a mechanism when they have been rendered obsolete by subsequent events.

3.2 Utilization of Access Correlations

Pre-computing decisions for all possible usage events is infeasible with regard to storage and computation time due to the generally vast number of such events. For selecting a limited set of decisions to pre-compute and yet achieving a high rate of successful lookups, we propose to use access correlations.

Access correlations on data result from access patterns encoded in a target program and from usage patterns established by the program's users. They capture which accesses to pieces of data are stochastically dependent. In this article, we focus on positive correlations, i.e., on cases in which the likelihood of accesses to two pieces of data occurring together during the run-time of the program is *higher* than the likelihood would be in case of stochastic independence. When the correlation between accesses to two pieces of data is sufficiently strong, we call the pieces of data correlated. An example of access correlations are correlations between accesses to disk blocks and files. Outside the domain of usage control, exploiting such correlations has already been proposed for accelerating file accesses through improved caching and prefetching (e.g., [16,23]). Access correlation between files have been successfully calculated, e.g., by treating the metadata of files as a multi-dimensional attribute space and marking files in close proximity as correlated [16].

SPEED^{AC} proposes that a mechanism utilizes access correlations as follows. Suppose a decision for a usage event that accesses a piece of data d has just been enforced. Then the mechanism selects, using some selection strategy, a subset D of all pieces of data correlated to d and pre-computes decisions for usage events on D. Later during the run-time, the mechanism employs some termination strategy to remove obsoleted pre-computed decisions again.

When access correlations are considered sufficiently strong as well as the selection strategy and termination strategy are scenario-specific and to be specified by concrete instances of SPEED^{AC}. When the chosen access correlations and strategies capture the actual program and user behavior well, the success rate of lookups is high and pre-computed decisions can be used often.

3.3   Perceivable Overhead

The perceivable overhead of a usage-control mechanism on a program is the additional delay, caused by the mechanism, that a user of the program experiences in her interaction with the program. The user might experience this delay, e.g., between a mouse click and the response by the GUI program or between a browser request to a web service and the resulting page being displayed.

Figures 1 and 2 depict the perceivable overhead caused with SPEED^{AC} for a single usage event. The overhead for a successful lookup includes the time for intercepting the event, for looking up a decision, and for enforcing a decision. The overhead for a failed lookup additionally includes the decision-making. The pre-computation of decisions is not part of the perceivable overhead, as it is performed while the target is running rather than while the target is blocked.

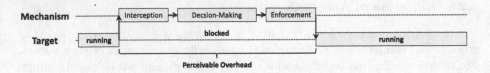

Fig. 3. Enforcement in current approaches

Traditional usage-control mechanisms (e.g., [1,11,13,20,27]) handle usage events as shown in Fig. 3: The mechanisms block the target program for interception, decision-making, and enforcement. They do not pre-compute decisions and perform their functionality sequentially to the target.

The perceived overhead in traditional mechanisms clearly is smaller than the perceived overhead caused by failed lookups in SPEED^{AC}: The latter comprises all tasks of the former and additionally includes the lookup. How the case of a lookup success compares to the traditional approach boils down to how the successful lookup compares to the decision-making. Decision-making can be significantly more time-consuming than a lookup in scenarios with complex usage constraints or in distributed systems, in which decision-making involves network communication. SPEED^{AC} reduces the perceived overhead if the time saved through successful lookups outweighs the overhead of failed lookups on average. We elaborate an example of such a setting in the remainder of this article.

4 Case Study

We use a case study to demonstrate how SPEED^{AC} can be realized in a concrete application scenario and how much acceleration can be achieved. In the application scenario, a distributed storage service offers storage space to its users. The service consists of multiple, spatially distributed servers through which the users can access the storage. Figure 4 depicts the possible interactions between users and the service. Through a network such as the Internet, each user can connect to any of the servers for storing and retrieving files from the service.

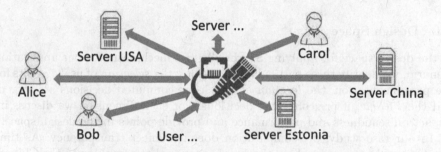

Fig. 4. Distributed storage service

The usage of the service shall be constrained according to a Chinese Wall policy [4] in order to technically counter conflicts of interest. That is, from a class of competing companies, each user may only access the files owned by one company. A mechanism can enforce this usage constraint by controlling the read and write events performed on behalf of users. We chose the Chinese Wall policy as an example of a business security requirement for which the computation of decisions in a distributed setting is non-trivial in general [26].

In the remainder of this article, we call two files *conflicting* if they are owned by competing companies. We lift this notion to usage (i.e., read and write events) by calling two usage events conflicting when the files accessed by the usage events are conflicting. The notion establishes an irreflexive and symmetric binary relation on usage events. By *equivalence classes* on usage events, we refer to the equivalence classes of the reflexive transitive closure of this relation.

5 Enforcement Mechanism

We design and implement a mechanism following SPEED^{AC} for the setting in Sect. 4. The core challenge is to design a mechanism that is effective *and* performant. Effectiveness demands *soundness*, i.e., that the mechanism assures the absence of policy violations, and *transparency*, i.e., that the mechanism permits all accesses that do not violate the policy [24].

Our mechanism is built on Gay, Hu, and Mantel's mechanism in [13]. To monitor and to intervene with a target program's execution, the mechanism encapsulates the target program into *enforcement capsules*, which we refer to as *nodes* in our setting. We re-use interception and enforcement, but design our own decision-making algorithm. The original decision-making algorithm works as follows: The mechanism maps usage events to nodes that are responsible for deciding on them. Effectiveness is achieved through the following requirement: Whenever two usage events conflict, the same node is responsible to decide on the events. We call this property on the responsibility distribution *properness*.

We first explore the design space for the decision-making algorithm. Second, we describe the proposed decision-making algorithm and the underlying design decisions. Finally, an overview over key implementation details is presented.

5.1   Design Space

In the design space for applying SPEED^{AC} to a mechanism for our application scenario, we identify three particular dimensions: the *selection* of usage events for the pre-computation, the *location* at which pre-computed decisions are stored, and the *lifetime* of pre-computed decisions. For each dimension we discuss its impact on soundness and performance and provide points in the design space.

In our case study, pre-computation does not affect transparency: As time advances and users access files, the set of accesses that comply with the Chinese Wall policy shrinks monotonically. Pre-computed decisions from earlier points in time, thus, do not violate transparency when they are enforced.

Selection. According to SPEED^{AC}, a mechanism can select the usage events for the pre-computation from the set of usage events that are correlated to the previously intercepted usage event. A greedy strategy is to select all correlated usage events. More cautious strategies are, e.g., to select at most a single correlated usage event and ensure that for each user only one pre-computed decision exists or to select a maximal set of correlated usage events such that for each user and equivalence class only one pre-computed decision exists.

The selection strategy can affect the soundness and performance of a mechanism. For instance, the greedy strategy might select two permissible but conflicting usage events for the pre-computation. If the mechanism would enforce the decisions for both events, it would violate the Chinese Wall policy. The more cautious strategies do not exhibit this property, as they prevent conflicting pre-computed decisions. Concerning performance, the greedy strategy yields a higher chance of successful lookups than the cautious strategies. However, it also increases the lookup and maintenance costs for pre-computed decisions.

Location. The location, i.e., the node at which a pre-computed decision is stored, is a dimension opened up by the distributed setting. A strategy to select the location can be static or dynamic. A static strategy does not adapt to system behavior but always uses the same node for each pre-computed decision. For example, the strategy could fix a node for each usage event based on the file location. A dynamic strategy selects the node based on observed system behavior. For example, the strategy could track where a usage event occurs most often and store the associated pre-computed decision there.

The locations of pre-computed decisions affect the mechanism's performance. A pre-computed decision can only be looked up efficiently, when it is stored at the node intercepting the respective usage event. Otherwise, comparatively expensive network communication is required. A dynamic strategy can increase efficient lookup chances but at the cost of additional bookkeeping. When some nodes are only temporarily reachable via the network, using these nodes to store pre-computed decisions also affects the soundness of the mechanism.

Lifetime. The lifetime of pre-computed decisions can be controlled by termination strategies. A termination strategy can be to terminate certain pre-computed decisions during the lookup, after the enforcement of a pre-computed decisions, during on-the-spot decision-making, and/or during the selection of events for pre-computation. The strategy can be to terminate pre-computed decisions that would conflict with newly selected usage events for the pre-computation. Conversely, the termination strategy can also be to keep once pre-computed decisions and rather select fewer events in the next pre-computation step.

The termination strategy can affect the soundness and performance of the mechanism. For instance, if conflicting pre-computed decisions are not terminated when a permitting decision is computed on the spot, then subsequently utilizing a pre-computed decision might violate the Chinese Wall policy. The performance impact of a termination strategy is in two directions. Firstly, terminating a pre-computed decision that is not located at the node that triggers the termination requires network communication and is therefore expensive. Secondly, terminating more or fewer decisions has the same impact on the performance as choosing fewer or, respectively, more events for the pre-computation.

5.2   Design for Effectiveness

In our mechanism responsible nodes compute and pre-compute decisions. Each node memorizes its permitted usage events to compute decisions in compliance with the Chinese Wall policy. A node decides to prevent a usage event only when a conflicting usage event was permitted previously. On enforcement of a permit decision the responsible node memorizes the permission of the usage event.

We let pre-computed decisions induce *temporary responsibility shifts*: a node storing the pre-computed decision becomes temporarily responsible for the equivalence class of the underlying usage event. Conversely, when the pre-computed decision is terminated, the responsibility shifts back to the originally responsible node. As a result, the responsibility distribution is proper also in presence of temporary responsibility shifts. A temporarily responsible node can only lookup the pre-computed decision, it can not compute decisions by itself.

Our mechanism employs a cautious selection strategy: it selects a maximal set of correlated usage events such that for each user and equivalence class only one pre-computed decision exists. The employed termination strategy is twofold:

it terminates pre-computed decisions after their enforcement and during on-the-spot decision-making. We fix a static strategy to select the location of pre-computed decisions: for each usage event a node is fixed based on the file location.

Augmenting the mechanism, i.e., [13] with SPEED^{AC} based on the design choices we made preserves the effectiveness of the mechanism. Our mechanism's on-the-spot decision-making preserves effectiveness due the responsibility distribution being proper even in presence of temporary responsibility shifts. As previously discussed, pre-computed decisions do not affect transparency of our mechanism. Enforcing a pre-computed decision u would only break soundness when u permits a usage event and a decision v permits a conflicting usage event. We distinguish the two cases that could lead to a policy violation. (1) between computation of u and its enforcement an on-the-spot decision v is enforced. However, the termination strategy prevents this by terminating u during computation of v. (2) between computation of u and its enforcement a pre-computed decision v is enforced. The selection strategy prevents this: if v is computed before u, u is never pre-computed. If u is computed before v, v is never pre-computed. Thus, only either u or v is enforced. Absence of both (1) and (2) preserves soundness.

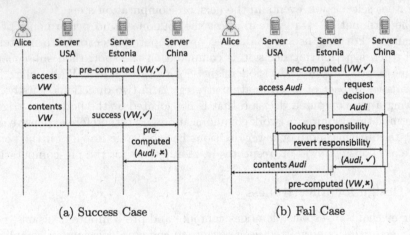

(a) Success Case   (b) Fail Case

Fig. 5. Decision-making examples

Example. Consider an audit process where *Alice* is hired to audit car manufacturing companies. The companies' data is distributed over servers *USA*, *Estonia*, and *China*. To avoid conflict-of-interest *Alice* is only given access to a single company's data. Server *China* is responsible to decide on usage events. Files of the companies *VW* and *Audi* are correlated in both directions.

A successful lookup is given in Fig. 5a. Initially, a pre-computed decision for access to *VW* arrives at node *USA*. An access to *VW* on server *USA* is performed by *Alice*. *USA* does a lookup of the pre-computed decision, terminates the decision, and reverts the responsibility shift through a notification to *China*.

The notification causes *China* to update the set of permitted events. Since *VW* and *Audi* are correlated, a decision for access to *Audi* is pre-computed.

A failed lookup is given in Fig. 5b. After the pre-computed decision for *VW* arrives at *USA*, Alice performs an access to *Audi* on *Estonia*. A decision for access to *Audi* is computed on-the-spot. The mechanism terminates the pre-computed decision for *VW* and reverts the responsibility shift. Correlation between *Audi* and *VW* causes a pre-computation of a decision for access to *VW*.

A key property of our design is that a lookup of a pre-computed decision requires no coordination among the nodes, as indicated also in our example. The coordination was done on the side during the pre-computation. That is, no network communication takes place for a usage event in case of a lookup success. We therefore expect that our mechanism exhibits a lower perceivable overhead compared to traditional approaches, which perform all coordination on the spot.

5.3 Implementation

We prototypically implemented our mechanism using the CliSeAu tool [13]. The implementation of the decision-making algorithm consists of 757 source lines of Java code (SLOC; empty/comment lines excluded) in 13 classes. This implementation is generic with respect to the target program. For a concrete target program to establish the distributed storage service, we selected CrossFTPServer.[2] The target-specific implementation consists of 81 SLOC in 3 classes.

We complement SPEED^{AC} for efficiency at the design level by efficient data structures at the implementation level that assure efficient lookup and maintenance of pre-computed decisions. This includes: pre-computed decision lookup, temporary responsibility shifts, and our usage event selection strategy. We realized all utilizing hash maps, which feature average $O(1)$ running time for all operations [8, p. 253]. By using representatives of equivalence classes as keys, we could efficiently implement lookup, deletion, and responsibility shifting.

To assure effectiveness of our mechanism we implemented JUnit tests and on top applied systematic manual testing. The JUnit tests cover functionality of decision-making, responsibility shifting, and bookkeeping for event selection. We complemented the tests with systematically testing an instantiation of our mechanism for CrossFTPServer. In a system setup with a concrete a policy, we manually accessed files to test the soundness and transparency of our implementation. In all cases we found our mechanism to be sound and to be transparent.

6 Performance Evaluation

We experimentally evaluate the performance of our mechanism by investigating its perceivable overhead. Through the evaluation, we assess whether and to which extent SPEED^{AC} reduces perceivable overhead compared to a traditional usage-control mechanism in our case study. As the reference point for the comparison, we employ the mechanism by Gay, Hu, and Mantel [13], which we call SOA (abbreviating State-Of-the-Art) in the following.

---

[2] http://www.crossftp.com/crossftpserver.htm.

6.1   Experimental Setup

For our experimental evaluation we employ a distributed file-storage system, a concrete instance of the system setting in Sect. 4. The system consists of 8 servers hosting a file structure modeled after the MSN BEFS trace [18], which captures the operation of a file server of Microsoft's Live services. We replay a post-processed MSN BEFS trace to simulate a system execution.

The MSN BEFS trace is a block I/O trace of a Microsoft's Live services server containing 8 physical disks. The trace captures operation during 6 hours. For our experiments, a file access event represents an aggregation of block accesses in close succession. We place the files for each disk on a separate server.

Our experiments use multiple synthesized Chinese Wall policies, i.e., 0.2, 0.5, 0.8, 0.8_d, 0.8_g. For each policy we target a rate of cooperative decision-making in SOA for a trace replay. Conflicting files distributed over nodes require SOA to cooperatively compute decisions. Our synthesis randomly selects files from different nodes and marks them as conflicting until we reach the targeted rate. The rate is represented by the name, e.g., 0.2 targets a rate of 20% in SOA. For policy 0.8_g an equivalence class contains files from at most seven nodes, for 0.8_d from at most four, and for the remaining ones from at most two.

Our experiments use a synthesized file-correlation that predicts, for each file access, the following file access in 80% of cases. Our synthesis follows the process: For each file f, the most frequent access following f is marked correlated until the hit-rate reaches 80%.[3]

In our experiments we employ 8 Lenovo ThinkCentre M93p as servers. Each is equipped with an Intel(R) Core(TM) i5-4590 CPU, 32 GB of RAM, and an 1000Mbit/s Intel I217-LM Ethernet adapter. As operating system Ubuntu Linux 14.04.2 LTS is run. The FTP server we employ is CrossFTPServer version 1.11. The JavaVM is OpenJDK version 7u79-2.5.5-0ubuntu0.14.04.2.

We conduct experiments for each Chinese Wall policies for both SOA and SPEED^{AC}. An experiment consists of 5 independent trace replays, i.e., we start fresh instances of all software. A trace replay measures the response time for 5552150 file accesses by 256 distinct users. We average the obtained results.

6.2   Perceivable Overhead

Our system exhibits an average response time of 2.03 ms without usage control enforcement. The perceivable overhead of SPEED^{AC} and SOA is obtained by subtracting 2.03 ms from their response times. Table 2 presents our results.

For SOA perceivable overhead is between 1.49 ms (for 0.2) and 2.62 ms (for 0.8 and 0.8_d) with an average of 2.27 ms. The perceivable overhead for SPEED^{AC} is between 0.96 ms (for 0.2) and 1.83 (for 0.8_g) with an average of 1.25 ms. The reduction is between 29.1% (for 0.8_g) and 61.5% (for 0.8), averaging at 44.9%.

---

[3]   The seemingly high hit-rate of 80% in fact constitutes a conservative choice: Hua et al. [16] obtained a 95.2% hit-rate on the same trace data.

Table 2. Perceivable overhead

Type	0.8	0.8_d	0.8_g	0.5	0.2	$\varnothing$
SOA	2.62 ms	2.62 ms	2.58 ms	2.02 ms	1.49 ms	2.27 ms
SPEED^{AC}	1.01 ms	1.48 ms	1.83 ms	0.99 ms	0.96 ms	1.25 ms
abs. speedup	1.61 ms	1.16 ms	0.75 ms	1.03 ms	0.53 ms	1.02 ms
rel. speedup	61.5%	44.3%	29.1%	51.0%	35.6%	44.9%

We particularly find the variability in the perceivable overhead among experiments 0.8, 0.8_d, and 0.8_g with SPEED^{AC} very interesting, given that these experiments feature nearly the same perceivable overhead with SOA. We identified the *size* of an equivalence class, i.e., its number of usage events, as cause. Our selection and termination strategies allow for fewer pre-computed decisions with increased size of equivalence classes. Further investigation showed that, indeed, fewer pre-compute decisions are enforced during 0.8_d and even fewer during 0.8_g.

Our results show a reduced perceivable overhead for SPEED^{AC} in all cases. On average SPEED^{AC} reduces perceivable overhead by 44.9% compared to SOA, i.e., the average perceivable overhead is reduced from 2.27 ms to 1.25 ms. We find this encouraging to depoly SPEED^{AC} for efficient usage control enforcement.

6.3 File-Correlation Effects

Our results made us curious about the effects of different file-correlation hit-rates on perceivable overhead. We conducted additional experiments to identify the relation between hit-rate and reduction of perceivable overhead.

A single run was conducted for 3 additionally synthesized file-correlations with hit-rates 50%, 35%, and 30%. As lowest hit-rate 30% captures only correlating the most frequent successive file access for each file. Our experiments use policy 0.8 due to high reductions for SPEED^{AC} in our previous experiments.

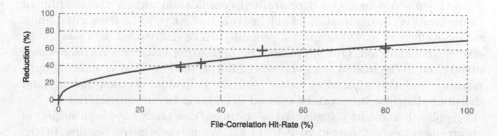

Fig. 6. Effects of file-correlation on perceivable overhead.

Figure 6 shows the results of our experiments. For 80% the reduction is taken from our previous experiment. Hit-rates 50%, 35%, and 30% show a reduction of

58%, 43%, and respectively 38%. Between hit-rates 50% and 80% the perceivable overhead reduction only differs by 3%. A curve, fitted on the obtained results, allows to anticipate reduction rates for hit-rates beyond the ones we employed.

To our surprise hit-rates above 50% only cause marginal additional reduction of perceivable overhead. On the other side of the spectrum, even hit-rates low as 30% result in a significant reduction of 38%. Thus, our results show SPEED^{AC} is useful even in settings with limited knowledge of access correlations.

7   Related Work

Optimizations for enforcement mechanisms, including usage-control mechanisms, are common and have been pursued in several directions. Mechanisms that utilize the inlining technique [10], e.g., based on aspect-oriented programming [21], use static program analysis to reduce the number of program instructions that are instrumented to invoke the mechanism (e.g., [1,13,27]). SASI [11] and Clara [3] expand the analysis to sequences of instructions in order to further reduce the number of invocations of the mechanism. Automata-theoretic techniques have been proposed for minimizing the composition of a program and an enforcement mechanism [6,22]. The optimizations performed by these approaches can be viewed as a form of statically pre-computed decisions ("permit") for security-irrelevant events.

Techniques for optimizing the performance of enforcement mechanisms themselves have also been proposed. JavaMOP [27] employs an optimization for enforcing properties on individual Java objects based on a decentralized indexing scheme that accelerates the lookup of mechanism state. JavaMOP furthermore optimizes the number of monitor state updates by exploiting the structure of the enforced property to achieve a low number of monitors. A dynamic optimization of an enforcement mechanism is proposed by the RV system [17], which at run-time collects dead monitors to reduce bookkeeping overhead.

Particularly for distributed enforcement mechanisms, architecture-based optimizations have been proposed. Gay et al. [13] and, subsequently, Decat et al. [9] propose to use a decentralized coordination among the distributed components of the mechanism for efficiently and effectively enforcing given properties. Kelbert et al. [20] employ a general-purpose distributed database for an efficient coordination. Our approach, i.e., optimizing via access correlations, is orthogonal to the related works presented in this section. That is, existing optimization techniques based on static analysis, individual decision-making, and distributed architectures can be utilized in addition to our optimization.

Application scenarios similar to the one in our case study have been subject of enforcement mechanisms before [9,13,14,26]. We use the scenario because, firstly, the Chinese Wall policy [4] is a classic business requirement and, secondly, we could use a publicly available mechanism as a reference for our evaluation [13]. Note, however, that SPEED^{AC} is a generally applicable approach for accelerating usage control. How special-purpose mechanisms, e.g., for DRM [28] or distributed access control [2], could be optimized is beyond the scope of the article.

8   Conclusion

We proposed SPEED^{AC}, an approach for accelerating distributed usage control enforcement by speculatively pre-computing decisions for usage events based on access correlations. In our case study, we developed a usage-control mechanism with SPEED^{AC} against conflicts of interest in distributed storage systems. The performance evaluation based on a real world data trace from Microsoft's Live services provides first evidence that our approach has the potential to significantly accelerate usage control. Concretely, our mechanism exhibited perceivable overheads that are up to 61.5% lower on average compared to not utilizing SPEED^{AC}. In absolute terms, the acceleration allowed us to reduce the perceivable overhead from 2.27 ms to 1.25 ms on average (see Table 2).

Our work constitutes a first step in this promising direction. Further investigation in this direction will provide a better understanding of the full potential of our approach. Questions for further investigations are: How can the approach be exploited to accelerate usage control even further and in other application scenarios, e.g., involving also dynamic policies? Does SPEED^{AC} influence how much information an attacker capable of measuring perceivable overhead can learn about processed secrets?

Acknowledgments. We thank the anonymous reviewers for their constructive comments. This work was partially funded by CASED (www.cased.de) and by the DFG as part of the project Secure Refinement of Cryptographic Algorithms (E3) within the CRC 1119 CROSSING.

References

1. Bauer, L., Ligatti, J., Walker, D.: Composing expressive runtime security policies. TOSEM **18**(3), 9:1–9:43 (2009)
2. Becker, M.Y., Sewell, P.: Cassandra: distributed access control policies with tunable expressiveness. In: POLICY, pp. 159–168. IEEE Computer Society (2004)
3. Bodden, E., Hendren, L.: The clara framework for hybrid typestate analysis. STTT **14**(3), 307–326 (2012)
4. Brewer, D.F., Nash, M.J.: The chinese wall security policy. In: IEEE S&P, pp. 206–214 (1989)
5. Brutlag, J.: Speed Matters for Google Web Search. (2009). https://services.google.com/fh/files/blogs/google_delayexp.pdf. Accessed 16 July 2017
6. Colcombet, T., Fradet, P.: Enforcing trace properties by program transformation. In: POPL, pp. 54–66. ACM (2000)
7. Colombo, M., Martinelli, F., Mori, P., Petrocchi, M., Vaccarelli, A.: Fine Grained Access Control with Trust and Reputation Management for Globus. In: Meersman, R., Tari, Z. (eds.) OTM 2007. LNCS, vol. 4804, pp. 1505–1515. Springer, Heidelberg (2007). doi:10.1007/978-3-540-76843-2_26
8. Cormen, T.H., Leierson, C.E., Rivest, R.L., Stein, C.: Introduction to Algorithms, 3rd edn. MIT Press, Cambridge (2009)
9. Decat, M., Lagaisse, B., Joosen, W.: Scalable and secure concurrent evaluation of history-based access control policies. In: ACSAC, pp. 281–290. ACM (2015)

10. Erlingsson, U.: The Inlined Reference Monitor Approach to Security Policy Enforcement. Ph.D. thesis, Cornell University (2004)
11. Erlingsson, Ú., Schneider, F.B.: sasi enforcement of security policies: a retrospective. In: NSPW, pp. 87–95. ACM (1999)
12. Forrest, B.: Bing and Google Agree: Slow Pages Lose Users. (2009). http://radar. oreilly.com/2009/06/bing-and-google-agree-slow-pag.html. Accessed 16 July 2016
13. Gay, R., Hu, J., Mantel, H.: CliSeAu: securing distributed java programs by cooperative dynamic enforcement. In: Prakash, A., Shyamasundar, R. (eds.) ICISS 2014. LNCS, vol. 8880, pp. 378–398. Springer, Cham (2014). doi:10.1007/ 978-3-319-13841-1_21
14. Gay, R., Mantel, H., Sprick, B.: Service automata. In: Barthe, G., Datta, A., Etalle, S. (eds.) FAST 2011. LNCS, vol. 7140, pp. 148–163. Springer, Heidelberg (2012). doi:10.1007/978-3-642-29420-4_10
15. Harvan, M., Pretschner, A.: State-based usage control enforcement with data flow tracking using system call interposition. In: NSS, pp. 373–380. IEEE Computer Society (2009)
16. Hua, Y., Jiang, H., Zhu, Y., Feng, D., Xu, L.: SANE: semantic-aware namespace in ultra-large-scale file systems. TPDS **25**(5), 1328–1338 (2014)
17. Jin, D., Meredith, P.O., Griffith, D., Rosu, G.: Garbage collection for monitoring parametric properties. In: PLDI, pp. 415–424. ACM (2011)
18. Kavalanekar, S., Worthington, B.L., Zhang, Q., Sharda, V.: Characterization of storage workload traces from production windows servers. In: IISWC, pp. 119–128 (2008)
19. Kelbert, F., Pretschner, A.: Data usage control enforcement in distributed systems. In: CODASPY, pp. 71–82. ACM (2013)
20. Kelbert, F., Pretschner, A.: A fully decentralized data usage control enforcement infrastructure. In: Malkin, T., Kolesnikov, V., Lewko, A.B., Polychronakis, M. (eds.) ACNS 2015. LNCS, vol. 9092, pp. 409–430. Springer, Cham (2015). doi:10. 1007/978-3-319-28166-7_20
21. Kiczales, G., Lamping, J., Mendhekar, A., Maeda, C., Lopes, C., Loingtier, J.-M., Irwin, J.: Aspect-oriented programming. In: Akşit, M., Matsuoka, S. (eds.) ECOOP 1997. LNCS, vol. 1241, pp. 220–242. Springer, Heidelberg (1997). doi:10. 1007/BFb0053381
22. Lemay, F., Khoury, R., Tawbi, N.: Optimized inlining of runtime monitors. In: Laud, P. (ed.) NordSec 2011. LNCS, vol. 7161, pp. 149–161. Springer, Heidelberg (2012). doi:10.1007/978-3-642-29615-4_11
23. Li, Z., Chen, Z., Srinivasan, S.M., Zhou, Y.: C-miner: mining block correlations in storage systems. In: FAST, pp. 173–186. USENIX (2004)
24. Ligatti, J., Bauer, L., Walker, D.: Edit automata: enforcement mechanisms for run-time security policies. Int. J. Inf. Secur. **4**(1–2), 2–16 (2005)
25. Lohr, S.: Bing and Google Agree: Slow Pages Lose Users (2012). http://www. nytimes.com/2012/03/01/technology/impatient-web-users-flee-slow-loading-sites. html. Accessed 16 July 2017
26. Martinelli, F., Matteucci, I.: Synthesis of local controller programs for enforcing global security properties. In: ARES, pp. 1120–1127. IEEE Computer Society (2008)
27. Meredith, P.O., Jin, D., Griffith, D., Chen, F., Roşu, G.: An overview of the MOP runtime verification framework. STTT **14**(3), 249–289 (2012)
28. Ongtang, M., Butler, K.R.B., McDaniel, P.D.: Porscha: policy oriented secure content handling in Android. In: Gates, C., Franz, M., McDermott, J.P. (eds.) ACSAC, pp. 221–230. ACM (2010)

29. Park, J., Sandhu, R.S.: The UCON$_{ABC}$ usage control model. TISSEC **7**(1), 128–174 (2004)
30. Shalom, N.: Amazon found every 100ms of latency cost them 1% in sales (2008). https://blog.gigaspaces.com/amazon-found-every-100ms-of-latency-cost-them-1-in-sales/. Accessed 16 July 2017
31. Singla, A., Chandrasekaran, B., Godfrey, B., Maggs, B.M.: The internet at the speed of light. In: HotNets, pp. 1:1–1:7. ACM (2014)

Emerging Security Areas

Generating Functionally Equivalent Programs Having Non-isomorphic Control-Flow Graphs

Rémi Géraud[1], Mirko Koscina[1,2(✉)], Paul Lenczner[1], David Naccache[1], and David Saulpic[1]

[1] École Normale Supérieure, 45 Rue d'Ulm, 75230 Paris Cedex 05, France
{remi.geraud,paul.lenczner,david.naccache,david.saulpic}@ens.fr
[2] Almerys, 46 Rue du Ressort, 63967 Clermont-Ferrand Cedex 9, France
mirko.koscina@almerys.com

Abstract. One of the big challenges in program obfuscation consists in modifying not only the program's straight-line code (SLC) but also the program's *control flow graph* (CFG). Indeed, if only SLC is modified, the program's CFG can be extracted and analyzed. Usually, the CFG leaks a considerable amount of information on the program's structure.

In this work we propose a method allowing to re-write a code P into a functionally equivalent code P' such that $\text{CFG}(P)$ and $\text{CFG}(P')$ are radically different.

1 Introduction

In the white-box security model, adversaries have access to a program's internals—assembly code, memory, etc. This model captures real-world attacks against low-end devices, as well as software disassembly and dynamic analysis. Such attacks may allow the adversary to extract secrets from the implementation, either in the form of tokens (passwords, etc.), intellectual property (algorithms, etc.), or may help uncover design flaws that may later be exploited. Reverse-engineering may also help the adversary recognize some trait that the program shares with other programs, e.g. in the case of malware analysis or intellectual property infringement. The general aim of obfuscation is to prevent reverse-engineering, by defeating automated methods and stave off human efforts to make sense of the code. Applications of RE-evasion techniques are many, and constitute for instance an essential building block of digital rights management (DRM) systems.

Historically, program identification focused on finding known code chunks called *signatures* in the binary. While this technique is still widely in use amongst intrusion and virus detection systems, such an approach requires both extensive, and up-to-date, databases (to account for the ever-growing corpus of threats) and a very efficient binary comparison method. At the same time, widely used packagers with self-modifying code capacity, now standard amongst virus designers, made the traditional signature-based approach less and less effective.

H. Lipmaa et al. (Eds.): NordSec 2017, LNCS 10674, pp. 265–279, 2017.
https://doi.org/10.1007/978-3-319-70290-2_16

Indeed, an increasing number of malicious programs re-write their executable code so as not to feature any recognizable code of significant length. In practice, it is not even necessary to resort to very complex re-writing mechanisms: the malicious code can simply add (or remove) useless instructions or instruction sequences (such as `nop`, and reversible register operations, e.g. `inc/dec`) to thwart a trivial comparison. While such variations can be accounted for, they require significantly more effort from the analyst, especially when scanning a large number of files.

An alternative, and certainly complementary approach to malware detection and analysis consists in running the program under certain controlled environment, or *sandbox*, in which every operation can be monitored and does not impact the "real" underlying system. Sandboxes typically implement a form of virtualized environment, and monitor access to resources, secrets, and peripherals to detect abnormal behavior. Naturally, the term "abnormal" is application-dependent, hence this approach assumes that characteristic behavioral features are known and are sufficiently distinguishable from those of uninfected software. Furthermore, running such a controlled environment is resource- and time-demanding. This limits the interest of sandboxing as a program identification tool.

Between these two approaches, recent research focused on methods for comparing programs using control-flow graph isomorphism [4,6,7]. The rationale is that the program's flow graph (CFG) wouldn't be altered significantly by the adjunction or removal of useless "decoy" operations, the kind of which thwarts direct comparison. CFG comparison techniques are also unaffected by straight-line code obfuscation techniques, e.g. when each function's code is completely rewritten. CFGs can be extracted statically to a large extent, and therefore constitute an attractive and resource-frugal alternative to full-blown virtualisation.

Defeating CFG analysis. In the malware-writing community, a typical anti-reverse engineering technique is the *trampoline*: instead of using typical control flow instructions such as `jmp` or `call`, the program makes heavy use of exception handling, that preempts the instruction pointer and runs the exception handler, which re-dispatches control flow to another program part (see e.g. [3]). After execution, each program part raises an exception, and falls back to an exception handler (hence the name, trampoline). There can be several trampolines, which may be created and moved at runtime, and code boundaries need not be rigid. This prevents disassemblers from reliably cross-referencing information, and makes it difficult to perform dynamic analysis as well, because it is typically impossible to run such code within a debugger.

However, because there is no classical call hierarchy, trampolines have to emulate the stack, and an analyst that recognizes the mechanism can easily reconstruct the control flow graph by following this pseudo-stack. Therefore, while the use of trampolines slows down analysis, it is by no means an efficient method anymore against trained reverse engineers, and the additional effort put into designing such code is not worth the marginal gain.

Recent work tried to automate the process, which strives to achieve a "flat" control flow graph, i.e. a graph with either a single central trampoline that dispatches execution, or a program that is fully unrolled and appears as a long straight-line code segment without internal structure [1,2,8,10,11,14,16]. However not only are such techniques not always applicable, but more importantly they tend to produce code that, while "flat", has salient signatures.

Our contribution. This paper addresses the question of rewriting a program in a way that hides its original control flow graph from static analysis (and, to a certain extent, from dynamic analysis as well), while preserving functionality. Straight-line code (SLC) obfuscation techniques can be used on top of our construction to destroy remaining signatures. Indeed SLC obfuscators have already been described in the literature and shown to effectively defeat classic code analysis techniques [15]. The rewriting is randomized, and produces different outputs every time. Unlike the trampoline construction, whose heavy use of exception handling is easily recognizable, and from there, traceable, our construction only uses common instructions and relies on a specific routing mechanism along execution—which is much harder to detect.

More formally, given a program P, we show how to obtain a functionally equivalent program P', such that the CFG of P' is essentially a random graph. This transformation is automatic, and we show how to implement a CFG-transcompiler for the x86-64 architecture, which is widely used and furthermore makes our implementation easier.

2   Control Flow Graph Transcompilation

2.1   Prerequisites

The control flow graph of a program is a graphical representation, based on nodes and edges, of the paths that might be traversed by the program during its execution.

Definition 1 (Control Flow Graph). *The (full) control flow graph of a program P is the graph whose nodes are the program's instructions and the edges are control flow transitions. The restricted control flow graph of P has for nodes straight-line blocks, i.e. a maximal sequence of code without departure or arrival of static jumps, and there is an edge from node x to node y (and we write $x \rightarrow y$) if either of the following conditions hold:*

- *The code of node y is located immediately after the node x, and both are separated by a conditional jump.*
- *The last instruction of the node x is either a conditional or a static jump, which is a call to the physical address of the beginning of the node y.*

In the following, unless specified otherwise, we always refer to the *restricted* control flow graph. This construction does not include information about dynamic jumps: In practice it is challenging to statically and reliably resolve dynamic jumps. The `ret` instruction, which we cannot ignore since it is often used to implement function calls, will be dealt with in a special way.

However, other dynamic and indirect control flow modifications (e.g. by direct alteration of the instruction pointer, or non-standard exception handling) are not considered in this work. On the one hand this is a limitation that may prevent some programs from undergoing the transformation that we propose. On the other hand, this may constitute an interesting countermeasure against code-reuse and hijack attacks that leverage such possibilities.

Let P be the program to be obfuscated. We denote by $G = (V, E)$ the CFG of P, where V and E correspond respectively to the nodes and edges of G. Let $G' = (V', E')$ be a given "final" target CFG.

Example 1. Consider the following program, implementing a simple double-and-add algorithm:

```
dbl_add(int, int):                  ; Compute ab from integer arguments a and b
        test    esi, esi
        mov     eax, 0              ; tmp = 0
        jle     .end                ; if b == 0, return tmp
.loop:
        lea     edx, [rax+rax]      ; tmp2 = 2 tmp
        add     eax, edi            ; tmp = tmp + a
        test    sil, 1
        cmovne  eax, edx            ; if b even set tmp = tmp2
        sar     esi                 ; shift b to the right
        jne     .loop               ; loop if b > 0
        rep ret
.end:
        rep ret
```

The CFG associated to this program is represented in Fig. 1, where the instructions' arguments have been removed for clarity. The associated restricted CFG is represented in Fig. 2.

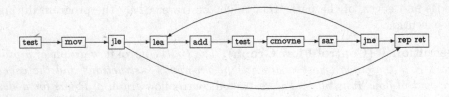

Fig. 1. Full CFG of the program of Example 1.

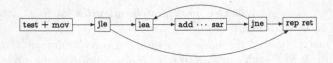

Fig. 2. Restricted CFG of the program of Example 1.

2.2   Overview of Our Approach

Our goal is to rewrite P into a program P' that achieves the same functionality as P, but whose CFG is $G' \not\cong G = (V, E) = \mathrm{CFG}(P)$. This is achieved in successive steps, illustrated in Figs. 3, 4, 5 and 6.

Step 1: Relabeling. We start from a morphism π between the two graphs, i.e. a function that is injective on nodes and preserves edges. If we fail to find enough nodes or edges to perform this operation, which happens with very low probability when the target graph is large enough, we simply start over with a new random graph G'. The process is illustrated in Fig. 3.

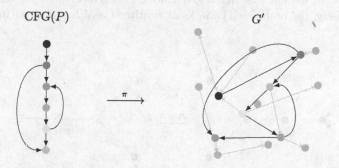

Fig. 3. Illustration of Step 1: Relabeling. The original nodes and edges from $\mathrm{CFG}(P)$ are assigned different colors, other nodes are in gray. (Color figure online)

Step 2: Breaking Edges. Then, additional nodes will be added by transforming the graph. The idea is to replace simple edges by paths in $G' = (V', E')$, i.e. for each edge $(a, b) \in E$, corresponding to an edge $(\pi(a), \pi(b)) \in E'$, we replace $(\pi(a), \pi(b))$ by a path $(\pi(a), f((a, b)), \pi(b))$, where f is a prescribed function. Such a function $f : E \to \mathrm{List}(V')$ must return paths already present in G', i.e. assuming that $f((a, b)) = (s_1, \ldots, s_n)$,

- $(\pi(a), s_1) \in E'$
- $(s_n, \pi(b)) \in E'$
- $\forall i \in \{1, \ldots, n-1\}, (s_i, s_{i+1}) \in E'$

We keep track of which edges were originally present and which edges were added at this step. The process is illustrated in Fig. 4.

Step 3: Identify Active and Passive Nodes. The previous step introduced "extra" operations between a and b. Since we wish to preserve the original program's functionality, we should make sure that only the original endpoints, a and b, are executed, while all the intermediary nodes are without effect when executed. We call a and b the *active* nodes, and the intermediary nodes (i.e. nodes that do not exist in the original CFG) are called *passive*.

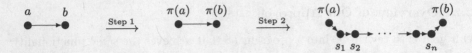

Fig. 4. Illustration of Step 2: Breaking edges. The original path $\pi(a) \to \pi(b)$ is extended by a path $f((a,b)) = (s_1, \ldots, s_n)$ of G'.

Remark 1. A node that is neither active nor passive in the control flow graph G can be considered either active or passive in G'.

Depending on the execution path taken, some nodes may be active or passive (e.g. Fig. 5). To decide whether a given node is active or passive, the program (more precisely, the node itself) checks at runtime the value of a routing variable (see below).

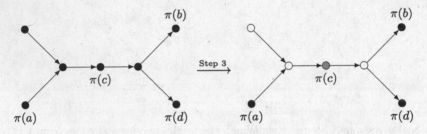

Fig. 5. Illustration of Step 3: Identifying active and passive nodes. Here two original sequences $\pi(a) \to \pi(b)$ and $\pi(c) \to \pi(d)$ cause some nodes to be passive (empty circle), active (filled black circle), or active depending on the execution path taken (grey circle).

Step 4: Routing. Finally, we transform each node so that the execution of passive nodes is without side effects (a process we call *passivation*), except continuing through the sequence of nodes until an active node is attained. To that end we introduce an additional "routing" variable that will be updated as the program is executed (e.g. Fig. 6).

Nodes consult the routing variable to know whether they are active or not; if not, they simply hand over execution to the next node in sequence (possibly after executing dummy instructions).

2.3   Contexts

During program execution, every node in the transformed program undergoes the following procedure:

1. Determine whether node is active or passive.
2. If active, restore the registers. Otherwise passivate itself.
3. Run the code.
4. Call the next node in the sequence.

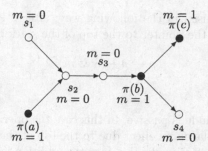

Fig. 6. Illustration of Step 4: The path is taken according the routing variable m. If the node is passive ($m = 0$), the path to be taken will be the subsequent node. In the case of a active node ($m = 1$), the next node will be defined by the current node

To allow this series of operations, we introduce the concept of *contexts*.

A context is a set of variables that save the node's state, in a way that can later be restored. Each traversed node is associated to a context, which is available just during the time that the node is being traversed.

Since passive nodes do not suffer side effects, they cannot in particular find the next node to be called; hence the next node is part of the context. If the node is active, it may ignore this part of the context and branch itself to another destination.

2.4   Node Passivation

Node passivation requires us to cancel the instruction(s) being executed, or compensate their effects in some way. We do this by using both the registers and the stack (it is not possible to rewrite registers that are in active mode), leveraging the specificities of the x86-64 architecture.

Register operations. Any register operation can be dealt with by using contexts, with the exception of the stack registers.

Stack operations. Stack operations are harder to compensate: the following instructions have an effect on the stack

```
PUSH, POP, PUSHA, POPA, PUSHAD, POPAD, PUSHF, POPF, PUSHFD,
POPFD
```

We control writing and reading in the stack by using a pointer to a "trash" address, stored as a fixed value. If a passive node attempts to write something in the stack, we redirect the address to the trash, nullifying the instruction's effects. The reading process is handled in the same way. If the node is active, the real address is used.

The context m is used in the following way: after a PUSH, we perform the following operation to the pointer to the top of the stack p:

$$p \leftarrow p + (8 \mathbin{\&} m)$$

where

- $m = 1 \cdots 1_2$ if the node is passive. In this case the operation will be compensated and will not have any effect due to the top of the stack not changing.
- $m = 0$ if the node is active. In this case the addition is useless and the PUSH works as intended.

MOV instruction. mov instructions from one register to another are already without effect, since register values are restored at the beginning of each active node, and are stored in the environment. However, mov instructions that involve a memory address require additional care, and we use the same technique as for the stack: the address is rewritten to the "bin" when the node is passive. This is followed by the transformation:

$$\text{address} = (\text{address} \mathbin{\&} (\neg m)) | (\text{trash\_address} \mathbin{\&} m)$$

This technique also hides the addresses that are really used during program execution.

Function calls. We will distinguish *library* function calls and calls to *internal* functions, that are defined in the code.

Library calls. In the case of library function calls, each of them is treated separately by using a specific context per function. Now, considering that it is impossible to handle all the functions at the same time, we propose to call the functions by using parameters that make them ineffective.

Example 2. In the following "Hello World" program, where we make ineffective the function printf by loading to EAX the address of an empty sentence (auxiliary parameter) and set the stack pointer to the address of EAX.

```
extern _printf
global _main

section .data
param1: db "Hello World",10,0
paramaux: db "",0              ; declaration of the empty sentence

section .text
_main:
    push param1
    lea eax, [paramaux]        ; paramaux address placed in EAX
    mov [esp], eax             ; pointer to the empty sentence
    call _printf
    add esp,4
    ret
```

2.5   Jumps and Internal Calls

Internal calls. Recall that we distinguish between a call to an address in a PUSH from a static jump. This makes the above transformation effective to handle these instructions. However, the RET instruction corresponds to a dynamic jump and is subtler to handle.

Let n be a node with a RET instruction in G, and assume that in G' the corresponding node $\pi(n)$ has two neighbors, f_1 and f_2. Their addresses are fixed, so that one can place, on the top of the stack, the address of the node that follows $\pi(n)$ (either f_1 or f_2).

Example 3. In the following example, we print on the screen the result returned by func1. In this case, we jump from func1 to func2 adding the desired address on the top of the stack by using a push operation. As a result, the program jumps to func2 instead of jumping back to the address after the call.

```
extern _printf
global _main

section .data
num DD 2,3
format: dd "num: %d" , 10, 0

section .text

_main:
  mov eax,0              ; eax = 0
  mov esi, [num]         ; edi = 2
  mov edi, [num+4]       ; esi = 3
  push esi               ; pass param 3 to .func1
  push edi               ; pass param 2 to .func1
  push eax               ; pass param 1 to .func1
  call .func1            ; jump to func1
  add esp,12             ; pop edi, esi and eax from the stack

  push eax
  push dword format
    call _printf         ; print eax in the screen
  add esp,8              ; pop stack 2*4-byte

.func1:
  push ebp
  mov ebp,esp            ; set stack base pointer
  sub esp, 4             ; creat space for one 4-byte local variable
  push edi               ; Save the values of the register that the function will use
  push esi
  mov eax,[ebp+8]        ; move param 1 to EAX
  mov edi,[ebp+12]       ; move param 2 to EDI
  mov esi,[ebp+16]       ; move param 3 to ESI

  mov [ebp-4],edi        ; var local = 2
  add [ebp-4],esi        ; var local = 5
  mov eax, [ebp-4]       ; EAX = 5

  pop esi                ; remove esi from the stack
  pop edi                ; remove edi from the stack
  mov esp,ebp
  pop ebp                ; takedown stack base pointer
  lea ecx,[.func2]
  push ecx               ; push func2 address on the top of the stack
  ret                    ; jump func2
```

```
.func2:
  push ebp
  mov ebp,esp         ; set stack base pointer
  sub esp, 4          ; creat space for one 4-byte local variable
  push edi            ; Save the values of the register that the function will use
  mov edi,[num]       ; edi = 2

  mov [ebp-4],eax     ; var local = 5
  add [ebp-4], edi    ; var local = 7
  mov eax,[ebp-4]     ; EAX = 7

  pop edi             ; remove edi from the stack
  mov esp,ebp
  pop ebp             ; takedown stack base pointer
  ret
```

2.6 Routing

Once we have passed through a passive node, without changing the environment, we must be capable to take the next desired branch. As each node is a maximum of two out-degree, all that we need is a boolean variable in the environment that will indicate to which child we must to go.

In practice, it is enough to maintain a global routing variable r. This allows the sequence of branches to follow (left or right) between two consecutive nodes. Hence, we modify r for each active node found and its i-th bit gives the direction of the i-th branch of the current path. We will denote by r_i the i-th bit of r.

Remark 2. Routing variables have a limited size if we use native types, it is straightforward to extend them but additional arithmetic is needed.

JUMP instruction. First of all, we need to transform a conditional jmp from P into a jmp that goes to the next node as determined by r_i. For simplicity, we assume that all conditional jumps test a "zero flag", which is set by a comparison just before the jump. For example, we have the node A (with children B and C) and the following program:

```
cmp (...)        ; comparison
je B             ; conditional jump to B
C                ; next node
```

As we know how to move from node A to node B or C in advance, we can save the routes in some constants A_to_B and A_to_C. For doing so we use the following code:

```
mov routing_variable, A_to_C   ; set routing variable
cmp (...)                      ; comparison
cmove routing_variable, A_to_B ; set routing variable iif comparison succeeds
```

This program then jumps according to the first value of the routing variable.

Note that, for passive nodes, routing variables are set to the (masked) trash address.

RET instruction. When a node is passive, we want to have two possible branches as in the case of the jump instruction. To achieve this we also store the constants A_to_B and A_to_C; and we will use the mask m as the context. We will go to node B if $r_i = 1$ and to node C if $r_i = 0$.

We want to put at the top of the stack the address to which we want to go. Hence, we just add the following line before the `ret`:

$$p \leftarrow (p \mathbin{\&} m)|((r \mathbin{\&} A\_to\_B)|(\neg r \mathbin{\&} A\_to\_C)) \mathbin{\&} \neg m$$

The transformation presented above allows us to modify the program's control flow graph. We are capable of transforming an arch into a path, and ensuring that the path's execution is identical to the effect of running the arch in the original graph.

3 Control Flow Graph Obfuscation

While the presented construction effectively transforms the program's CFG, the resulting construction has a strong signature, and it is easy to reverse the process to obtain the initial graph. It is indeed enough to run the program and identify nodes that change the routing variable. These nodes are the active ones, and it is possible to reconstruct the original control flow graph.

In this section we propose several ways to obfuscate the transformed program and make this reconstruction harder. First, we will "force" the execution of the program in order to recover successfully the initial control flow graph, we then hide the nodes' activity, including the operations on the routing variable, which is a signature of an active node.

3.1 Forcing Execution

For now, we know that the routing variable suffices to determine the next active node. We will modify its definition and use it to hide the control flow from static analysis. The routing variable is now maintained as a sequence of bits $(r_1, \ldots, r_n)$.

Upon transitioning to node i, we apply to the routing bit r_i a random permutation f_i of $\{0, 1\}$.

Example 4. For example, if one seeks to obtain at the end of the function a bit equal to 1, the following operations can be used:

$$
\begin{aligned}
&r \leftarrow 0 &&\text{Null routing variable}\\
&a \leftarrow \mathrm{rand}() &&\text{Introduce randomness}\\
&t \leftarrow 5a\\
&t \leftarrow t + r \times a\\
&r' \leftarrow t/a \bmod 2
\end{aligned}
$$

At the end of the code execution we obtain $r' = 1$. If we declare $r = 1$ instead, we get $r' = 0$.

We can easily generate the random flips f_i by using an arithmetic operation and its inverses. As determining the value of a variable is undecidable, running the program is the most natural way to get information about the execution paths taken.

3.2 Node Hiding

The same way that routing bits are masked, we can hide the value of the bit indicating whether a node is active or passive. However by doing so node i only hides the status of node $i+1$. The mask's value can also be changed by choosing a random number between m and $\neg m$, and updating the formula accordingly.

3.3 Route Hiding

Updates of the routing variable are crucial, as they immediately reveal active nodes. To hide the information about the routing changes, we extend each path beyond the active node, and introduce a weak form of "onion" routing, where the next node is determined at runtime. The rationale is that determining whether a node is active or inactive will require recovering the full route leading to this particular node.

We introduce two additional variables per node, called *path* and *next path*. The *next path* variable is masked (XORed) with a value that depends on the node. Upon execution of an active node, the values of these two new variables are further modified.

If the next node is C, the route from B to C is stored in *next path*, and masked by being XORed with the constants of every intermediary node between A and B.

The number of hops is counted. Upon arriving at the final hop B of the path from A to B, we swap *next path* and *path*.

The route hiding process is illustrated in Fig. 7.

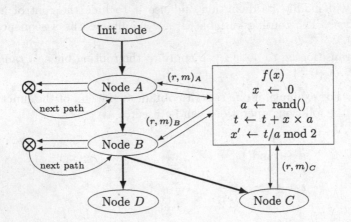

Fig. 7. Diagram of hiding process for nodes and routes

4   Security

Intuitively, the security of our construction depends on the hardness of identifying active nodes. This can be formalized as an adversarial game, whereby a more precise security notion can be given:

CFG-FullRecovery Game:
1. The challenger provides a program CFG $G = (V, E)$
2. The adversary chooses a set $N \subseteq V$

The adversary wins the game if the nodes in N are the active ones.

To get a grasp on how hard this game is, assume that we choose N at random in V, where there are exactly $|N|$ active nodes:

$$\Pr\left[N \text{ is exactly the active nodes} \mid N \subseteq_R V\right] = \frac{1}{\binom{|V|}{|N|}} = \frac{|N|!(|V| - |N|)!}{|V|!}.$$

If one node out of two is active, and there are more than 42×2 nodes in V, this probability is negligible. Thus we may hope, for realistically large programs, to resist adversaries for which there is no better way to choose N than selecting a random subset of V.

However, in practice, adversaries may succeed in recovering smaller portions of the CFG. This corresponds to the following game:

CFG-OneRecovery Game:
1. The challenger provides a program CFG $G = (V, E)$
2. The adversary chooses a node $n \in V$

The adversary wins the game if n is active and n is not the first node of G (which is always active).

The success probability if n is chosen at random is

$$\Pr\left[n \text{ is active} \mid n \in_R N\right] = \frac{|N|}{|V|}$$

where again N is the set of (actually) active nodes. In the balanced case, where $2|N| = |V|$ this probability is exactly one half. When that is the case, and V is large enough, security in this second game implies security in the first game.

As discussed above, static analysis cannot in general determine the variables' values in a given node (by Rice's theorem [13]). Given that the difference between active and passive nodes is only semantic, for a general program determining whether a given node is active is undecidable.

Hence, our obfuscation scheme should be secure against static analysis, for large enough values of N and few enough active nodes.

4.1   Security Against Dynamic Analysis

Dynamic analysis is performed by running and monitoring the program. As mentioned previously, the first node is always active. The second node can be determined as follows: Execution continues until the *next path* variable is updated.

At that point, we know that there is an active node between the current node B and the first node A.

The analyst then performs the following operation: For each node n between A and B in the CFG, replace n by another operation, and run the program up to B. There are at most $|V|$ nodes to test. A node is active if, when modified, the program's state at B has changed.

As each test is required to continue running the program until B, which can take up to $|V|$ steps, it is then possible to determine the next active node in $O(|V|^2)$. By running this procedure iteratively for all nodes, we reconstruct the list of active nodes, i.e. the original CFG, in $O(|V|^3)$ operations.

5   Implementation

Given as input a program CFG, we construct a "target" CFG to which the original program is mapped.

1. *Graph generation.* We generate a random graph with n edge and maximum out-degree two, using a variant of the Tarjan-Eswaran algorithm [5, 12].
2. *Linearisation.* This graph is linearised, so that it corresponds to a CFG. For this purpose we use the scheme presented by Leroy for the CompCert compiler [9]. We then select a random morphism π between the initial graph and the new graph that we are creating.
3. *Transformation.* We begin the transformation by identifying the active and passive nodes, the edges for paths are then changed by neutralizing (passivation) the instructions using: registers and stacks operations, transforming the jumps and internal call, and defining the route to follow according to the routing variable. Finally, we remove the signature of the new graph hiding the routing variable and node's status (active or passive) by randomizing their values and adding the variables *path* and *next path*. To mask the variable *next path* we XOR it with the node's values.

The source code of this implementation is available from the authors upon request.

6   Conclusion

This paper presents a control flow graph trans-compilation algorithm allowing to transforms a program into a new functionally equivalent program. This algorithm uses common instructions such as register and stack operations, and a random routing variable, such that the resulting CFG is entirely different from the original one. We let as a future work the study of the obfuscation performance regarding the time expended in the transformation process for different code size and if the obfuscation can be improved if we apply on P the same transformation more than once.

References

1. Cappaert, J., Preneel, B.: A general model for hiding control flow. In Proceedings of the tenth annual ACM workshop on Digital rights management, pp. 35–42. ACM, 2010
2. Chow, S., Gu, Y., Johnson, H., Zakharov, V.A.: An approach to the obfuscation of control-flow of sequential computer programs. In: Davida, G.I., Frankel, Y. (eds.) ISC 2001. LNCS, vol. 2200, pp. 144–155. Springer, Heidelberg (2001). doi:10.1007/3-540-45439-X_10
3. Davi, L.V.: Code-Reuse attacks and defenses. Ph.D. thesis (2015)
4. Dullien, T., Rolles, R.: Graph-based comparison of executable objects (English version). In: SSTIC, vol. 5, pp. 1–3 (2005)
5. Eswaran, K.P., Tarjan, R.E.: Augmentation problems. SIAM J. Comput. $5(4)$, 653–665 (1976)
6. Flake, H.: Structural comparison of executable objects. In: DIMVA 2004, 6–7 July, Dortmund, Germany, pp. 161–173 (2004)
7. Kruegel, C., Kirda, E., Mutz, D., Robertson, W., Vigna, G.: Polymorphic worm detection using structural information of executables. In: Valdes, A., Zamboni, D. (eds.) RAID 2005. LNCS, vol. 3858, pp. 207–226. Springer, Heidelberg (2006). doi:10.1007/11663812_11
8. László, T., Kiss, Á.: Obfuscating C++ programs via control flow flattening. Annales Universitatis Scientarum Budapestinensis de Rolando Eötvös Nominatae, Sectio Computatorica **30**, 3–19 (2009)
9. Leroy, X.: The CompCert C verified compiler: documentation and user's manual. Ph.D. thesis, Inria (2015)
10. Linn, C., Debray, S.: Obfuscation of executable code to improve resistance to static disassembly. In: Proceedings of the 10th ACM Conference on Computer and Communications Security, pp. 290–299. ACM (2003)
11. Popov, I.V., Debray, S.K., Andrews, G.R.: Binary obfuscation using signals. In: USENIX Security (2007)
12. Raghavan, S.: A note on Eswaran and Tarjan's algorithm for the strong connectivity augmentation problem. In: Golden, B., Raghavan, S., Wasil, E. (eds.) The Next Wave in Computing, Optimization, and Decision Technologies, vol. 29. Springer, Boston (2005). doi:10.1007/0-387-23529-9_2
13. Rice, H.G.: Classes of recursively enumerable sets and their decision problems. Trans. Am. Math. Soc. **74**(2), 358–366 (1953)
14. Schrittwieser, S., Katzenbeisser, S.: Code obfuscation against static and dynamic reverse engineering. In: Filler, T., Pevný, T., Craver, S., Ker, A. (eds.) IH 2011. LNCS, vol. 6958, pp. 270–284. Springer, Heidelberg (2011). doi:10.1007/978-3-642-24178-9_19
15. Schrittwieser, S., Katzenbeisser, S., Kinder, J., Merzdovnik, G., Weippl, E.: Protecting software through obfuscation: Can it keep pace with progress in code analysis? ACM Computing Surveys (CSUR) **49**(1), 4 (2016)
16. Wang, C., Hill, J., Knight, J., Davidson, J.: Software tamper resistance: obstructing static analysis of programs. Technical Report CS-2000-12, University of Virginia, 12 2000 (2000)

Proof of a Shuffle for Lattice-Based Cryptography

Nuria Costa[1(✉)], Ramiro Martínez[2], ánd Paz Morillo[2]

[1] Scytl Secure Electronic Voting, Barcelona, Spain
nuria.costa@scytl.com
[2] Universitat Politècnica de Catalunya, Barcelona, Spain

Abstract. In this paper we present the first proof of a shuffle for lattice-based cryptography which can be used to build a universally verifiable mix-net capable of mixing votes encrypted with a post-quantum algorithm, thus achieving long-term privacy. Universal verifiability is achieved by means of the publication of a non-interactive zero knowledge proof of a shuffle generated by each mix-node which can be verified by any observer. This published data guarantees long-term privacy since its security is based on perfectly hiding commitments and also on the hardness of solving the Ring Learning With Errors (RLWE) problem, that is widely believed to be quantum resistant.

Keywords: Mix-nets · Evoting · Post-quantum cryptographic protocol · RLWE encryption · Proof of a shuffle

1 Introduction

In the last years, several countries have been introducing electronic voting systems to improve their democratic processes: electronic voting systems provide more accurate and fast vote counts, reduce the logistic cost of organizing an election and can offer specific mechanisms for voters with disabilities to be able to cast their votes independently. In particular, internet voting systems provide voters with the chance to cast their votes from anywhere: their homes, hospitals, or even from foreign countries in case they are abroad at the time of the election.

Anonymity and verifiability are two fundamental requirements for internet voting systems that seem to be contradictory. Anonymity requires that the link between the vote and the voter who has cast it must remain secret during the whole process, while verifiability requires that all the steps of the electoral process - vote casting, vote storage and vote counting - can be checked by the voters, the auditors or external observers.

The different techniques used by the actual internet voting systems to achieve anonymity can be classified in three categories:

Blind signature: this method allows the voter to obtain a message signed by an authorized entity in such a way that this entity gets no information at all about the message. Consequently, votes are anonymized before being sent.

H. Lipmaa et al. (Eds.): NordSec 2017, LNCS 10674, pp. 280–296, 2017.
https://doi.org/10.1007/978-3-319-70290-2_17

Homomorphic tallying: the votes are encrypted using a homomorphic cryptosystem and during the tallying phase they are aggregated. The resulting ciphertext is decrypted and the ballot count results are obtained. This anonymizes the votes at the end of the election since no vote is individually decrypted.

Mixing: the ciphertexts are permuted and re-encrypted in such a way that the correlation between the input and output of the mixing process is hidden and it is not possible to trace it back. This operation is called a *shuffle* and it is executed in a mixing network (*mix-net*) composed of mixing nodes each one performing in turns the same operation. This is done in order to be able to preserve the privacy of the process even if some nodes are dishonest: as long as one of the mix nodes remains faithful and does not reveal the secret permutation or re-encryption values, unlinkability is preserved. Notice that this method requires to provide a *proof of a shuffle* so that it can be checked that the contents of the output are the same as the contents of the input.

On the other hand, in order to build verifiable systems one key instrument is the Bulletin Board: a public place where all the audit information of the election (encrypted votes, election configuration, ...) is published by authorized parties and can be verified by anyone: voters, auditors or third parties. However, once published in the Bulletin Board, it is not possible to ensure that all the copies are deleted after the election and the audit period ends, and long-term privacy may not be ensured by encryption algorithms used nowadays, for example due to efficient quantum computers. Learning how a person voted some years ago may have political, as well as personal implications (e.g. in case of family coercion).

Everlasting privacy is a recent research topic meaning that even if a computationally unbounded adversary exists, the voter's privacy is preserved. Several solutions have been proposed in order to address the problem exposed above and the majority of them use Pedersen commitments [31] to protect the information that is going to be published in the Bulletin Board. These commitments perfectly hide the committed message since its privacy does not depend on any computational assumption whose strength may be eroded in the future. Nevertheless most of these proposals require an anonymous channel to send additional information (for instance, the encrypted openings of the commitments) to the server either during the voting phase [13,28] or during the authentication phase [23,24]. Since these anonymous channels are difficult to implement, mix-nets are frequently used as an alternative. There are some proposals to construct universally verifiable mix-nets with everlasting privacy [6], where a mixing of commitments instead of ciphertexts is performed.

Nevertheless, our goal is to achieve long-term privacy, in which the voter's privacy is preserved against a polynomially bounded quantum capable adversary. Lattice-based cryptography [27] is maybe the most promising approach to get cryptosystems that will remain secure in the post-quantum era, and so it has become a very active area of research in the last years. The security of lattice-based cryptography is based on the worst-case problem meaning that breaking a lattice-based cryptosystem implies finding an efficient algorithm for solving any instance of the underlying lattice problem, for instance, the Shortest Vector

Problem (SVP), the Closest Vector Problem (CVP) or the Shortest Independent Vector Problem (SIVP). There are several proposals to build lattice-based cryptosystems such as public key encryption schemes, digital signatures schemes, hash functions, Identity Based Encryption schemes or ZK proofs.

Since mix-nets are of paramount importance in an online voting scenario and lattice-based cryptography seems to be one of the main alternatives to achieve post-quantum security, we consider it necessary to have a system with a mix-net capable of shuffling lattice-based encryptions. As far as we know there is only one proposal of an e-voting protocol that uses lattice-based cryptography [9]. In the cited paper, the authors present an e-voting scheme that uses LWE-based fully homomorphic encryption in order to provide a homomorphic tally system. Nevertheless, to the best of our knowledge, there is no proposal for a lattice based e-voting scheme using mix-nets in the literature.

1.1   Related Work

The first mix-net was introduced by Chaum [8] in 1981 where the plaintext is encrypted as many times as mixing nodes using RSA onions with random padding, and during the mixing process each node decrypts the outer layer and removes the random padding, so the last node obtains the original message. In 1993, Park *et al.* noticed that Chaum's mix-net required a ciphertext size proportional to the number of mixing nodes and proposed a re-encryption mix-net [30] where instead of concatenating, they re-randomized the ciphertexts using a homomorphic cryptosystem like ElGamal. In this system decryption occurs after shuffling is finished, however they also proposed a different mix-net in the same paper [30] where each node performs partial decryption besides the shuffling. Two years later, Sako and Kilian [35] defined the property of *universal verifiability* and proposed the first universally verifiable mix-net, that provides a zero-knowledge proof of correct mixing that any observer can verify. Achieving efficient mixing proofs was the challenge of the late 1990s, where two solutions were proposed for an efficient universally verifiable mix-net [1,26]. In 2001, Furukawa and Sako [16] proposed a proof of correct mixing more efficient than the previous ones, in this scheme each node uses a matrix to do the ciphertexts permutation and proves that this matrix is a permutation matrix. In the same year, Neff [29] introduced the fastest, fully-private, universally verifiable mix-net shuffle proof known so far, optimized and generalized by Groth in [18]. In 2004, Golle *et al.* [17] proposed a mix-net with universal re-encryption, that does not require that each mix node knows the public key of the ciphertexts they are mixing. This can be done with homomorphic cryptosystems like ElGamal. In the same year, Wikström [40] gave the first mix-net definition and implementation in the UC framework [7] as well as a simpler and efficient construction [41].

Adida and Wikström introduced a different mix-net approach [2,3] motivated by the complexity of using mix-nets in elections. They proposed an offline precomputation technique in order to reduce the online computation complexity. However, the scheme [2] was quite inefficient while the construction in [3] was very efficient but reduced to a relatively small number of senders. In 2010 Terelius and Wikström [38] proposed a provably secure technique to prove the correctness of a cryptographic shuffle using simple shuffle arguments and two years later, Bayer and Groth, proposed an honest verifier zero-knowledge argument for the correctness of a shuffle of homomorphic encryptions that, compared with previous work, matched the lowest computation cost for the verifier. Nevertheless, as these non-interactive proofs are known in the random oracle model, several works have studied how to construct NIZK shuffle arguments in the Common Reference String (CRS) model without using random oracles [14,15,19,22]. However, given that these CRS-based proposals are constructed for bilinear groups, we are going to use the approach presented in [38,42] to build our proof of a shuffle. In [42] Wikström presented a mix-net based on homomorphic cryptosystems using the idea of permutation matrices. In the proposal, a proof of a shuffle is split in an offline and online phase that reduces significantly the computational complexity in the online part. More precisely, in the offline part the mixing node computes a commitment to the permutation matrix and proves in zero knowledge that it knows an opening for that commitment. In the online part, the node computes a commitment-consistent proof of a shuffle to demonstrate that the committed matrix has been used to shuffle the input.

To the best of our knowledge, the concept of using mix-nets for lattice-based cryptography is very new in the research literature, and as such, there are not many proposed schemes. There have been proposals for a lattice based universal re-encryption for mix-nets [36,37] but none of them proposes a proof of a shuffle, which is essential for verifiable protocols.

1.2 Our Contribution

We propose the first universally verifiable mix-net for a post-quantum cryptosystem. The mix-net receives at its input a set of messages encrypted using a RLWE encryption scheme [25] whose security is based on the hardness of solving the Learning With Errors problem over rings (RLWE problem) [33]. In the proposal, we show how to permute and re-encrypt RLWE encryptions and we also give the first proof of a shuffle that works for a lattice-based cryptosystem. This proof is based on what is proposed in [42] but it is not a direct adaptation of it, since we introduce a new technique to implement the last part of the proof that differs from what is presented in that article.

We split the proof of a shuffle into two protocols following Wikström's technique. In the offline part, the permutation and re-encryption parameters used to shuffle the ciphertexts are committed and it is demonstrated using zero knowledge proofs that these values meet certain properties and that the openings for the commitments are known. The zero-knowledge proofs used in this part satisfy special soundness and special honest verifier zero-knowledge [11]. The first property means that given two accepting conversations with identical first messages but different challenges, it is possible to extract a valid witness. Regarding the second property, it means that for a given challenge the verifier can be simulated.

In the online part, instead of computing a commitment-consistent proof of a shuffle, each mix node should compute a commitment to its output using the commitments calculated in the offline protocol taking advantage of the homomorphic property of both the commitment and encryption schemes. Finally, the node should reveal the opening of the output commitment in order to demonstrate that it has used the committed permutation and re-encryption values to do the shuffle. It is important to notice that we are not opening directly the commitments to the secret permutation neither to the secret re-encryption values but the commitments to a linear combination of them. The openings revealed by each node perfectly hide the secret values and no information is leaked that could compromise the privacy of the process. Commitments used to construct the proof are generalized versions of the Pedersen commitment, which is perfectly hiding and computationally binding under the discrete logarithm assumption and it is widely used to provide everlasting privacy. The reason why we use this commitment is for efficiency and simplicity, nevertheless since our protocol only requires a commitment that allows us to prove linear relations between committed elements, the protocol presented in this paper could be modified in order to use the commitment scheme proposed by Benhamouda *et al.* in [5]. This would allow us to construct a mix-net totally based on post-quantum cryptography. As this is a non-trivial modification we first show how to mix RLWE ciphertexts using Pedersen commitments and how to do it universally verifiable.

The organization of this paper is as follows. In Sect. 2 we define the notation and review the cryptographic background that is necessary to understand the mix-net proposal. In Sect. 3 we give the details about the shuffle of RLWE encryptions, and finally in Sect. 4 we conclude the paper.

2 Preliminaries

In this section we present the notation that we are going to use throughout the paper and we also give some details about the cryptographic background required for the latter sections.

Standard notation regarding vectors and matrices will be used. Vectors will be represented by boldface lowercase roman letters (such as $\mathbf{v}$ or $\mathbf{w}$) and matrices will be represented by boldface uppercase roman letters (such as $\mathbf{M}$ or $\mathbf{A}$). Let $\langle \cdot, \cdot \rangle$ denote the standard inner product in $\mathbb{Z}_q^N$, given two vectors $\mathbf{v}, \mathbf{w} \in \mathbb{Z}_q^N$ $\langle \mathbf{v}, \mathbf{w} \rangle$ is defined by means of $\sum_{i=1}^N v_i w_i$. When working with lattices we are going to follow the notation proposed in [25].

2.1 Ideal Lattices

A lattice is a set of points in an n-dimensional space with a periodic structure. Given m-linearly independent vectors $\mathbf{a_1}, \ldots, \mathbf{a_m} \in \mathbb{R}^n$, the rank m lattice generated by them is the set of vectors:

$$\mathcal{L}(\mathbf{a_1}, \ldots, \mathbf{a_m}) = \left\{ \sum_{i=1}^m x_i \mathbf{a_i} : x_i \in \mathbb{Z} \right\}$$

We denote the basis of the lattice as $\mathbf{A} = (\mathbf{a_1}, \ldots, \mathbf{a_m})$, i.e., the matrix whose columns are $\mathbf{a_1}, \ldots, \mathbf{a_m}$. We are going to work with lattices that are *full-rank* $(n = m)$, that is, the number of linearly independent vectors in the basis of the lattice is equal to the number of dimensions in which the lattice is embedded.

Definition 1. *An ideal lattice is a lattice defined by a basis $\mathbf{A}$ constructed with a vector $\mathbf{a} \in \mathbb{Z}^n$ iteratively multiplied by a transformation matrix $\mathbf{F} \in \mathbb{Z}^{n \times n}$ defined from a vector $\mathbf{f} \in \mathbb{Z}^n$ as follows.*

$$
\mathbf{F} = \begin{bmatrix} 0 & \cdots & 0 & -f_0 \\ & & & -f_1 \\ & I & & \vdots \\ & & \cdots & -f_{n-1} \end{bmatrix}
$$

The basis is defined as: $\mathbf{A} = \mathbf{F}^*\mathbf{a} = [\mathbf{a}, \mathbf{Fa}, \ldots, \mathbf{F}^{n-1}\mathbf{a}]$.

Lattices that follow this particular structure have been named ideal lattices because they can be equivalently characterized as ideals of the ring of modular polynomials $R = \mathbb{Z}[x]/\langle f(x)\rangle$ where $f(x) = x^n + f_{n-1}x^{n-1} + \cdots + f_0 \in \mathbb{Z}[x]$. That means that working on the polynomials domain modulo $f(x)$ is equivalent to working on the ideal lattice domain characterized by $\mathbf{F}$. We will use the ring where $f(x) = x^n + 1$, as proposed by [25]. When working in this ring, where we know that $f(x)$ is a cyclotomic polynomial for n a power of 2, one obtains the family of the so called *anti-cyclic integer lattices*, i.e., lattices in $\mathbb{Z}^n$ that are closed under the operation that cyclically rotates the coordinates and negates the cycled element. The vector $\mathbf{f}$ corresponding to $f(x) = x^n + 1$ is $\mathbf{f} = (1, 0, \ldots, 0)$ and therefore the basis $\mathbf{A}$ is:

$$
\mathbf{A} = \begin{pmatrix} a_1 & -a_n & -a_{n-1} & \cdots & -a_2 \\ a_2 & a_1 & -a_n & \cdots & -a_3 \\ a_3 & a_2 & a_1 & \cdots & -a_4 \\ \vdots & \vdots & \vdots & \ddots & \vdots \\ a_n & a_{n-1} & a_{n-2} & \cdots & a_1 \end{pmatrix} \tag{1}
$$

Notice that using ideal lattices we are able to express a rank n ideal lattice with only n values, rather than $n \times n$ as is the case for general lattices, which allows a more compact representation that requires less storage space.

Given a prime q, let R_q be $\mathbb{Z}_q[x]/\langle f(x)\rangle$. Henceforth we will write a either as a polynomial $a = a_1 + a_2x + a_3x^2 + \ldots + a_nx^{n-1} \in R_q$ or as a vector with coefficients $(a_1, a_2, a_3, \ldots, a_n) \in \mathbb{Z}_q^n$. Notice that given two polynomials $a \in R_q$ and $p \in R_q$, the product $a \cdot p$ in R_q is equivalent to the product of the matrix $\mathbf{A}$ with the vector $\mathbf{p} = (p_1, \ldots, p_n)$. Working with the polynomial representation in R_q allows a speedup in operations commonly used in lattice-based schemes: polynomial multiplication can be performed in $\mathcal{O}(n \log n)$ scalar operations, and in parallel depth $\mathcal{O}(\log n)$, using the Fast Fourier Transform (FFT).

2.2   RLWE Encryption Scheme

Let R_q be the ring of integer polynomials $R_q = \mathbb{Z}_q[x]/\langle x^n + 1\rangle$ where n is a power of 2, q is a prime; and let χ_σ be a discretized Gaussian distribution with standard deviation $\sigma = \alpha q/\sqrt{2\pi}$.

Definition 2 (RLWE distribution). *Given the "secret" $s \in R_q$ and an error distribution χ_σ, the RLWE distribution $A_{s,\chi}$ over $R_q \times R_q$ consists of samples of the form $(a, b = a \cdot s + e) \in R_q \times R_q$ where $a \leftarrow R_q$ is chosen uniformly, and e is the error polynomial sampled from the error distribution χ_σ.*

Definition 3 (*Search* RLWE). *Given many samples $(a_i, b_i = a_i \cdot s + e_i) \in R_q \times R_q$ from the RLWE distribution $A_{s,\chi}$ the goal is to recover the "secret" $s \in R_q$ with high probability.*

Definition 4 (*Decision* RLWE). *Given a vector (a, b) the goal is to efficiently distinguish if it has been sampled uniformly at random from $R_q \times R_q$ or from a RLWE distribution $A_{s,\chi}$.*

Hardness of RLWE. For any large enough q, solving certain instantiations of the *search* RLWE problem is at least as hard as quantumly solving a corresponding poly(n)-approximate Shortest Vector Problem (*approx*-SVP) on any ideal lattice. On the other hand, solving *decision* RLWE in any cyclotomic ring (for any poly(n)-bounded prime $q = 1 \mod n$) is as hard as solving *search* RLWE. See [25] for the details about the hardness of RLWE and the quantum reduction from worst-case *approx*-SVP on ideal lattices to the *search* RLWE.

RLWE parameters. How to choose secure parameters for lattice based cryptosystems is still an open question, nevertheless there are some parameters' proposals in the literature that take into account various security levels or attacker types [34], that consider the requirements of security reductions [32] or that consider an upper bound on the decryption error probability [20].

A RLWE encryption scheme is a triplet (KeyGen, Enc, Dec) which operates on rings such as R_q for which the RLWE is difficult to be solved. The original definition of the algorithm requires choosing *small* elements in R_q from an error distribution at several points. For practical purposes we will construct these *small* elements by taking their coefficients from an error distribution χ_σ which will be a discrete Gaussian distribution with parameter σ, as defined above. The RLWE encryption scheme that we are going to use is that proposed in [25] which defines the following algorithms:

KeyGen(): choose a uniformly random element $a \in R_q$ as well as two random *small* elements $s, e \in R_q$ from the error distribution. Output $sk = s$ as the secret key and the pair $pk = (a, b = a \cdot s + e) \in R_q \times R_q$ as the public key.

Encrypt(pk, z, r, e_1, e_2): to encrypt an n-bit message $z \in \{0,1\}^n$, we view it as an element of R_q by using its bits as the 0–1 coefficients of a polynomial. The encryption algorithm then chooses three random *small* elements $r, e_1, e_2 \in R_q$ from the error distribution and outputs the pair $(u, v) \in R_q \times R_q$, as the encryption of z: $(u, v) = (r \cdot a + e_1 \mod q, \ b \cdot r + e_2 + \lfloor \frac{q}{2} \rfloor z \mod q) \in R_q \times R_q$

Decrypt$(sk, (u, v))$: the decryption algorithm simply computes $v - u \cdot s = (r \cdot e - s \cdot e_1 + e_2) + \lfloor \frac{q}{2} \rceil z \mod q$.

For an appropriate choice of parameters (namely q and σ) the coefficients of $r \cdot e - s \cdot e_1 + e_2$ have magnitude less than $q/4$, so the bits of z can be recovered by rounding each coefficient of $v - u \cdot s$ back to either 0 or $\lfloor \frac{q}{2} \rceil$, whichever is closest modulo q. Notice that in the RLWE encryption scheme presented above the secret s is taken from the error distribution, as well as r, e_1 and e_2. This is done in order to build efficient encryption schemes and it is demonstrated in [4] that the hardness of the underlying problem is not affected by this change.

Security. The RLWE encryption scheme is semantically secure given the pseudorandomness of the RLWE samples [25]. Notice that both the public key (a, b) and the ciphertexts are RLWE samples, where the encrypted messages can be seen as the pairs $(a, u), (b, v) \in R_q \times R_q$ (ignoring the message component $\lfloor \frac{q}{2} \rceil z \mod q$) with secret r. Then, these values are pseudorandom. Consequently, the encryption of a message using the RLWE encryption scheme is indistinguishable from an element sampled uniformly at random from $R_q \times R_q$ as long as the number of samples containing the same secret r is polynomial on the security parameter. The number of nodes in a mix-net is fixed and every re-encryption is computed using different parameters r, so the security is not compromised.

For our proposal we will need not only to encrypt messages but also to re-encrypt them. Since an RLWE encryption scheme is an additive homomorphic cryptosystem, we can re-encrypt a message just adding to the original ciphertext the encryption of the neutral element, that is, the encryption of a polynomial whose coefficients are 0. Notice that semantic security in an RLWE encryption scheme implies semantic security under re-encryption.

Reencrypt$((u, v), r', e'_1, e'_2)$: to re-encrypt an n-bit message z, the algorithm chooses three random *small* elements $r', e'_1, e'_2 \in R_q$ from the error distribution and outputs the pair $(u', v') = (u, v) + \mathsf{Encrypt}(pk, 0, r', e'_1, e'_2) \in R_q \times R_q$.

Decrypting this re-encrypted ciphertext we would obtain $v' - u' \cdot s = (r + r') + (e_2 + e'_2) - s \cdot (e_1 + e'_1) + \lfloor \frac{q}{2} \rceil z$. The plaintext is preserved but the error terms may grow after every homomorphic operation. In order to avoid decrypting errors the number of mixing nodes must be taken into account when choosing the parameters q and σ, such that the error is still small compared to q even after as many re-encryptions as mixing nodes we are planning to use.

2.3   Zero Knowledge Proofs

A zero-knowledge proof is a protocol between two parties, the prover $\mathcal{P}$ and the verifier $\mathcal{V}$, where the first tries to convince the second that it knows some secret w that satisfies a public relation $(x, w) \in R$ (where x would be some public information), for instance, that $\mathcal{P}$ knows the discrete logarithm of a public element. This proof is done in such a way that the prover does not reveal any information beyond the fact that a certain statement is true.

Definition 5. *A two party protocol $(\mathcal{P}, \mathcal{V})$ is a Σ-protocol [12] for relation R if it is a three round public-coin protocol of the form:*

1. *The prover $\mathcal{P}$ sends a message t to the verifier $\mathcal{V}$.*
2. *$\mathcal{V}$ sends a random string e to $\mathcal{P}$.*
3. *$\mathcal{P}$ sends a response s to $\mathcal{V}$. The verifier decides to accept or reject the proof based on the protocol transcript (t, e, s).*

and the following requirements hold:

- *Completeness: if an honest prover $\mathcal{P}$ knows w satisfying the public relation $(x, w) \in R$, then $\mathcal{V}$ always accepts.*
- *Special soundness: given two accepted protocol transcripts (t, e, s) and (t, e', s') where $e \neq e'$, there exists a Probabilistic Polynomial-Time (PPT) algorithm which outputs w such that $(x, w) \in R$.*
- *Special honest-verifier zero-knowledge: given a pair (x, e) there exists a PPT algorithm that outputs a valid protocol transcript (t, e, s) with the same probability distribution as transcripts between the honest $\mathcal{P}$ and $\mathcal{V}$.*

The underlying structure behind the zero-knowledge proofs constructed in our proposal is that of a Σ-protocol. As Terelius and Wikström mention in their article [38], we need to prove knowledge on how to open commitments such that the committed values satisfy a public polynomial relation.

$$\Sigma\text{-proof}[\mathbf{e} \in \mathbb{Z}_q^N, s \in \mathbb{Z}_q | a = \mathsf{Com}(\mathbf{e}, s) \wedge f(\mathbf{e}) = e')] \tag{2}$$

We refer the reader to [38] for more details on how this can be done.

2.4   Pedersen Commitments

Let p and q be large primes, $\mathbb{Z}_p^*$ a group of integers modulo $p = 2q + 1$ and $G_q \subset \mathbb{Z}_p^*$ a subgroup of order q where the discrete logarithm assumption holds. Given two independent generators $\{g, g_1\}$ of G_q, to commit to a message $x \in \mathbb{Z}_q$ using the Pedersen commitment scheme [31], choose a random $\alpha \xleftarrow{\$} \mathbb{Z}_q$ and output $\mathsf{Com}(x, \alpha) = g^\alpha g_1^x$. In order to open this commitment simply reveal the values α and x. This scheme is *perfectly hiding* and *computationally binding* as long as the discrete logarithm problem is hard in G_q.

In our proposal we are going to work with the *extended version* of the Pedersen commitment scheme, that allows to commit to more than one message at once. Given $N + 1$ independent generators $\{g, g_1, \ldots, g_N\}$ of G_q and a randomness $\alpha \xleftarrow{\$} \mathbb{Z}_q$, the commitment to N messages $\mathbf{x} = (x_1, \ldots, x_N) \in \mathbb{Z}_q^N$ is computed as:

$$\mathsf{Com}(\mathbf{x}, \alpha) = g^\alpha \prod_{i=1}^N g_i^{x_i}$$

We use this *extended version* of the Pedersen commitment to commit to a matrix $\mathbf{M} \in \mathbb{Z}_q^{N \times N}$. In order to do that just compute a commitment to each of its columns $(\mathbf{m_1}, \ldots, \mathbf{m_N})$ where $\mathbf{m_j} = (m_{1j}, m_{2j}, \ldots, m_{Nj})$ for $j = 1, \ldots, N$. This means that a matrix commitment is a vector whose components are the commitments to the matrix columns:

$$\mathsf{Com}(\mathbf{M}, \alpha_1, \alpha_2, \ldots, \alpha_N) = (\mathsf{Com}(\mathbf{m_1}, \alpha_1), \ldots, \mathsf{Com}(\mathbf{m_N}, \alpha_N)) \qquad (3)$$

Due to the homomorphic property of the Pedersen commitment we can compute a commitment to the product of a matrix $\mathbf{M}$ by a vector $\mathbf{x}$ from the commitment to the matrix $\mathsf{Com}(\mathbf{M}, \boldsymbol{\alpha}) = (c_{\mathbf{m_1}}, \ldots, c_{\mathbf{m_N}})$.

$$\prod_{j=1}^{N} c_{\mathbf{m_j}}^{x_j} = \prod_{j=1}^{N} \left(g^{\alpha_j} \prod_{i=1}^{N} g_i^{m_{i,j}} \right)^{x_j} = g^{\langle \boldsymbol{\alpha}, \mathbf{x} \rangle} \prod_{i=1}^{N} g_i^{\langle (m_{i,1}, \ldots, m_{i,N}), (x_1, \ldots, x_N) \rangle} \qquad (4)$$

3   Shuffling Ring-LWE Encryptions

In this section we first present an overview of the mixing protocol and then we explain in more detail how is it proved that the committed matrix is a permutation matrix, the random re-encryption paramenters are *small* and that all these values have been used to perform the shuffle.

Let $M_1, \ldots, M_k$ be the mix-nodes that participate in the mixing protocol and let N be the number of encrypted messages at the input of each node.

3.1   Protocol Overview

Given the the ring R_q and the encryption scheme presented in Sect. 2, and the matrices $\mathbf{A}$ and $\mathbf{B}$ constructed from vectors $\mathbf{a}$ and $\mathbf{b}$ (RLWE public key) following Eq. 1; we can express a ciphertext $(u^{(i)}, v^{(i)}) \in R_q^2$ as a vector of $2n$ elements $(\boldsymbol{u}^{(i)}, \boldsymbol{v}^{(i)}) = (u_1^{(i)}, \ldots, u_n^{(i)}, v_1^{(i)}, \ldots, v_n^{(i)}) \in \mathbb{Z}_q^{2n}$ and its re-encryption as:

$$\left(\boldsymbol{u}'^{(i)} \ \boldsymbol{v}'^{(i)} \right)^T = \left(\boldsymbol{u}^{(i)} \ \boldsymbol{v}^{(i)} \right)^T + \begin{pmatrix} \mathbf{A} \\ \mathbf{B} \end{pmatrix} (\boldsymbol{r}'^{(i)})^T + \left(\boldsymbol{e}_1'^{(i)} \ \boldsymbol{e}_2'^{(i)} \right)^T \quad \forall i \in [1, \ldots, N]$$

Following this notation and given a permutation π characterized by the matrix $\mathbf{M}$ and a set of re-encryption parameters $\left(\boldsymbol{r}'^{(i)}, \boldsymbol{e}_1'^{(i)}, \boldsymbol{e}_2'^{(i)} \right)$ for each one of the messages, we can express the shuffling of N RLWE encryptions as:

$$\begin{pmatrix} u_1''^{(1)} \cdots u_n''^{(1)} \ v_1''^{(1)} \cdots v_n''^{(1)} \\ \vdots \ \ddots \ \vdots \ \vdots \ \ddots \ \vdots \\ u_1''^{(N)} \cdots u_n''^{(N)} \ v_1''^{(N)} \cdots v_n''^{(N)} \end{pmatrix}_{N \times 2n} = \begin{pmatrix} m_{11} \cdots m_{1N} \\ \vdots \ \ddots \ \vdots \\ m_{N1} \cdots m_{NN} \end{pmatrix}_{N \times N} \begin{pmatrix} u_1^{(1)} \cdots u_n^{(1)} \ v_1^{(1)} \cdots v_n^{(1)} \\ \vdots \ \ddots \ \vdots \ \vdots \ \ddots \ \vdots \\ u_1^{(N)} \cdots u_n^{(N)} \ v_1^{(N)} \cdots v_n^{(N)} \end{pmatrix}_{N \times 2n}$$

$$+ \begin{pmatrix} r_1'^{(1)} \cdots r_n'^{(1)} \\ \vdots \ \ddots \ \vdots \\ r_1'^{(N)} \cdots r_n'^{(N)} \end{pmatrix}_{N \times n} \begin{pmatrix} a_1 \cdots a_n & b_1 \cdots b_n \\ -a_n \cdots a_{n-1} & b_2 \cdots b_{n-1} \\ \vdots \ \ \vdots \ \ \vdots \ \ \vdots \\ -a_2 \cdots a_1 & -b_2 \cdots b_1 \end{pmatrix}_{n \times 2n} + \begin{pmatrix} e_{1,1}'^{(1)} \cdots e_{1,n}'^{(1)} \ e_{2,1}'^{(1)} \cdots e_{2,n}'^{(1)} \\ \vdots \ \ddots \ \vdots \ \vdots \ \ddots \ \vdots \\ e_{1,1}'^{(N)} \cdots e_{1,n}'^{(N)} \ e_{2,1}'^{(N)} \cdots e_{2,n}'^{(N)} \end{pmatrix}_{N \times 2n}$$

$$\left(\mathbf{U''\ V''}\right) = \mathbf{M}\left(\mathbf{U\ V}\right) + \mathbf{R'}\left(\mathbf{A}^T\ \mathbf{B}^T\right) + \left(\mathbf{E'_1}\ ,\mathbf{E'_2}\right) \tag{5}$$

A mix-net node should prove that it knows the matrices $\mathbf{M}, \mathbf{R'}, \mathbf{E'_1}, \mathbf{E'_2}$ such that the output of the node $\left(\mathbf{U''\ V''}\right)$ is the input $\left(\mathbf{U\ V}\right)$ re-encrypted and permuted, without revealing any information about $\mathbf{M}, \mathbf{R'}, \mathbf{E'}_1$ and $\mathbf{E'}_2$.

$$\Sigma\text{-proof}\left[\begin{array}{c|c} \begin{array}{c} \pi \\ r'^{(1)},\ldots,r'^{(N)} \\ e_1'^{(1)},\ldots,e_1'^{(N)} \\ e_2'^{(1)},\ldots,e_2'^{(N)} \end{array} & \begin{array}{c} \left(\left(u''^{(1)},v''^{(1)}\right),\ldots,\left(u''^{(N)},v''^{(N)}\right)\right)^T \\ = \\ \mathsf{Re\text{-}encrypt}\left(\left(u^{\pi(1)},v^{\pi(1)}\right),r'^{(1)},e_1'^{(1)},e_2'^{(1)}\right)^T \\ \ldots \\ \mathsf{Re\text{-}encrypt}\left(\left(u^{\pi(N)},v^{\pi(N)}\right),r'^{(N)},e_1'^{(N)},e_2'^{(N)}\right)^T \end{array} \end{array}\right]$$

Following Wikström's proposal we are going to split the proof into two protocols.

Offline Phase

1. The mix-node M_j chooses a random permutation π_j characterized by the matrix $\boldsymbol{M}_j \in \mathbb{Z}_q^{N\times N}$, computes a matrix commitment $\mathsf{Com}(\mathbf{M}_j, \boldsymbol{\alpha}_{m_j})$ and publishes it. It also proves knowledge of the committed permutation.
2. M_j chooses randomly the re-encryption parameters: $\mathbf{R}'_j \in \mathbb{Z}_q^{N\times n}$, $\mathbf{E}'_{1j} \in \mathbb{Z}_q^{N\times n}$ and $\mathbf{E}'_{2j} \in \mathbb{Z}_q^{N\times n}$. It computes the corresponding matrix commitments, publishes them and prove that the committed elements are *small*.

Online Phase

1. Given a list of N input ciphertexts, the mix-node M_j permutes and re-encrypts the list using Eq. 5.
2. In order to prove that the committed matrices have been used to perform the mixing, M_j computes the commitment to its output using those commitments calculated during the online phase, and finally reveals its opening.

3.2   Proof of Knowledge of Permutation Matrix

The permutation matrix is characterized by the following theorem.

Theorem 1. *Given a matrix $\boldsymbol{M} \in \mathbb{Z}_q^{N\times N}$ and a vector $\boldsymbol{x} = (x_1,\ldots,x_N) \in \mathbb{Z}_q^N$ of N independent variables, $\boldsymbol{M}$ is a permutation matrix if and only if $\boldsymbol{M1} = \boldsymbol{1}$ and $\prod_{i=1}^N x_i = \prod_{i=1}^N x'_i$ where $\boldsymbol{x'} = \boldsymbol{Mx}$.*

We refer the reader to [38] for the details about the theorem's proof.

Given a commitment to a matrix $\mathsf{Com}(\mathbf{M}, \boldsymbol{\alpha}_m) = (c_{m_1},\ldots,c_{m_N})$ and a vector $\mathbf{x} = (x_1,\ldots,x_N)$, we can compute a commitment to the product of the matrix by a vector $\mathsf{Com}(\mathbf{Mx}, k)$ using Eq. 4, where $k = \langle\boldsymbol{\alpha}_m, \mathbf{x}\rangle$. In the special case where the vector $\mathbf{x} = \mathbf{1}$ the identity above is $\mathsf{Com}(\mathbf{1}, t)$ where $t = \sum_{i=1}^N \alpha_{m_j}$. Another important observation is that given a vector $\hat{\mathbf{r}} = (\hat{r}_1,\ldots,\hat{r}_N)$ we can express a commitment to the product of the elements of $\mathbf{x}'$ in a recursive way $\hat{c}_i = g^{\hat{r}_i} \hat{c}_{i-1}^{x'_i}$ for $i = 1, \ldots, N$ and $\hat{c}_0 = g_1$.

Applying the second condition for a permutation matrix $\left(\prod_{i=1}^{N} x_i = \prod_{i=1}^{N} x'_i \right)$, it is possible to obtain a commitment $\hat{c}_N$ such that $\hat{c}_N = g^{\hat{r}} g_1^{\prod_{i=1}^{N} x'_i} = g^{\hat{r}'} g_1^{\prod_{i=1}^{N} x_i}$, and prove that two different valid openings, $(\hat{r}, \prod_{i=1}^{N} x'_i)$ and $(\hat{r}', \prod_{i=1}^{N} x_i)$, are known. Due to the binding property of the commitments we know that if someone is able to open a commitment to two different openings, this means that either both openings are the same or the discrete logarithm, $g_1 = g^z$ where $z = (\hat{r} - \hat{r}') / \left(\prod_{i=1}^{N} x_i - \prod_{i=1}^{N} x'_i \right)$, can be computed.

Observe that using the Schwartz-Zippel lemma we can prove the polynomial equality $\prod_{i=1}^{N} x_i = \prod_{i=1}^{N} x'_i$ holds with overwhelming probability just verifying that the equation holds for a point $(\lambda_1, \ldots, \lambda_N)$ randomly chosen from $\mathbb{Z}_q^N$.

Given these preliminaries we can construct a Σ-proof to prove that the mix-net node knows an opening for the commitment and that the element committed is a permutation matrix. Since this proof follows the approach given by Wikström, we left the details for the full version [10].

3.3   Proof of Knowledge of Small Exponents

The second step in the offline part will be to prove that the random values used to re-encrypt are small. Remember that in order to re-encrypt a message, the following randomness is used: $\mathbf{r}'^{(i)} = \left(r_1'^{(i)}, \ldots, r_n'^{(i)} \right)$, $\mathbf{e}_1'^{(i)} = \left(e_{1,1}'^{(i)}, \ldots, e_{1,n}'^{(i)} \right)$ and $\mathbf{e}_2'^{(i)} = \left(e_{2,1}'^{(i)}, \ldots, e_{2,n}'^{(i)} \right)$ for $i \in [1, \ldots, N]$. In our case, we would require that the coefficients of these vectors belong to $[-\beta + 1, \beta - 1]$ where $\beta = 2^k$. In order to prove this we are going to use the strategy proposed in [21] by Ling et al. As it is explained in [5] the probability of obtaining an element from the error distribution with norm larger than β is negligible (notice that β will depend on the parameters of the encryption). Even when this restriction on the re-encryption elements norm is applied, the RLWE samples remain pseudorandom. This prevents a corrupted node from modifying the plaintext of the ciphertexts, while an honest node can still use the pseudorandomness to hide the relation between its input an output. We decompose $r_j'^{(i)}, e_{1,j}'^{(i)}$ and $e_{2,j}'^{(i)}$ as

$$ r_j'^{(i)} = \sum_{l=0}^{k-1} r_{j,l}'^{(i)} 2^l \qquad e_{1,j}'^{(i)} = \sum_{l=0}^{k-1} e_{1,j,l}'^{(i)} 2^l \qquad e_{2,j}'^{(i)} = \sum_{l=0}^{k-1} e_{2,j,l}'^{(i)} 2^l $$

with $r_j'^{(i)}, e_{1,j,l}'^{(i)}, e_{2,j,l}'^{(i)} \in \{-1, 0, 1\}$ and we prove that these elements have one of the possible values in the set $\{-1, 0, 1\}$ using an OR-proof. Afterwards, using the commitment to every bit of the decomposition we obtain a commitment to the coefficients, and consequently a commitment to each of the corresponding matrix columns. The protocol used to demonstrate that a value belongs to a specific set, $x \in \{-1, 0, 1\}$, is based on a zero knowledge proof that proves that the element x has one of the values in the set without revealing which one it is.

$$\Sigma\text{-proof}\left[\,x\,\middle|\,x\in\{-1,0,1\},c=g^r h^x\,\right]$$

Informally, the proof consists of computing three proofs simultaneously, for $x=-1, x=0$ and $x=1$, where two of them will be simulated and only that which corresponds to the real value of x will be the *real* proof. As this is a standard proof [11] the details are omitted here and both the proof and the demonstration of its properties are given in [10].

3.4   Opening the Commitments

Given the commitments to the permutation matrix and to the re-encryption matrices, the only thing that is left to prove is that these matrices have been used during the mixing process. This is an operation that should be done online since we need the list of encrypted messages to compute the proof. In order to do that we propose a methodology that differs from what Wikström proposes.

Given the commitments $c_{m_j} = \mathsf{Com}\left(m_j, \alpha_{m_j}\right)$, $c_{r'_j} = \mathsf{Com}\left(r'_j, \alpha_{r'_j}\right)$, $c_{e'_{1,j}} = \mathsf{Com}\left(e'_{1,j}, \alpha_{e'_{1,j}}\right)$ and $c_{e'_{2,j}} = \mathsf{Com}\left(e'_{2,j}, \alpha_{e'_{2,j}}\right)$ and Eq. 5, we can compute the following commitments to matrix products and sums,

$$c_{\mathbf{e}'_{1,k}}\left(\prod_{j=1}^{N} c_{\mathbf{m}_j}^{u_k^{(j)}}\right)\left(\prod_{j=1}^{n} c_{\mathbf{r}'_j}^{a_{k,j}}\right) = \mathsf{Com}\left(\hat{\mathbf{u}}''_{\mathbf{k}}, \alpha_{e'_{1,k}} + \langle\boldsymbol{\alpha}_M, \hat{\mathbf{u}}_{\mathbf{k}}\rangle + \langle\boldsymbol{\alpha}_{r'}, (a_{k,1},\ldots,a_{k,n})\rangle\right)$$

$$c_{\mathbf{e}'_{2,k}}\left(\prod_{j=1}^{N} c_{\mathbf{m}_j}^{v_k^{(j)}}\right)\left(\prod_{j=1}^{n} c_{\mathbf{r}'_j}^{b_{k,j}}\right) = \mathsf{Com}\left(\hat{\mathbf{v}}''_{\mathbf{k}}, \alpha_{e'_{2,k}} + \langle\boldsymbol{\alpha}_M, \hat{\mathbf{v}}_{\mathbf{k}}\rangle + \langle\boldsymbol{\alpha}_{r'}, (b_{k,1},\ldots,b_{k,n})\rangle\right)$$

naming $\hat{\mathbf{u}}_{\mathbf{k}}, \hat{\mathbf{v}}_{\mathbf{k}}, \hat{\mathbf{u}}''_{\mathbf{k}}, \hat{\mathbf{v}}''_{\mathbf{k}}$ the corresponding k-column of each matrix $\mathbf{U}, \mathbf{V}, \mathbf{U}'', \mathbf{V}''$. The only thing that the mix node should do in order to prove that it has used the appropriate values during the shuffling, is to open the commitments above revealing the openings.

$$\left(\alpha_{e'_{1,k}} + \left\langle\boldsymbol{\alpha}_M, \left(u_k^{(1)},\ldots,u_k^{(N)}\right)\right\rangle + \langle\boldsymbol{\alpha}_{r'}, (a_{k,1},\ldots,a_{k,n})\rangle\right) \quad \forall k \in [1,\ldots,n]$$

$$\left(\alpha_{e'_{2,k}} + \left\langle\boldsymbol{\alpha}_M, \left(v_k^{(1)},\ldots,v_k^{(N)}\right)\right\rangle + \langle\boldsymbol{\alpha}_{r'}, (b_{k,1},\ldots,b_{k,n})\rangle\right) \quad \forall k \in [1,\ldots,n]$$

The verifier has to check that these values are appropriate openings of the commitments in order to verify the node has used the committed matrices $\mathbf{M}, \mathbf{R}', \mathbf{E}'_1$ and $\mathbf{E}'_2$ to shuffle the encrypted messages (at its input).

As we have seen above, given the commitments to $\mathbf{M}, \mathbf{R}', \mathbf{E}'_1$ and $\mathbf{E}'_2$ we can compute the commitment to the matrix of permuted votes $\mathbf{M}\left(\mathbf{U}\,\mathbf{V}\right)$ and the re-encryption matrix $\left(\mathbf{R}'\left(\mathbf{A}^T\,\mathbf{B}^T\right) + \left(\mathbf{E}'_1\,\mathbf{E}'_2\right)\right)$. Notice that the $2n$ linear combinations of the values $\alpha_{m_j}, \alpha_{r'_j}, \alpha_{e'_{1,j}}, \alpha_{e'_{2,j}}$ that the mix node reveals, allow us to open the commitments to the sum of these matrices, but not to each matrix separately. Given that $\boldsymbol{\alpha}_M$, and $\boldsymbol{\alpha}_r$ appear on all the openings that we reveal we have to double check if they could leak any information about any relations between the α's that (in a post-quantum scenario) may reveal information about the permutation and the re-encryption elements. This is not the case because all the $\alpha_{e_1',j}$ and $\alpha_{e_2',j}$ are uniformly and independently chosen from $\mathbb{Z}_q$. All the linear combinations that we reveal have a different $\alpha_{e_i',j}$, and this implies that the combinations are also uniformly and independently distributed, and thereby it is impossible to isolate any of the α. The full protocol and a discussion about its properties are given in the full version of the paper [10].

4   Conclusions

We have proposed the first universally verifiable proof of a shuffle for a lattice-based cryptosystem. The messages at the input of the mix-net are encrypted using an RLWE encryption system and then they are shuffled by the mixing nodes. In order to prove the correctness of this shuffle each node must provide a proof of a shuffle, demonstrating that the protocol has been executed correctly without leaking any secret information. Our proposal follows the idea presented in [42] but introduces two significant differences: during the offline part the random elements used to re-encrypt the ciphertexts are committed using the generalized version of Pedersen commitment and it is proved that these elements belong to a certain interval using OR-proofs. On the other hand, during the online part each node computes a commitment to its output using the homomorphic properties of both the commitment scheme and the encryption scheme. Opening this commitment the mix node proves that it has used the values committed during the offline part to compute its output. Revealing this opening does not give any information about the secret information required to do the shuffling.

It is worth noticing that shuffling the votes is not enough to guarantee the voters' privacy, as the system can be insecure, for instance, due to malleability attacks [39]. To avoid this kind of attack additional security proofs might be provided before the mixing process starts.

Regarding efficiency, the number of OR-proofs to be computed by each mix node is proportional to knN, where N is the number of encrypted messages received by the node, n is the dimension of the lattice and k is the number of bits of each element of the re-encryption matrices. There are some techniques that allow to reduce the computational cost of these proofs and we leave for a future work to explore these improvements. We refer the reader to [42] for the details about the efficiency of the ZKP for a permutation matrix.

References

1. Abe, M.: Mix-networks on permutation networks. In: Lam, K.-Y., Okamoto, E., Xing, C. (eds.) ASIACRYPT 1999. LNCS, vol. 1716, pp. 258–273. Springer, Heidelberg (1999). doi:10.1007/978-3-540-48000-6_21
2. Adida, B., Wikström, D.: How to shuffle in public. In: Vadhan, S.P. (ed.) TCC 2007. LNCS, vol. 4392, pp. 555–574. Springer, Heidelberg (2007). doi:10.1007/978-3-540-70936-7_30

3. Adida, B., Wikström, D.: Offline/Online mixing. In: Arge, L., Cachin, C., Jurdziński, T., Tarlecki, A. (eds.) ICALP 2007. LNCS, vol. 4596, pp. 484–495. Springer, Heidelberg (2007). doi:10.1007/978-3-540-73420-8_43

4. Applebaum, B., Cash, D., Peikert, C., Sahai, A.: Fast cryptographic primitives and circular-secure encryption based on hard learning problems. In: Halevi, S. (ed.) CRYPTO 2009. LNCS, vol. 5677, pp. 595–618. Springer, Heidelberg (2009). doi:10.1007/978-3-642-03356-8_35

5. Benhamouda, F., Krenn, S., Lyubashevsky, V., Pietrzak, K.: Efficient zero-knowledge proofs for commitments from learning with errors over rings. In: Pernul, G., Ryan, P.Y.A., Weippl, E. (eds.) ESORICS 2015. LNCS, vol. 9326, pp. 305–325. Springer, Cham (2015). doi:10.1007/978-3-319-24174-6_16

6. Buchmann, J., Demirel, D., Graaf, J.: Towards a publicly-verifiable mix-net providing everlasting privacy. In: Sadeghi, A.-R. (ed.) FC 2013. LNCS, vol. 7859, pp. 197–204. Springer, Heidelberg (2013). doi:10.1007/978-3-642-39884-1_16

7. Canetti, R.: Universally composable security: a new paradigm for cryptographic protocols. In: Proceedings of the 42nd IEEE Symposium on Foundations of Computer Science, FOCS 2001, Washington, USA, pp. 136–145. IEEE Computer Society (2001)

8. Chaum, D.L.: Untraceable electronic mail, return addresses, and digital pseudonyms. Commun. ACM $24(2)$, 84–90 (1981)

9. Chillotti, I., Gama, N., Georgieva, M., Izabachène, M.: A homomorphic LWE based E-voting scheme. In: Takagi, T. (ed.) PQCrypto 2016. LNCS, vol. 9606, pp. 245–265. Springer, Cham (2016). doi:10.1007/978-3-319-29360-8_16

10. Costa, N., Martínez, R., Morillo, P.: Proof of a shuffle for lattice-based cryptography. IACR Cryptology ePrint Archive (2017)

11. Cramer, R., Gennaro, R., Schoenmakers, B.: A secure and optimally efficient multi-authority election scheme. In: Fumy, W. (ed.) EUROCRYPT 1997. LNCS, vol. 1233, pp. 103–118. Springer, Heidelberg (1997). doi:10.1007/3-540-69053-0_9

12. Damgard, I.: On σ-protocols. Lecture on Cryptologic Protocol Theory, Faculty of Science, University of Aarhus (2010)

13. Demirel, D., Henning, M., van de Graaf, J., Ryan, P.Y.A., Buchmann, J.: Prêt à voter providing everlasting privacy. In: Heather, J., Schneider, S., Teague, V. (eds.) Vote-ID 2013. LNCS, vol. 7985, pp. 156–175. Springer, Heidelberg (2013). doi:10.1007/978-3-642-39185-9_10

14. Fauzi, P., Lipmaa, H.: Efficient culpably sound NIZK shuffle argument without random oracles. In: Sako, K. (ed.) CT-RSA 2016. LNCS, vol. 9610, pp. 200–216. Springer, Cham (2016). doi:10.1007/978-3-319-29485-8_12

15. Fauzi, P., Lipmaa, H., Zając, M.: A shuffle argument secure in the generic model. In: Cheon, J.H., Takagi, T. (eds.) ASIACRYPT 2016. LNCS, vol. 10032, pp. 841–872. Springer, Heidelberg (2016). doi:10.1007/978-3-662-53890-6_28

16. Furukawa, J., Sako, K.: An efficient scheme for proving a shuffle. In: Kilian, J. (ed.) CRYPTO 2001. LNCS, vol. 2139, pp. 368–387. Springer, Heidelberg (2001). doi:10.1007/3-540-44647-8_22

17. Golle, P., Jakobsson, M., Juels, A., Syverson, P.: Universal re-encryption for mixnets. In: Okamoto, T. (ed.) CT-RSA 2004. LNCS, vol. 2964, pp. 163–178. Springer, Heidelberg (2004). doi:10.1007/978-3-540-24660-2_14

18. Groth, J.: A verifiable secret shuffe of homomorphic encryptions. In: Desmedt, Y.G. (ed.) PKC 2003. LNCS, vol. 2567, pp. 145–160. Springer, Heidelberg (2003). doi:10.1007/3-540-36288-6_11

19. Groth, J., Lu, S.: A non-interactive shuffle with pairing based verifiability. In: Kurosawa, K. (ed.) ASIACRYPT 2007. LNCS, vol. 4833, pp. 51–67. Springer, Heidelberg (2007). doi:10.1007/978-3-540-76900-2_4
20. Lindner, R., Peikert, C.: Better key sizes (and attacks) for LWE-based encryption. In: Kiayias, A. (ed.) CT-RSA 2011. LNCS, vol. 6558, pp. 319–339. Springer, Heidelberg (2011). doi:10.1007/978-3-642-19074-2_21
21. Ling, S., Nguyen, K., Stehlé, D., Wang, H.: Improved zero-knowledge proofs of knowledge for the ISIS problem, and applications. In: Kurosawa, K., Hanaoka, G. (eds.) PKC 2013. LNCS, vol. 7778, pp. 107–124. Springer, Heidelberg (2013). doi:10.1007/978-3-642-36362-7_8
22. Lipmaa, H., Zhang, B.: A more efficient computationally sound non-interactive zero-knowledge shuffle argument. In: Visconti, I., Prisco, R. (eds.) SCN 2012. LNCS, vol. 7485, pp. 477–502. Springer, Heidelberg (2012). doi:10.1007/978-3-642-32928-9_27
23. Locher, P., Haenni, R.: Verifiable internet elections with everlasting privacy and minimal trust. In: Haenni, R., Koenig, R.E., Wikström, D. (eds.) VOTELID 2015. LNCS, vol. 9269, pp. 74–91. Springer, Cham (2015). doi:10.1007/978-3-319-22270-7_5
24. Locher, P., Haenni, R., Koenig, R.E.: Coercion-resistant internet voting with everlasting privacy. In: Clark, J., Meiklejohn, S., Ryan, P.Y.A., Wallach, D., Brenner, M., Rohloff, K. (eds.) FC 2016. LNCS, vol. 9604, pp. 161–175. Springer, Heidelberg (2016). doi:10.1007/978-3-662-53357-4_11
25. Lyubashevsky, V., Peikert, C., Regev, O.: On ideal lattices and learning with errors over rings. J. ACM 60(6), 43:1–43:35 (2013)
26. Markus, J., Ari, J.: Millimix: mixing in small batches. Technical report (1999)
27. Micciancio, D., Regev, O.: Lattice-based cryptography. In: Bernstein, D.J., Buchmann, J., Dahmen, E. (eds.) Post-Quantum Cryptography, pp. 147–191. Springer, Heidelberg (2009). doi:10.1007/978-3-540-88702-7_5
28. Moran, T., Naor, M.: Split-ballot voting: everlasting privacy with distributed trust. In: Proceedings of the 14th ACM Conference on Computer and Communications Security, CCS 2007, pp. 246–255. ACM (2007)
29. Andrew Neff, C.: A verifiable secret shuffle and its application to e-voting. In: Proceedings of the 8th ACM Conference on Computer and Communication Security, CCS 2001, pp. 116–125, NY, USA (2001)
30. Park, C., Itoh, K., Kurosawa, K.: Efficient anonymous channel and all/nothing election scheme. In: Helleseth, T. (ed.) EUROCRYPT 1993. LNCS, vol. 765, pp. 248–259. Springer, Heidelberg (1994). doi:10.1007/3-540-48285-7_21
31. Pedersen, T.P.: Non-interactive and information-theoretic secure verifiable secret sharing. In: Feigenbaum, J. (ed.) CRYPTO 1991. LNCS, vol. 576, pp. 129–140. Springer, Heidelberg (1992). doi:10.1007/3-540-46766-1_9
32. Regev, O.: On lattices, learning with errors, random linear codes, and cryptography. In: Proceedings of the Thirty-seventh Annual ACM Symposium on Theory of Computing, STOC 2005, pp. 84–93, New York, NY, USA. ACM (2005)
33. Regev, O.: The learning with errors problem. In: IEEE 25th Annual Conference on Computational Complexity (CCC), pp. 191–204 (2010)
34. Rückert, M., Schneider, M.: Estimating the security of lattice-based cryptosystems. IACR Cryptology ePrint Archive, Report 2010/137 (2010). http://eprint.iacr.org/2010/137
35. Sako, K., Kilian, J.: Receipt-free mix-type voting scheme. In: Guillou, L.C., Quisquater, J.-J. (eds.) EUROCRYPT 1995. LNCS, vol. 921, pp. 393–403. Springer, Heidelberg (1995). doi:10.1007/3-540-49264-X_32

36. Singh, K., Pandu Rangan, C., Banerjee, A.K.: Lattice based universal re-encryption for mixnet. J. Int. Serv. Inf. Secur. (JISIS) **4**(1), 1–11 (2014)
37. Singh, K., Pandu Rangan, C., Banerjee, A.K.: Lattice based mix network for location privacy in mobile system. Mob. Inf. Syst. **1–9**, 2015 (2015)
38. Terelius, B., Wikström, D.: Proofs of restricted shuffles. In: Bernstein, D.J., Lange, T. (eds.) AFRICACRYPT 2010. LNCS, vol. 6055, pp. 100–113. Springer, Heidelberg (2010). doi:10.1007/978-3-642-12678-9_7
39. Wikström, D.: The security of a mix-center based on a semantically secure cryptosystem. In: Menezes, A., Sarkar, P. (eds.) INDOCRYPT 2002. LNCS, vol. 2551, pp. 368–381. Springer, Heidelberg (2002). doi:10.1007/3-540-36231-2_29
40. Wikström, D.: A universally composable mix-net. In: Naor, M. (ed.) TCC 2004. LNCS, vol. 2951, pp. 317–335. Springer, Heidelberg (2004). doi:10.1007/978-3-540-24638-1_18
41. Wikström, D.: A sender verifiable mix-net and a new proof of a shuffle. In: Roy, B. (ed.) ASIACRYPT 2005. LNCS, vol. 3788, pp. 273–292. Springer, Heidelberg (2005). doi:10.1007/11593447_15
42. Wikström, D.: A commitment-consistent proof of a shuffle. In: Boyd, C., González Nieto, J. (eds.) ACISP 2009. LNCS, vol. 5594, pp. 407–421. Springer, Heidelberg (2009). doi:10.1007/978-3-642-02620-1_28

An Analysis of Bitcoin Laundry Services

Thibault de Balthasar[1] and Julio Hernandez-Castro[2(✉)]

[1] Chainalysis Inc., 43 West 23rd Street, 2nd Floor, New York, NY 10010, USA
thibault@chainalysis.com
[2] School of Computer Science, University of Kent, Cornwallis South,
Canterbury CT2 7NF, UK
jch27@kent.ac.uk

Abstract. This work briefly (An extended version can be found at https://kar.kent.ac.uk/id/eprint/63502) examines some of the most relevant Bitcoin Laundry Services, commonly known as tumblers or mixers, and studies their main features to try to answer some fundamental questions including their security, popularity, transaction volume, and generated revenue. Our research aims to inform both legitimate users and Law Enforcement about the characteristics and limitations of these services.

Keywords: Bitcoin · Tumbler · Alphabay · Helix · Anonymity · Cybercrime

1 Introduction to Tumblers

Bitcoin offers pseudo-anonymity [10] because all transactions are visible and traceable, but no names are stored in the Blockchain. Bitcoin laundry services are open, like most modern technologies, to dual use. They are employed by regular users who do not engage in any illicit activities and simply want to improve on the anonymity features of Bitcoin. On the other hand, they can also be used by cyber criminals for laundering their ill-gotten gains before exchanging them into traditional currencies such as Dollars, Euros or Sterling. It is also common for stolen Bitcoins (i.e. after a wallet compromise or a hack) and for ransom money to be processed by one or more tumblers to reduce its traceability. In either scenario, Bitcoin laundry services play a central role in the Bitcoin economy, but they have been relatively poorly studied [8,9,11], and their operation is not that well understood. We will try to address this in this work, by focusing on a small number of very well-known Bitcoin tumblers that vary widely in their characteristics and sophistication.

Methodology. For this purpose, multiple transactions have been carried out involving the mixers under study, transactions that have been later carefully studied for finding patterns, regularities and correlations with a set of tools we have developed. Using our own tools, and together with other commercially

© Springer International Publishing AG 2017
H. Lipmaa et al. (Eds.): NordSec 2017, LNCS 10674, pp. 297–312, 2017.
https://doi.org/10.1007/978-3-319-70290-2_18

available ones[1], it becomes possible to demonstrate that these services suffer from serious limitations that expose their users to traceability and, sometimes, even de-anonymisation attacks.

Attacker Model. The attacker model we will consider in this paper is based around the concept of taint analysis. The objective of taint analysis is to link multiple Bitcoin addresses. Typically at least one is known to contain stolen Bitcoins, or Bitcoins that are otherwise clearly linked with a criminal activity, so establishing this link will show the latter addresses (ones that have received funds from it) are tainted and, for example, money from them should not be accepted by reputable merchants or at legitimate exchanges. To break this link or taint cyber-criminals use mixers, so our aim at attacking a mixer is first and foremost to be able to characterize all (or a sizable proportion) of the Bitcoins that have gone through it. Of course, this taint can also be interpreted in terms of anonymity levels, when tainted addresses and wallets can be linked back to individuals. Apart from this, we will try to find how exactly these mixers work and establish clusters or other patterns between input and output addresses so that, to a certain extent, we can 'reverse' the operation of a tumbler and, at least probabilistically, trace back and deanonymise it.

2   Results

We present in the following our most relevant results in terms of security and privacy characteristics of the mixers we have studied.

2.1   DarkLaunder, Bitlaunder and CoinMixer

Darklaunder, Bitlaunder and CoinMixer are probably the weakest mixers of all tested in this work. We analyse these jointly because we have reasons to believe they share a common owner and are almost identical in their functioning and features. So, albeit in the following we will mostly refer to Darklaunder many of our findings also apply to Bitlaunder and CoinMixer, which will be explicitly mentioned only to highlight any differences. Darklaunder is available on both the clearnet[2] where it makes usage of CloudFlare (a widely used proxy service) and on the darknet[3]. This duality is uncommon in good mixers, as is the use of CloudFlare.

The service offers two types of laundering: the **quick** one is claimed to take between one and six hours to process, and has a 2% fixed fee. The **secure** one is said to be dealt with by hand and to be more secure. In this case there is

---

[1] In addition to a large set of python scripts developed by the authors, we have also been given access to some of the proprietary Chainalysis tools.

[2] At https://darklaunder.com, last accessed on 17/02/2017.

[3] At http://wwxoxavgqbhthyz7.onion, last accessed on 17/02/2017.

a 3% fixed fee. For both, the lowest accepted sum is 0.01 BTC[4]. According to the service's FAQ, there is an upper limit of 1,000 BTC. To be able to use the service, registration is mandatory and a username, name, password and email address have to be provided. This is common in other mixers, but not a good practice regarding privacy. To launder Bitcoin, the user has to make a deposit on a given address. When withdrawing, the only choices are the amount of Bitcoin to withdraw and the destination address. Despite their claim that it does not keep any personal information, we have found it stores data about their user's previous transactions with the service, including their exact date and time and the involved IPs and Bitcoin addresses. All these weaknesses could be also found in Bitlaunder. Since there is precedent of authorities arresting owners of laundering services[5], and the service retains full historical transaction data, this mixer can not be considered secure. In addition, PHP errors creep around frequently during its usage.

Security Analysis. On top of its bad design, the service is also subject to other critical problems. First, it is possible to find the IP address of the server hosting the mixer. This makes easy to establish a link to an individual's name and address, and to other mixers he owns and operates. Since the server is using CloudFlare, which is only an HTTP proxy, the emails sent by the service (in response to customer's questions) do not go through it. By analyzing the header of these emails it is possible to find that the mail server is located at the address mail.darklaunder.com, which points to the IP address 94.23.45.166. We can, therefore, access the website directly now without going through CloudFlare. Furthermore, the SSL certificate used by the service is quite weak: It is using the SHA-1 algorithm, that is deprecated [3,4], with a 2048-bit key. Finally, the service certificate is self-signed, and has expired. It was signed in August 2015, which suggest the service has been probably first online around this time. The HTTP server used is Nginx 1.0.14, which is a legacy version as the latest one at the time of writing is 1.10. There are multiple CVEs affecting the server version, as shown at cvedctails.com [5], notably CVE-2013-4547, CVE-2013-0337, CVE-2012-2089 and CVE-2012-1180. An additional serious security issue is that the server is allowing SSL v3, which is vulnerable to multiple attacks [6,7]. Bitlaunder suffers from many of the previously described problems.

Transactions with the Service. At total of 61 transactions were carried out with Darklaunder. At the beginning, the transactions were processed correctly even if the time needed to get the money back was longer than expected, usually between 8 and 10 h. From the 29th test on, transactions took more than 20 h to withdraw. From the 45th, it took between one and seven days to get the withdraw (sometimes, due to multiple failures during the laundering process). Eleven

---

[4] During our tests we encountered some issues, and the contact support stated that the minimum value was 0.5 BTC. This is strange, since despite this message the mixer eventually worked after some time with the initial 0.01 BTC.

[5] For example in the case of coin.mx [1,2].

transactions have also been made with Bitlaunder but no delays were encountered, probably because they were requested to be more evenly spaced on time. For both services, the fees taken have always been exactly as announced, but once Darklaunder returned the money twice (so we received double the money we sent!) and another time, the service returned slightly less: 10% of the total sum was missing. These errors suggest that, at some point, the algorithm in charge of withdrawing the money was suffering from flaws. Another important mistake is that the service is using counters as transaction IDs, so the total number of transactions can be simply read. Furthermore, several issues with the laundering algorithm can be detected after analyzing our database of transactions. First and foremost, the independent accounts we have used happen to have common transactions. Also, when the service takes money from the wallet, the transaction used involves multiple input wallets and they have always exactly two outputs, one of them, as we will see later, being a central address. This is quite a poor practice since a malicious user may simply engage in making transactions on a regular basis to find the addresses of other users, thus partially de-anonymising the service (Fig. 1).

Fig. 1. Darklaunder: withdrawing to multiple addresses

Tracking the Money. Using a script to trace the money, some common paths between the addresses used have emerged. In particular, we can detect a path between wallets and return addresses, showing the anonymity offered by the service is poor.

Figure 2 shows the output of the program we developed to follow the money, where we can see the results when tracking the wallet generated in the first transaction with the service. The watch-list is made of the addresses given to the service to get the money back. We can see that, in this case, the money has been redistributed to three known addresses generated in the next tests (these three addresses are the only ones that belong to us within four levels of tracking for all the tests, however, with a deeper tracking it is possible to find even more).

Drawing and Analysing the Transactions. We will begin the analysis of the service by using Fig. 3, which is the graph generated after analysing the transactions we performed with Darklaunder and Bitlaunder, to one level of depth.

The image allows to quickly visualise the very high centralization of the service, which is a poor characteristic regarding anonymity. All the wallets are

```
Analyzing wallet 1 of 61 (1CJULzTjfbpQWKrVsHocG9imKcC1a6Wk7T)
------------------------------------------------------------
1A1yFjzA1jMZim358Uew1gqWLzKMuE772k IS IN WATCHLIST :
2016-05-03T20:31:04Z 1HRpW9rPBFUYUmxB3nhyTPK6PmRAY7cZjj
2016-05-03T20:19:35Z 1MGj1xzFtZvV7EBL7TLU6RBccAk6QxocRM
2016-04-18T19:06:07Z 15uyvmNQtLPyzeNcBCvuvgH4f7MUN6XFKF
None 1CJULzTjfbpQWKrVsHocG9imKcC1a6Wk7T
------------------------------------------------------------
1LwAB8s9ytMoNMvLevzrXvEEX1tWjFMc9b IS IN WATCHLIST :
2016-04-21T20:28:26Z 1FyPcniC17LWxGj3crtxhY8wUQwJfjGChb
2016-04-21T20:22:03Z 1JqjRynSnMd8ogQDi7pmZi6LGQdDkMB9iu
2016-04-18T19:06:07Z 15uyvmNQtLPyzeNcBCvuvgH4f7MUN6XFKF
None 1CJULzTjfbpQWKrVsHocG9imKcC1a6Wk7T
------------------------------------------------------------
1D8mC6ywLiCSAZCwdvRLxn4GqEaKLEvQpm IS IN WATCHLIST :
2016-04-21T20:28:26Z 1FyPcniC17LWxGj3crtxhY8wUQwJfjGChb
2016-04-21T20:22:03Z 1JqjRynSnMd8ogQDi7pmZi6LGQdDkMB9iu
2016-04-18T19:06:07Z 15uyvmNQtLPyzeNcBCvuvgH4f7MUN6XFKF
None 1CJULzTjfbpQWKrVsHocG9imKcC1a6Wk7T
```

Fig. 2. Darklaunder: output of the program following the money

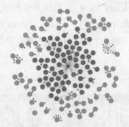

Fig. 3. Darklaunder and Bitlaunder transactions, at depth one

sending their funds to address 15u...FKF[6]. This central address has been used for the first time on the 18th October 2015 - which matches nicely with our estimate of the creation time of the service - and has been continuously used since.

Figure 4 shows the number of operations of the 15u...FKF address since its creation.

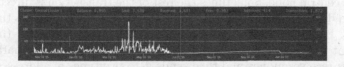

Fig. 4. Number of transactions by 15u...FKF, from October 2015 to February 2017

The address has a total of 1,635 operations. The number of credit transactions (934) is roughly equal to the number of debit transactions (719). However,

---

[6] 15uyvmNQtLPyzeNcBCvuvgH4f7MUN6XFKF.

this address received money from 4,277 addresses but sent money to only 1,327 addresses. The total in and out by day from the creation of the address to the 26th February 2017 tends to confirm the hypothesis that the address has only be used as a gateway. We can observe that the credit and debit per day are approximately equal, leaving the address with only a few bitcoin in reserve. Our last 2016 test transaction with the service was on the 5th May 2016, and at this time the wallet[7] was still using the same central address. However, another transaction has been carried out on the 27 June, and we can see that the wallet[8] has used another address to get back the money: 13K...isR[9]. This address has been created on the 15 June 2016, which matches with the moment when 15u...FKF's traffic started to decrease. In just 10 days, the new address made 110 transactions, sent 187.812357 BTC and received 190.612357 BTC.

By analysing money in, out and the total credit by day of address 13K...isR, we can see a very similar behaviour to that shown in Fig. 5. We can also observe that the percentages of credit and debit transactions are similar for the two addresses. Considering these elements, we can guess that the service periodically switches its *central* address. This is a good security practice, but by itself not sufficient to provide enough anonymity. The characteristics we underlined above may allow to easily detect these new addresses, thus completely defeating its security aims.

Fig. 5. Transactions by 13K...isR from June 2016 to March 2017

The interactions involving the central address follow recognisable patterns, as shown in Fig. 6. We can see that in each case wallets send bitcoins to the central address (label 1) but sometimes they also send to another addresses (labels 2 and 3). These secondary addresses will receive bitcoins from other transactions involving the withdraw addresses, and will then send it to other addresses and back to the central node. Sometimes the rest of the transaction is directly sent to the central address, as with node 26. Node 6 on the graph represents the address 15v...j2N[10]. Looking at its transactions is particularly interesting: there are a total of 51 at the time of writing, and the pattern followed is very characteristic. The address receives money and then sends it to two types of addresses; most of the bitcoins go to the central node, but a few of them go to another address (not the same every time) which is probably there to confuse a potential attacker.

---

7 1GgfvBoVpeJLKdVkqMehbFPrm4VjoqUP7.
8 1MC8VD89moVwXL4s213vNpdbrmUZQZf1DV.
9 13KtxHChVmGu43A19narE3hbKGCUBGAisR.
10 15vXhKcnNZo6su5PkKeZQPavvFhjVG3j2N.

The fact that the money of all wallets is sent to the central node is a terrible weakness, since it allows to find the wallet addresses with great ease. In addition, performing most of the withdrawal transactions within only one or two levels of the central node is also extremely poor.

Fig. 6. Interaction with central address

When observing the withdrawal transactions, a specific pattern is also interesting to notice: almost every withdrawal is at a distance of one address, as we can see on Fig. 7 which has been adapted from real data. Address 1 (that has not been analysed) establishes a link between two withdrawals, and Address 2 send the funds to the addresses used to withdraw.

Fig. 7. Interaction between withdrawal transactions.

Using Walletexplorer, we can find that Address 2 belongs to localBitcoins.com while Address 1 behaves similarly to Addresses 3 and 4. We can see that the debit transactions follow two distinct patterns. Either the central address gives money to localBitcoins, or it makes a peeling-chain (which consist in dividing the sums again and again) and eventually sends money to localBitcoins after a small number of transactions. Using the information gathered so far, we are now able to understand the complete workflow of the service, that we display in Fig. 8.

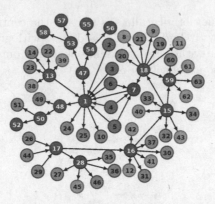

Fig. 8. Darklaunder workflow

The wallets are used to credit the central address (1) but at the same time they can also make use of a change address (7). This change address will receive credit at the end of the withdraw chain and send it to the central address. The central address sends Bitcoins to localBitcoins directly but also, sometimes, starts a peeling chain where bitcoins can be sent to localBitcoins.com during the process (49, 51) or later (55, 56, 57, 58). Then, localBitcoins sends back money to the service (17, 28) which starts a new withdraw chain. Using the Chainalysis tool, a graph of the exchanges has been drawn (Fig. 9).

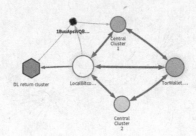

Fig. 9. Interaction of the service with external clusters

So we can conclude that the laundering algorithm itself is quite poor. The service is characterised by a heavy centralization because a central address is gathering all the Bitcoins from the customer's wallets and receives the rest of the money at the end of the withdraw chain. Furthermore, it is easy to find a direct route (only a few levels deep) from the central address to some of the wallets. Tracking is further facilitated because a significant number of transactions have multiple input addresses. Finally, the scarcity of traffic makes for an even easier address identification (Fig. 10).

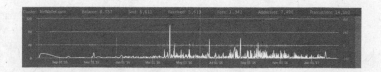

Fig. 10. Estimated darklaunder transaction volume

2.2   Helix

The Service. Helix is accessible only using Tor[11] and offers two different services: a standard version and a light version. The two versions only differ in that the light one allows to withdraw to up to five addresses, and to choose to receive multiple transactions and/or within a random time delay of a few hours while the standard version requires registration and allows to manage a wallet and to automatically mix money send to the wallet to a defined address. Both standard and light services are taking 2.5% off fees, and only allow withdrawals of 0.02 BTC or more.

Analysis of the Transactions. A total of 34 transactions were carried out with this service. The money always returned on time, and to the right number of addresses. On the more negative side, the page which displays the status of the laundering process has been observed to remain active a few days after the mixing has finished, when it is claimed to be available only for 24 h. Furthermore, a major problem has been found in the pattern of transactions: Regardless of whether we ask for multiple transactions or to use multiple addresses, our tests suggests that there will always be 5 transactions done in total. Some of our wallets and return addresses (issued from different tests) have also been observed taking part in the same transactions. Finally, our tests revealed that it always takes between one and two minutes to make a transaction. The average time is ninety seconds and the average duration for all the transactions to perform is five minutes and fifty seconds. This allows for a trivial timing-based attack.

Analysis of the Addresses. Our analysis has started by drawing a graph of the exchanges we carried out, at a depth level of 2 from our wallet and return addresses, as shown in Fig. 11.

First, it is possible to observe that the green addresses (which are the addresses where the coins have been returned) are very close to each other. Sometimes even present in the same transaction. This can be explained by the fact that the service is using a peeling chain to fund its customers. An interesting fact is that the transactions in these chains have always a single input but can have between two and five outputs, thus allowing withdrawals to multiple customers at the same time. Three addresses involved in a big amount of transactions are shown in the graph. The one in the center is identified by

---

[11] Main address is at http://grams7enufi7jmdl.onion/helix/light.

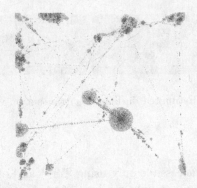

Fig. 11. Helix light exchanges at depth 2 - August 2016 (Color figure online)

Chainalysis's tool as part of LocalBitcoins.com. This cluster receives money from multiple points in peeling chains; This suggests it is widely used by customers of the service. The two other addresses are identified by the tool as part of the same cluster, which will be named C1[12] and studied later. Finally, the graph shows that multiple addresses are receiving coins from multiple wallet addresses (in red).

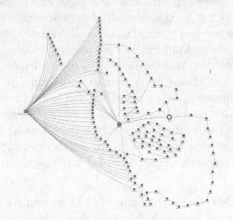

Fig. 12. Helix light withdrawal pattern - February 2017 (Color figure online)

Figure 12 has been drawn using Reactor. It represents the return transactions made by Helix. The point on the left is a custom cluster, made of all the return addresses used to perform the tests. The red point on the right is the Helix cluster, and the big point in the middle is cluster C1. It is possible to observe that C1 is receiving multiple transactions originated from an important number of addresses in the graph. This cluster seems to be receiving only coins from the

---

[12] 1MiaNEG1jqoAeLPSE8JuZ8ync1e6i1y6ho.

Helix's peeling chains. The money sent by C1 can not be linked to Helix, but it is possible to formulate two hypothesis: Either the owner of the service is using the cluster to recover some money, or this is a very special customer making an extensive usage of the service (190 BTC have been received).

The second interesting point is that (even though not all the chains are shown in the graph) all the money that have been sent to the return addresses goes through the Helix cluster, after a few transactions on the peeling chain. This allows to guess the algorithm used by the service: it generates multiple wallets and recovers their money using a few transactions. This money is then directly sent to a peeling chain for the customers withdrawal.

Assessment of the Service. The Chainalysis's tool suggests that at least 216,000 BTC have been mixed using this service until early 2017 showing it is widely[13] used. However, our findings indicate that it does not offer adequate anonymity. We observed that wallets and withdrawals of multiple customers are present on the same transaction, or very close to each other, so that it is easy to identify them. For example, by processing regular and small-amount transactions with the service we can gain valuable insights that can help us compromise its security and anonymity at a very low cost.

2.3   Alphabay

The Service. Alphabay was accessible only through Tor[14] and required to open a customer account to use it. The registration was straightforward, and only a username and password were required. A wallet address is automatically generated by the service for the user, and that address changes every time a deposit is made. However, the address was still usable seven days after the change. On top of that, if the generated address is not used after ten days, it will be deleted. Each deposit to the service must be at least of 0.01 BTC, and it is possible to withdraw money for a fee of 0.001 BTC. The service offered the possibility to withdraw to one to five addresses, in an interval of time between one and twenty-four hours. An option labeled *Sent a single transaction* suggest that the service was capable of returning money in multiple transactions.

Transaction Analysis. We performed 35 transactions with the service before proceeding with a first analysis of the tumbler. Multiple problems were detected as this early point. First, the service was taking more than what was claimed in fees (0.007 BTC instead of 0.001). In addition, the money was moved from the user's wallet before the withdrawal was carried out. During the tests, we also noticed that the service was **never** returning the money in multiple transactions

---

[13] The number of bitcoins in circulation at the time of writing is approximately 16.2 million, according to Blockchain.info.

[14] The main address was at pwoah7foa6au2pul.onion, but many others existed to cope with frequent DDoS attacks.

(a feature that is proposed in the form) and that it did not returned money to multiple addresses if the sum to withdraw is less than (#Addresses ·0.01 BTC). During our tests, we also notice that the service **never** returned money by doing multiple transactions to a single address. This was still true as of our last test on the 16th February, 2017. Another important problem was in the history of withdrawals, as IDs are used and they are simply incremental counters that leak the number of transactions. Using this information we can, for example, estimate the number of transactions to be around 33.76 per minute between the 4 and the 6 August 2016. Recent cluster size estimation tends to suggest the number of transactions did not changed a lot a year later. Finally, we can also detect some specific patterns concerning the number of input and outputs in the transactions performed by Alphabay, which can allow for simple heuristics to recognize them. On a more positive note, the money is always returned on time.

Analysis of the Addresses. By drawing the exchanges of the addresses on two depth levels, we can observe that while the money on the wallet addresses is going to addresses with a lot of traffic, the withdrawals are performed within a basic peeling chain. The peeling chain is a pattern of use widely present in the Bitcoin network; for example, various services often use it to withdraw their customers. Basically, the chains starts with an address receiving a decent amount of coins. This address will then send the coins to two (or more) addresses. One of these addresses will belong to the service and will then send coins to two (or more) addresses until there is no money left. In our example (Fig. 13), we can see a withdraw chain started by the service (in blue) with 50 BTC. Another common characteristic is that the nodes of a peeling chains have only two transactions. One credit and one debit. In the case that a peeling chain is used by a service, it can happen that the orange nodes are not to withdraw to a customer but just a redirection of some part of the money to another peeling chain also owned by the service. For example, 1dj6nAA7Sp456Ph9EvM8LYnvb6aYX9NPQ is the start of a classical peeling chain.

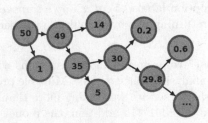

Fig. 13. How a peeling chain works (Color figure online)

If we look closely at Fig. 14, the first thing we will notice is that three addresses on the graph are involved in a lot of transactions. These addresses

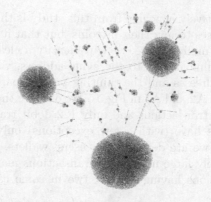

Fig. 14. Alphabay exchanges at depth 2.

will be discussed later and will be named A1[15] (the one on the left), A2[16] (the one at the bottom) and A3[17] (the one on the upper right). We will first focus on the exchanges carried out by our wallets: We can observe that, every time, the service is making transactions with the wallet following the exact same pattern. Money is moved with a transaction having only one input and one output, and goes to an address which has multiple transactions. We can observe a lot of transactions on this destination address, so we used Chainalysis's tools to gather more information.

There are multiple types of clusters we can observe in Fig. 15:

1. Some are identified by Chainalysis as belonging to well-known services, such as BTC-e or localBitcoins.com
2. Some are not identified as known services, and only receive money from addresses identified as part of the Alphabay cluster, or from addresses matching the wallet pattern. Then, they send money to an unique service using different addresses (for example 19Gc...X1d[18]). In this case, we hypothesize that Alphabay is using these services to mix the bitcoins.
3. Some are matching pattern 2, but do not send money directly to services and instead start peeling chains (e.g.: 1JJww8DFoAp5whSu4oV 89yZyY8MPVomsiz). The peeling chain is probably used to withdraw money off the service.
4. Some are matching the pattern 2, but sending money directly to multiple services. We do not have a good enough explanation for these cases.
5. A few clusters are sending and receiving money to/from multiple services. In this case, they probably belong to services that are not detected by the tool yet.

---

[15] 1HBsi9dDzHQecyy4xtRnvqjiT1KvLUwRcH.
[16] 16ZZ6svbB36o5Q2gLtAMHMiKJXtbs6nvuF.
[17] 14cGaFD4iUyqX9NQaB1ff8uLUb42qd5deM.
[18] 19Gc23Ggr58ZRhemmx7rtZnqTj6tasX1d.

What we can tentatively conclude from this study is that Alphabay is using other third party services to mix their bitcoins, but that it probably also makes some custom in-house mixing. Here, we can identify a clear flaw: multiple customer wallets are sending money to the same address, which could make the detection of wallet addresses and the tainting process particularly easy. When studying addresses A1, A2 and A3 in Fig. 15, we observe that these are receiving a total of 8,582 credit transactions and only 122 debit transactions. Moreover, the credit transactions have (with a few exceptions) only one input and one output. This suggests we are dealing with users' wallets. We noticed that the clusters have a relatively large number of transactions matching the same pattern: Deposit transactions having one (or two in some cases) inputs and two outputs.

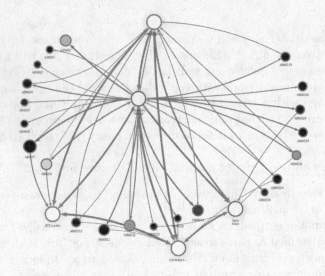

Fig. 15. Alphabay Wallet exchanges

We then analyzed return addresses. We can see in Fig. 15 that these addresses are often linked together and, sometimes, have associations to addresses exchanging with many more addresses than usual. In the first case, we observed that the addresses are part of a peeling chain. These peeling chains are most of the time tainted with Alphabay addresses, but sometimes we can track-back their origin to services such as localBitcoins. The second case is a direct withdraw from a known service (most of the time, localBitcoins). Our address is never alone on the outputs, and most of the time another output in the withdraw transaction leads to Alphabay. On both cases, we can say that the withdraw is not secure. On the first case, since the peeling chain is highly tainted by Alphabay, it would be easy to identify withdraw chains tainted by the service. Even without knowing which cluster Alphabay is, engaging in regular, low-cost transactions with it should be enough for taint purposes. In the second case, we would need to know Alphabay's addresses to be able to find the withdraw addresses. This would be a little more involved but not too difficult. In any case, this is not the most common way the money is withdrew.

3   Conclusions

Bitcoin mixers are quite popular nowadays, but even the most well-known and established ones seem to have serious security and privacy limitations, as exposed in this work. Together with the major players, a myriad of smaller laundry services such as Bitlaunder, Darklaunder and Coinmixer exist, and we have shown some of them offer an appalling service that can seriously compromise the security and privacy expectations of any legitimate user. Unfortunately also the major players such as Alphabay and Helix present significant deficiencies. Our findings show that devising and implementing a secure mixer is far from an easy task, and as such it is plagued with a multitude of opportunities to get things wrong and compromise the service. This is refreshing news for Law Enforcement, who will be able to taint Bitcoin transactions and even back-track them by using our findings and some readily available technology. But at the same time this is worrying news to any legitimate Bitcoin user that simply wants to use these services for the purpose of increasing its anonymity. More study needs to be done on the advantages and shortcomings of the different algorithms employed by these tumblers, as a well-founded theoretical analysis of a highly secure and privacy-aware protocol for providing the required mixing services is unfortunately still lacking. Whether these mix services will continue to be popular and profitable in the near future, when alternative cryptocurrencies that offer improved anonymity and untraceability properties such as Monero or Zcash become widely accepted, is still an open question.

Acknowledgements

 This project has received funding from the European Union's Horizon 2020 research and innovation programme, under grant agreement No. 700326 (RAMSES project). One co-author also wants to thank EPSRC for project EP/P011772/1 on the EconoMical, PsycHologicAl and Societal Impact of RanSomware (EMPHASIS) which partly supported this work.

References

1. Higgins, S.: Coin.mx Execs Arrested for Operating Illegal Bitcoin Exchange (2015). http://www.coindesk.com/coin-mx-arrested-operating-illegal-bitcoin-exchange/
2. United States District Court Southern District of New York Sealed Indictment (2015). http://bit.ly/2aC9Mpl
3. Stevens, M., Karpman, P., Peyrin, T.: Freestart collision for full SHA-1. https://eprint.iacr.org/2015/967.pdf

4. Prince, M.: SHA-1 Deprecation: No Browser Left Behind. https://blog.cloudflare.com/sha-1-deprecation-no-browser-left-behind

5. Nginx CVE for version 1.0.14 (2013). CVEdetails.com

6. Barnes, R.: The POODLE Attack and the End of SSL 3.0 (2014). https://blog.mozilla.org/security/2014/10/14/the-poodle-attack-and-the-end-of-ssl-3-0/

7. Möller, B., Duong, T., Kotowicz, K.: This POODLE Bites: Exploiting The SSL3. Fallback (2014). https://www.openssl.org/~bodo/ssl-poodle.pdf

8. Meiklejohn, S., Pomarole, M., Jordan, G., Levchenko, K., McCoy, D., Voelker, G.M., Savage, S.: A fistful of bitcoins: characterizing payments among men with no names. In: Proceedings of the 2013 Conference on Internet Measurement Conference, pp. 127–140. ACM (2013)

9. Moser, M., Bohme, R., Breuker, D.: An inquiry into money laundering tools in the Bitcoin ecosystem. In: eCrime Researchers Summit (eCRS), 2013, pp. 1–14. IEEE (2013)

10. Bitcoin Organisation: Protect your Privacy (2016). https://bitcoin.org/en/protect-your-privacy

11. Bonneau, J., Narayanan, A., Miller, A., Clark, J., Kroll, J.A., Felten, E.W.: Mixcoin: anonymity for bitcoin with accountable mixes. In: Christin, N., Safavi-Naini, R. (eds.) FC 2014. LNCS, vol. 8437, pp. 486–504. Springer, Heidelberg (2014). https://doi.org/10.1007/978-3-662-45472-5_31

Author Index

Printed in the United States
By Bookmasters